Environmental Science
Appreciation & Perception

ENVIRONMENTAL SCIENCE

APPRECIATION AND PERCEPTION

Arvind Kumar

2026

Daya Publishing House®

A Division of

Astral International Pvt. Ltd.

New Delhi – 110 002

First Published, 2008

Reprinted, 2026

ISBN: 978-81-7035-489-5 (HB)

Published by : **Daya Publishing House®**
A Division of
Astral International Pvt. Ltd.
– ISO 9001:2015 Certified Company –
4736/23, Ansari Road, Darya Ganj,
New Delhi-110 002
Ph. 011-43549197, 23278134
E-mail: info@astralint.com
Website: www.astralint.com

Printed at : Replika Press Pvt. Ltd.

Preface

Environmental Science is a young albeit maturing discipline. The new understanding that complexity is vital to the integrity and stability of natural systems, allows ecologists to argue, more coherently - why must we preserve the diverse elements and species that coexist in a healthy, sustainable and well-functioning ecological community? As we enter the "**Century of the Environment**" it needs to be realized that one crucial function of Environmental Science is to provide an unbiased, scientific basis on which policy makers can decide about how best to treat our natural environment.

Increasing human population strains the resources and availability of energy and materials. The environment gradually degrades by the processes of waste dispersal, nutrient cycling and spread of crop pests and parasites. The quality of air, water and soil deteriorates as a result of human activities.

Environmental disruption may be considered under three steps (i) mankind's misdeeds or insult to the environment; (ii) the response of the environment to the insult and (iii) the consequential damage to the welfare of society, or the environmental cost. Examples of insults are the discharges of wastes into air, water and land; the production of heat; noise; the removal of natural vegetation and fauna; and such physical activities as drilling damming, mining, pumping and dredging.

Some common examples of the environmental responses are accumulation of materials in undesirable forms, concentrations and in the wrong place; alterations in the distribution, abundance and proportions of organisms, decline in primary productivity; changes in predator-prey and host-parasite relations, and such geologic catastrophes as earthquakes, upheavals, erosions and landslides. So, to have a clear cut regarding different aspects of Environmental Science is pertinent. Keeping this in view, the present endeavour has been undertaken to have clear cut perception of various environmental disasters.

This book entitled "*Environmental Science: Appreciation and Perception*" is the unique compilation of esteemed articles of internationally acknowledged experts in the field of Environmental sciences with the intention of providing a sufficient depth of the subject to satisfy the needs at a level which will be comprehensive and interesting. This book will be useful to the students, research scholars, scientists in the field of Environmental management and ecoplanners, politicians and other people with similar interest.

I consider myself extremely fortunate in having got the invaluable support and erudite suggestions of eminent persons like Prof. (Dr.) Christian Ulrichs of Germany, Dr. M. A. Kabir Chowdhury of Malaysia, Prof. Dr. Ahmed H. Al-Harbi of Saudi Arabia, Prof. Tej Kumar Shrestha of Nepal, Prof. P. V. Dehadrai, Former Director, CIFRI, Dr. Dilip Kumar, Director, CIFE, Mumbai, Prof. K. C. Pandey, Former V. C. (Lucknow), Prof. A. R. Yousuf of University of Kashmir, Srinagar (J & K), Professor N. C. Datta of Calcutta University, Professor S. K. Konar of Kalyani University, Professor D. K. Belsare of Bhopal University, Professor Professor U. C. Goswami of Gauhati University, Professor P. C. Mishra of Sambalpur University, Professor P. Natarajan of Kerala University, Professor P. S. Murthy of Bangalore University, Professor Ajit Varma of JNU, Professor A. L. Bhatia of Jaipur University, Professor A. K. Mittal of BHU, Professor S. P. Hosmani of Mysore University, Professor K. Kapoor of Udaipur University, Professor K. B. Reddy of Nagarjuna University, Professor M. Vikram Reddy of Pondicherry University, Professor B. K. Tiwari of NEHU, Professor G. Tripathi of Jodhpur University, Professor R. Ramalingam of Annamalai University, Professor Sharif U. Ahmad of Nagaland University, Professor G. K. Kulkarni of Aurangabad University, Professor S. U. Mehram of Nagpur University. Professor S. K. Battish of PAU, Professor B. M. Sharma of Manipur University, Professor B. D. Joshi of Hardwar University, Professor G. C. Pandey of Faizabad University, Professor K. C. Sharma of Ajmer University, Professor M. Raziuddin of Hazaribag University, Professor U. S. Bagde of Mumbai University, Professor Gurdeep Singh of I. S. M., Dhanbad, Dr. P. K. Goel of Karad, Professor S. P. Roy and Shri Tribhuwan Poddar of Bhagalpur University, Bhagalpur for encouragements.

I also express my deep sense of gratitude to my parents whose blessings have always prompted me to pursue academic activities deeply. I am also thankful to my wife, *Kumari Bimla* and my two lovely sons, *Kumar Pallav Shivshankaran* and *Kumar Prasun Ramakrishnan* whose natural smiles extended to me relief all through this tiresome endeavour.

Last but not the least, I am also thankful to Mr. Anil Mittal, Proprietor, Daya Publishing House, New Delhi for taking keen interest in bringing out of this book. Finally, I will always remain a debtor to all my well-wishers for their blessings, without which this book would not have come into existence.

Professor A. Kumar

Contents

Chapter 1

Bioreclamation of Water as well as Soil Resource with Special Reference to Phytoremediation

Arvind Kumar

Environmental Science Research Unit, Post Graduate Department of Zoology, S.K.M. university, Dumka – 814 101, Jharkhand

E-mail: arvind6dmk@sancharnet.in

Introduction

In broad terms, a resource, or natural resource is any thing by an organism or group of organisms. In other words, a resource is something useful, but for humanity what is useful or useless can change because of technology economics, and the environmental effects of getting and using a resource.

Technology cannot bring back an extinct animal resource or a paved over wilderness area, but it can extend the supply of some resources by improving them, using them more efficiently, or recycling them. However, while many matter resources such as copper, lead and silver, can be recycled, we can never recycle energy resources. Once a fossil fuel resource, such as coal, oil or natural gas, is burnt, it is gone forever as a useful energy source.

In addition to technology, resource use is tied to economics. Something is useful as a resource only if it can be made available at a reasonable cost. For example, once we deplete the easily available supplies of a resource, we have to look harder and dig deeper to find remaining supplies. If the costs of findings and making a scarce resource available rise, the resource will eventually become too expansive for most people.

The sum of all physical, chemical, biological and social factors which compose the surroundings of man is referred to as environment and each element of these surroundings constitutes a resource on which man draws in order to develop a better life. Thus, any part of our natural environment – such as land, water, air, minerals, forest, wildlife, fish or even human population – that man can utilize to promote his welfare may be regarded as a natural or environmental resource.

Humans live on a planet that is dominated by water. More than 70 % of the Earth's surface is covered with it. Scientists estimate that the hydrosphere contain about 1.36 billion cubic kilometers of this substance mostly in the form of liquid that occupies topographic depressions on the earth. Freshwater is a primary resource for all terrestrial life on this planet. Water is important for the facilitation of most biotic and abiotic environmental processes. Humans use water for basic survival and require water for use in industry, agriculture, transportation, and electrical power generation. As the world's human population and industrial activity increase, so does the need for water.

According to the uses water resources may be classified into three main groups:

1. Water for consumptive uses
2. Water for partially consumptive uses
3. Water for non-consumptive uses.

Water for Consumptive Uses

Irrigation, urban and rural water supplies are grouped in this category, since, in these situations, water is directly consumed by living beings.

Water for Partially Consumptive Uses

Use of water for domestic and industrial purposes, thermal and nuclear power generation etc. are partially consumptive, as a part of the water can be renewed for reuse after proper conditioning.

Water for Non-consumptive Uses

This includes the quantity required for hydro-power generation, navigation, pollution control, recreation, preservation of fish and wild life etc.

According to the placement, water resources of India are divisible into two distinct categories:

1. Surface water Resources
2. Ground water Resources.

The precipitation that falls on land is the ultimate source for both the categories of water resources. When rain falls, a sizable portion is intercepted by the vegetation, temporarily detained in surface depressions. When the available interception or the depression storages are completely exhausted and when the rainfall intensity at the soil surface exceeds the infiltration capacity of the soils, the over land flow begins. This water reaches the rivers, streams, lakes, surface reservoirs, dams etc.. Our irrigation mainly depends upon surface water resources. Runoff water from streams and rivers is stored in reservoirs or is diverted directly through canal system for irrigation.

Goethe rightly said that, "Everything Originated in the water, and everything is sustained by water". Water is needed to fulfill diverse requirements in so many diverse ways. It is vital to life, for all physiological activities of plants and is essential for animals as well. On an average it constitutes 80 per cent protoplasm ranging from 8 per cent in dry dormant seeds to 95 per cent jelly fish. It serves not only to quench the thirst but also to meet the food requirements because it is an essential raw material in the process of photosynthesis through which green plants make food that is used by all trophic levels directly or indirectly. Ambient humidity which is indeed the invisible form of water that surrounds us is necessary to prevent desiccation of terrestrial life forms. Thus, it can be said that water is a biologically essential and it serves as milieu internal as well as external.

Besides temperature, water is the other such key factor that influences the global ecology. Covering about 70 per cent of land surface and circulating in hydrologic cycle, water influences weather and climate on any region and thus, its flora and fauna. Water is an essential requirement of all living beings and indispensable for many human related activities. In nature it transports eroded materials from mountains and forests to the plains and the sea. Man uses water to carry away hi source of recreation. Apart from all these the role of water in agriculture is very significant. Agriculture is dependent on water, may it be rainfall, or surface or ground water irrigation. It eventually is a more important commodity in a country like India because her largely agrarian economy depends critically on water. Huge amount of protein rich food in the form of fish, shell fish, and prawn is obtained from the inland water.

With increase in population there is an increase in waste. It is causing stress on environment management of waste, causing pollution of water, land and air. Water being the best and convenient dilution medium is the worst affected resource by pollution. To a common man pollution of water means only the drainage of industrial effluents or domestic sewage in natural waters. It may astonish a lay reader that even pure distilled water when released in sea may act as a pollutant since it alters the normal salinity of sea water which adversely affects the marine life. Besides chemical alterations, changes in physical condition of water also have grave effects on the biota of system as during thermal pollution caused by the discharge of heated effluents in a water body. Pathogens, chiefly entering natural waters through municipal wastes in excreta and other organic filth, also render the water unsuitable for human use.

In fact, it is not only man who has to use the water resources but equally important to land animals and more to those fish and other aquatic animals who live in it. Thus, pollution of a body of water does prove detrimental not only to the human society but also to regional ecology and the ecology of every aquatic ecosystem. It is, therefore, essential that the water should not be treated as a simple repository of waste disposal, and if inevitable, the wastes only after proper treatment should be released in it under strict ecological consideration (Kumar *et al.*, 2002). The qualitative degradation of water is also in a way quantitative depletion of usable resources. Therefore, recycling of waste water after due treatment would relieve the water scarcity to a great extent, especially in regions carried by water shortage. Current waste disposal philosophy is to treat all waste as resource materials some for recycling, some for conversion to fertilizer or as a source of energy and the balance for land or pond reclamation. A waste is not a waste. It is wealth if properly recycled. The aquatic macrophytes and macrophytes are considerable ecological and economic importance. They contribute significantly to the productivity of water, bodies, mobilize mineral elements from the bottom sediment and provide shelter to aquatic invertebrates and fishes. Excessive use of chemical fertilizer causes ecological imbalance in long run. To overcome this in the long run recycling of weed will help to a great extent (Verma, 2003). Greater attention need to given in aquaculture to management for sustainability (Dash and Chatterjee, 2003).

There are reports of widespread arsenic contamination in several parts of the world, of which the scenario of arsenic contamination of ground water delta basin seems to the worst one affecting millions of people (Sanyal, 2000 and Sanyal and Nasar, 2002). There have been reports of elevated levels of arsenic build-up in agricultural produce obtained with contaminated ground water used as the irrigation source (Sanyal and Nasar, 2002 and ICAR, 1998-2001).

Among the possible mitigation options/interventions that are being examined in this regard. Phytoremediation tends to offer a potentially useful avenue to address the problem of contaminated agricultural soils and crops.

The fact that the standard for the safe limit of arsenic concentration in drinking water has recently been reduced from 50 to 10 ppd by the USEPA (Ma *et al.*, 2004) makes it all the more urgent to develop reliable and cost-effective technologies capable of reducing arsenic in ground water to environmentally acceptable levels. The phytoremediation techniques seem to offer viable options, and are discussed herein.

Different Techniques of Phytoremediation

Phytoremediation is defined as the use of vegetation or other macro and microscopic biota for *in situ* treatment of contamination soils, sediments and water where green plants degrade, assimilate, metabolize (Thuraisingham *et al.*, 2004). Or detoxify inorganic and organic chemicals in soils to environmentally acceptable levels; in other words, it is equivalent to a bioremediation process that uses various types of plants to remove, transfer, stabilize, and/or destroy contaminants in the soils and groundwater.

There are several ways and means to regulate phytoremediation technique.

Rhizosphere Biodegradation

In this process, the plant releases natural substance through its roots, supplying nutrients to microorganisms in the soils. The microorganisms in turn, facilitate biological degradation through rhizosphere interactions. Substantial progress has been made towards an understanding of arsenic transformation processes in soils. However, adequate information is not available addressing directly the issue of arsenic in the rhizosphere. It is only rather recently that Fitz and Wenzel (2002) developed a model which correlates the fate of arsenic in the soil- rhizosphere plant system.

Phyto Stabilization

In this process chemical compounds produced by the plants immobilize contaminants rather than degrade them. This is essentially a process of stabilizing contaminated land or the pollutants present in the soil and in doing so it prevents or reduces erosion, water flow and flow of pollutant. In this case metal tolerant plant species that do not take up large quantities of metals are often used. As for example, apples, cabbage, carrots, tomato, wheat, etc. are tolerant to arsenic (Adriano, 1986) and may possibly be used for this purpose.

Phytoaccumulation

Phytoaccumulation or hyper-accumulation of heavy metals is likely to involve several steps, including metal transport across root cell membranes, xylem loading and translocation. The following are the mechanisms of metal hyper-accumulation, especially for arsenic.

1. Metal uptake
2. Metal transformation from root to shoot
3. Mechanisms of metal tolerance in hyper-accumulator plants which include-
   (*a*) Cellular and sub-cellular compartmentalization.
   (*b*)Complexation with ligands – which results in decreased free metal ion activity and thus decreased toxicity.

Phytovolatilization or Biotransformation

In this process, plants and microorganisms take up waste containing toxic contaminants and release the same into the air through volatilization for this, arsenic culture solution are utilize by certain bacterial streams and further methylation of arsenic takes place, thereby rendering the more toxic forms (inorganic ones) to less toxic methylated forms (Wong *et al.*, 1977).

Hydroponic System for Treating Water Streams (Rhizofiltration)

Rhizofiltration similar to the phyto accumulation, but the plants used for clean-up are raised in green house with their roots in water. This system can be used for *ex situ* groundwater treatment. Groundwater is pumped to the surface to irrigate these plants. Typically hydroponic systems utilize an artificial soil medium, such as sand mixed with perlite or vermiculite. As the roots become saturated with contaminants, they are harvested and suitably disposed off.

Phytodegradation

In this process, plants (Ma *et al.*, 2004) actually metabolize and destroy the contaminants within the plant tissues.

Hydraulic Control

Generally, the use of phytoremediation is limited to the sites with lower contaminant concentrations and contamination in shallow soils, streams and groundwater. In this process, trees indirectly remediate by controlling groundwater movement. Trees act as natural pumps when their roots reach very close to the water table and establish a dense root mass that takes up large quantities of water. Using poplars and will ours developed in the horticultural researcher conservation plant breeding programme at New Zealand (2000), hybrids with a propensity for arsenic accumulation have been identified. A poplar tree, for example pulls out of the ground 30 gallons of water per day, while a cotton wood can absorb up to 350 gallons of water per day, thereby drawing a substantial amount of water along with the toxicant.

Notwithstanding the fact phytoremediation is a "green technology" which could be cost effective and ecofriendly *in situ* remediation technique, nevertheless it suffers from many limitations, yet to overcome, while contemplating the implementation of such phytoremediation options some of these are mentioned hereunder:

1. Lack of sustained, intensive research on phytoremediation of plants and microorganisms.
2. Some hyperaccumulator plant species may not lead to effective detoxification.
3. Such remediation processes are not generally stable on a long-term perspectives.
4. Phytoremediation may even generate the secondary source of taken while decomposition takes place.
5. The expertise and the knowledge base available in the field do not seem to be adequate, while the cost effectiveness of many such phytoremediation techniques is yet to be fully assessed.

Lastly we can say that management water implies making the best use of available water resources for human benefit, while, not only preventing and controlling its depletion and degradation but also developing it in view of the present and future needs. Water, like forest, is a multipurpose resource and it is important to see that its various uses should not conflict with each other and it can be enjoyed in its totality by man and others. Thus, its right allocation, as well as quantitative and qualitative conservation is the primary tasks before water managers.

References

Adriano, D.C., 1986. In: *Trace Elements in the Terrestrial Environment*. Springler Verlag, New York, USA.

Dash, G. and Chatterjee, N.R., 2003. Strategies for environmental management in aquaculture. In: *Aquatic Ecosystem*, (Ed.) A. Kumar. A.P.H. Pub. Crop., New Delhi, pp. 319–322.

Fitz, W.J. and Wenzel, W.W., 2002. *J. Biotechnol.*, pp. 99.

ICAR, 1998–2001. Final Report on Status, causes and Impacts of Arsenic Contamination in Groundwater in Parts of W.B. *Vis-à-vis Management of Agricultural Systems.*

Ma, L.Q., Tu, M.S., Fayiaga, A.O., Stamps, R.H. and Zilliux, E.J., 2004. *The Association for Environmental Health and Sciences.*

Sanyal, S.K., 2000. In: *SATSA Mukhapatra: Annual Technical Issue*, 4 : 41 – 69.

Sanyal, S.K. and Nasar, S.K.T., 2002. In: *Water Resources Engineering for Disaster Mitigation*, 1 : 216–222.

The Horticultural and Food Research Institute of New Zealand Limited. 2000.

Thuraisingham, R., Mueller, R. and Arguello, R., 2004. *The Association for Env. Hlth. and Sci.*

Verma, J.P., 2003. Recycling of aquatic vegetation and waste material for enrichment of nutrient status of pond. In: *Environmental Challenges for 21st Century*, (Ed.) A. Kumar. APH Publ. Corp., New Delhi.

Wong, P.T.S., Chau, Y.K., Luxon, L. and Bengert, G.A., 1977. *J. Environ. Qual.*, 13: 499–504.

Chapter 2

Toxicological Effects Caused by Mercury Contained SWE of a Chlor-alkali Industry on a Nitrogen Fixing BGA and its Detoxification

R.K. Behera, Alaka Sahu and A.K. Panigrahi

Environmental Science Research Division, Department of Botany, Berhampur University, Berhampur – 760 007, Orissa

ABSTRACT

The mercury containing solid waste extract of a chlor-alkali industry was tested for possible use of BGA as a nitrogen fixer and an agent for detoxification of a contaminated environment. Significant increase in nitrogen fixation and removal of mercury by way of volatilisation by the blue-green alga was marked at sub-lethal concentration of solid waste extract (SWE), showing stimulation and detoxification. At higher concentrations of the SWE both nitrogen fixation and removal of mercury from the contaminated environment was affected. The SWE of the chlor-alkali industry should be diluted to its sub-lethal concentration and can be used in the crop fields, the tolerant BGA can be charged into the field and allowed to grow for 20 days and then any crop can be cultivated, which will show better yield/production. This alga can be used as a biofertiliser cum detoxifier in mercury contaminated environments.

Introduction

Algal assays have proven to be very sensitive indicators of contaminant stress. Tewari *et al.* (1990) studied the growth and biochemical composition of two marine macro algae exposed to chlor-alkali industry effluent. The role of blue-green algae as biological inputs in agriculture has been well documented and substantiated (Venkataraman and Rajyalakhmi, 1972). Contamination of the aquatic and terrestrial environment by Industrial discharges occur both as effluents and solid wastes and cause hazardous effect on living organisms (Agarwal and Kumar, 1978). Although the toxic effect of the effluent have been studied on different algae (Adhikary and Sahu, 1985; Shaw, 1987, Tewari *et al.*, 1990; Rath, 1991 and Sahu (2000), much work has yet to be done on the effects of solid wastes and its leached chemicals on crop field inhabiting blue-green algae. The solid waste under present study contains a significant quantity of mercury (Shaw *et al.*, 1990). It has been reported that mercurial compounds were toxic to algae (den Doorendejong, 1965), mercuric ion prolonged the lag phase of growth and at low concentrations of mercurial compounds stimulatory effects were reported (Shaw, 1987; Sahu, 1987 and Rath, 1991). Nitrogen fixing blue-green algae make a major contribution to the fertility of paddy fields (Venkataraman, 1972). The atmospheric nitrogen fixed by the algae is released to the external medium as extra-cellular nitrogenous chemicals. Hence, the extra cellular and the cellular nitrogen content of algae, can be an index of nitrogen fixing ability of the organism. Sahu (2000) suggested that *Westiellopsis* can fix nitrogen equally well under aerobic, micro-aerobic and anaerobic conditions so that they can tolerate the very wide range of oxygen tensions found in rice fields. Therefore, any adverse effect on these organisms, caused by indiscriminate use of heavy metals or by discharge of industrial wastes might influence the total productivity and nitrogen fixing ability of the algae, which helps for increasing the fertility of the paddy fields. Shaw *et al.* (1990) studied the effect of effluent of a chlor-alkali industry on the nitrogen fixation capability of the blue-green alga, *Westiellopsis prolifica*, Janet.

The present piece of work was described to study the effect of the leached chemicals of the solid waste of a chlor-alkali industry containing mercury on the nitrogen fixing blue-green alga, abundantly available in nearby crop fields, the *Anabaena cylindrica*. The observed luxuriant growth of the alga in paddy fields might be used as a biological agent to detoxify the contaminated environments simultaneously fixing atmospheric nitrogen and acting as a biofertilizer.

Materials and Methods

Anabaena cylindrica, Lemm. is photo-autotrophic, unbranched, filamentous, heterocystous, blue-green alga belonging to the family Nostocaceae and can grow well in Allen and Arnon's (1955) nitrogen free medium with trace elements of Fogg (1949) as modified by Pattnaik (1964). The solid waste was collected from the huge dumps near a chlor-alkali industry was brought to the laboratory, air dried, powdered and seived. The solid waste (SW) powder was used to prepare the solid waste extract (SWE), by taking 5 kgs of SW and 5 liters of distilled water and stirred for 48 hours. It was allowed to stand for 24h and the supernatant was decanted and used for the experimental purpose. The SW and SWE was analysed to understand its contents. Table 2.1 indicates the nature of SW and SWE.

Estimation of Cellular and Extra-cellular Nitrogen Content

It was measured in terms of cellular and extra-cellular Kjeldahal nitrogen, which were estimated by Kjeldahal Nesslerization method of Herbert *et al.* (1971). The data were expressed as µg of nitrogen/ 100 ml algal culture. Residual mercury was estimated by digesting the samples following Wanntorp and Dyfverman (1955) as modified by Sahu (1987) and measuring in a cold vapour atomic absorption spectrophotometer, Mercury Analyser (ECIL).

Results

The control set showed 100 per cent survival. The toxicity study revealed that 10 per cent survival at 1.53 per cent SWE, 50 per cent survival at 0.82 per cent SWE, 90 per cent survival at 0.46 per cent

SWE, and 100 per cent survival at 0.21 per cent SWE was marked. Out of the above concentrations, LC_{00} or PS_{100} as safe MAC value of 0.2 per cent SWE was selected as 'A' and LC_{90} or PS_{10} value of 1.5 per cent SWE was selected as 'B' for conducting future experiments.

Table 2.1: Physico-chemical Analysis of Solid Waste (SW) and Solid Waste Extract (SWE)

(a) Physical Properties	*Solid Waste*	*Solid Waste Extract*
Texture	Clay	
Colour	Grayish white	Pale yellow
Temperature	26°±2°C	24°±2°C
Specific gravity	2.8	–
Water holding capacity	39 per cent by volume	–
Air content	27 per cent by volume	14 per cent by volume
(b) Chemical Properties	*(mg kg^{-1} dry wt.) Solid Waste*	*(mg l^{-1}) Solid Waste Extract*
pH of solid waste	4.6±0.4	4.8±.0.2
Phosphate	59.00±10.93	42.6±8.1
Chloride	11614±165.2	18.2±1.4
Calcium	126±18.33	108±26
Magnesium	75.00±10.22	51.24±16.87
Sodium	1157±96.44	5.2±0.8
Potassium	96.0±10.25	106.4±12.6
Total nitrogen	11.6±4.5	0.24±0.06
Mercury	1149±248.5	12.86

(Values are mean of five samples ± standard deviation).

In case of the control set, the cellular nitrogen content increased from 6.2±0.8 μg/100 ml culture to 14.1±2.6 μg/100 ml culture within 15 days of exposure. The cellular nitrogen content increased to 25.4±4.8 μg/100 ml culture on 15th day of recovery. The cellular nitrogen increase showed a positive correlation with the exposure period (Table 2.2). In concentration 'A' (0.2 per cent SWE), significant higher values, when compared to control was recorded. The cellular nitrogen content increased from 6.2±0.8 to 17.6±2.2 μg/100 ml culture within 15 days of exposure. The cellular nitrogen content in concentration A was much higher than the control value at all exposure periods. When exposed alga was transferred to toxicant free medium, significant increase in cellular nitrogen content was recorded, where maximum value of 30.2±3.6 μg/100 ml culture was recorded (Table 2.2). In concentration B (1.5 per cent SWE), the cellular nitrogen content increased insignificantly from 6.2±0.8 to 7.1±1.1 on 3rd day and 6.9±0.9 μg/100 ml culture on 6th day of exposure. These values were less than the respective control values and concentration A set values. After 9th day of exposure, cellular nitrogen content significantly declined to 0.6±0.1 μg/100 ml culture on 15th day of exposure. When the exposed alga was transferred to toxicant free medium, no recovery was marked. In case of concentration A, at all exposure periods, rise in cellular content level was marked, when compared to the control value. On 15th day of exposure, a maximum of 24.8 per cent increase over the control value was marked. In concentration B, significant depletion in cellular nitrogen content was recorded. A maximum of 95.74

per cent decrease over the control value was recorded on 15th day of exposure. In case of concentration A, significant recovery was marked and the 7th and 15th day recovery values were much more than the control value. But in case of concentration B, significant depression and no recovery was recorded. The correlation coefficient analysis between days of exposure and cellular nitrogen content shows the existence of a significant positive correlation in control ($r = 0.979, p \leq 0.001$) and in concentration ($r = 0.969, p \leq 0.01$). A significant negative ($r = -0.905, p \leq 0.05$) correlation in concentration B was marked.

Table 2.2: Changes in Cellular Nitrogen Content (µg 100 ml culture) of Control and Exposed Blue-green Alga, at Different Exposure and Recovery Periods, and at Different Concentrations of the Toxicant. Data are mean of 5 samples±standard deviation.

Concentration of the Toxicant, % (v/v)	*Exposure in Days*						*Recovery in Days*	
	0	*3*	*6*	*9*	*12*	*15*	*7*	*15*
Control	6.2±0.8	7.8±1.1	8.5±1.6	9.9±0.6	11.2±1.9	14.1±2.6	18.6±3.4	25.4±4.8
A (0.2%)	6.2±0.8	9.1±0.4	9.8±1.1	10.8±0.7	13.9±1.2	17.6±2.2	24.2±4.2	30.2±3.6
B (1.5%)	6.2±0.8	7.1±1.1	6.9±0.9	4.2±0.4	2.1±0.2	0.6±0.1	0.8±0.2	0.4±0.1

The extra-cellular nitrogen was almost zero in the inoculation day, as the selected nutrient media is free from nitrogen, which was more suitable for the growth of the blue-green alga. The exuded extra-cellular nitrogen content in the control set showed an increasing trend, with the increase in exposure period. The value increased to 16.3±2.4 on 3rd day, 31.6±8.2 on 6th day, 54.3±3.5 on 9th day 86.7±11.3 on 12th day and 122.7±18.9 µg/100 ml culture on 15th day of exposure. The extra-cellular nitrogen content in the recovery flask, increased to 218.9±16.8 µg/100 ml culture on 15th day of recovery. This value is the total amount of extra-cellular nitrogen fixed by the blue-green alga, which has been exuded to the medium. In concentration A (0.2 per cent SWE), at all exposure periods, the extra-cellular exudate nitrogen content was more than the control value. The value increased from 22.8±3.8 to 156.8±18.8 µg/100 ml culture on 15th day of exposure and the value reached maximum to 254.2±11.6 µg/100 ml culture on 15th day of recovery (Table 2.3). In case of concentration B (1.5 per cent SWE), an insignificant amount of nitrogen was fixed on 3rd day 10.8±1.4, on 6th day 12.6±2.2 and on 9th day 8.2±1.1 µg/100 ml culture. These values were much less when compared to the control values. With the increase in exposure period, the extra-cellular nitrogen availability in the flask reduced and on 15th day of exposure, no extra-cellular nitrogen was recorded in the culture flask (Table 2.3). When the exposed alga was transferred to toxicant free medium for recovery, no recovery was marked. In concentration A, higher fixation of nitrogen was marked on 3rd day. At all exposure periods, higher values were recorded. But in the early phases, upto 9th day of exposure, maximum fixation was recorded. On 3rd day 38.18 per cent, 6th day 29.43 per cent, on 9th day 28.17 per cent, on 12th day 16.95 per cent and on 15th day 27.79 per cent increase over the control value was recorded. In case of concentration B, the extra-cellular nitrogen content decreased and the per cent decrease increased with the increase in exposure period. Hundred per cent inhibition was recorded on 15th day of exposure. When the exposed alga was transferred to toxicant free medium, no recovery was marked in case of concentration B. In concentration A, higher values and significant recovery was recorded. The obtained values were much more than the control values. The correlation coefficient analysis between days of exposure and extra-cellular nitrogen content indicated the existence of a positive correlation in control ($r = 0.981, p \leq 0.01$) and in concentration A ($r = 0.988, p \leq 0.001$). A non-significant negative correlation ($r = -0.299$, p = NS) in concentration B was marked.

Table 2.3: Changes in Extra-cellular Nitrogen Content (μg/100 ml culture) of Control and Exposed Blue-green Alga, at Different Exposure and Recovery Periods, and at Different Concentrations of the Toxicant. Data are mean of 5 samples±standard deviation.

Concentration of the Toxicant, % (v/v)	*Exposure in Days*						*Recovery in Days*	
	0	*3*	*6*	*9*	*12*	*15*	*7*	*15*
Control	NT	16.5±2.4	31.6±8.2	54.3±3.5	86.7±11.3	122.7±18.9	194.3±14.2	218.9±16.8
A (0.2%)	NT	22.8±3.8	40.9±9.5	69.6±11.5	101.4±18.2	156.8±18.8	204.4±28.2	254.2±11.6
B (1.5%)	NT	10.8±1.4	12.6±2.2	8.2±1.1	1.4±0.2	NT	NT	0.8±0.1

NT: Not traceable.

Table 2.4 shows the residual mercury accumulation in SWE exposed blue-green alga at different exposure period and recovery period. In concentration A (0.2 per cent SWE), the exposed alga could accumulate 0.682±0.065 μg of Hg/100 ml culture within 3 days of exposure. With the increase in exposure period, the residual mercury accumulation increased showing a positive correlation. The alga could accumulate 1.324±0.065 μg of Hg/100 ml culture within 15 days of exposure. When the exposed alga was transferred to toxicant free medium, on 7th day, 1.106±0.066 μg of Hg/100 ml culture was noted and on 15th day 0.864±0.071 μg of Hg/100 ml culture was recorded. Within a period of 15 days during recovery period, 0.46 μg of Hg/100 ml culture was excreted from the exposed alga. Within 7 days, 0.218 μg of Hg/100 ml culture excreted from the exposed alga. In concentration B (1.5 per cent SWE), the exposed alga could accumulate 0.914±0.084 μg of Hg/100 ml culture within 3 days of exposure and the residual accumulation of mercury increased significantly and a maximum of 1.336±0.114 μg of Hg/100 ml culture was recorded on 15th day of exposure. When the exposed alga, was transferred to toxicant free medium, on 7th day 1.208±0.092 μg of Hg/100 ml culture was recorded and on 15th day 1.106 μg of Hg/100 ml culture was recorded. When the exposed alga was transferred to toxicant free medium, within 7th day, 0.128 μg of Hg/100 ml culture has been excreted and within 15 days of recovery 0.23 μg of Hg/100 ml culture was excreted. A maximum of 16.47 per cent and 34.74 per cent residual mercury got excreted during recovery period in concentration and 9.58 per cent of residual mercury and 17.22 per cent residual mercury got excreted from the exposed alga in concentration B set.

Table 2.4: Showing the Mercury Content in the Medium, Accumulation in the Alga After 15 Days of Exposure, Amount of Volatilised Mercury within 15 Days of Exposure (μg/100 ml culture), Total Mercury Estimated Budget and Unseen/Non-recordable Mercury in the Exposed Systems. (Sensitivity: 88.65 per cent; Accuracy: 99.8 per cent)

Concentration of SWE in % (V/V)	*Hg Content in the Medium*	*Residual Hg in the BGA After 15 Days of Exp.*	*Hg in the Medium After 15 Days of Exposure*	*Volatised Hg Within 15 Days of Exposure*	*Total Hg Removal from the medium*	*Total Hg Recorded in the Experiment*
	A	*B*	*C*	*D*	*E (B + D)*	*F (C + E)*
A (0.2%)	3.065±0.002	1.324	0.005	1.732	3.056	3.061
B (1.5%)	6.122±0.041	1.336	0.165	4.702	6.038	6.203

Discussion

The review made by Whitton (1970) on impact and effect of heavy metals on algae added a lot of information to the literature of algal toxicology. Algae, have been shown to concentrate heavy metals to a larger extent (Jannett and Wixson, 1975). Agarwal and Kumar (1978) showed decrease in growth of *Chlorella* sp. when exposed to mercurial effluent and solid wastes indicating toxic nature of mercury on the organism. A liquid industrial waste may affect the algal growth in any of three ways: stimulation, inhibition and stimulation at lower concentrations but inhibition at higher concentrations (Sahu, 1987 and Rath, 1991). Some suggested uptake and metabolisation of the constituent as the probable mechanism for growth stimulation. The observed stimulation in the algal growth at low concentrations of Cd, Pb and Ni. Neither Cd, Pb and Ni have been reported to be essential micronutrients for algae (O'Kelley, 1974) nor the pure solution of these heavy metals are expected to contain any growth regulator. Thus, an ideal explanation for stimulation of growth at lower concentrations of the toxicant is yet to be ascertained. The solid waste extract under present investigation contained huge amount of mercury. Compounds based on mercury are toxic to algal organisms (Rath *et al.*, 1983). Organic forms of Hg is more toxic than inorganic form (Rath, 1984 and Sahu, 1987). Further inorganic forms are active only when these are present in free ionic state. Possible formation of complex of heavy metals with calcium and phosphate ions might be responsible for reduction in the toxicity, because of the complex formation they might not be getting an entry into the cell either through membrane transport or by surface adsorption which are the two known mechanisms for uptake of the substances (Rai *et al.*, 1981). Further, different heavy meals interact among themselves either to reduce or to increase the toxicity of each other. Rath (1984), Shaw (1987) and Mohapatra (1992) reported the dual behaviour of mercurial compounds on the growth of BGA and confirmed the dichotomous behaviour of the toxicants on living systems (Sahu, 1987 and Rath, 1991). However, this piece of work strongly agrees with the findings of the above authors. Probably the result indicated a new line of thinking, which can become a possibility in case of heterogeneous toxicants, where synergistic and antagonistic effects were expected. Here, it can be presumed that chemicals present other than mercury probably act as a masking agent on mercury, reducing the toxicity in turn, showing variation in the observed data. More work is essential on different live systems to confirm the synergistic and antagonistic characteristic features of the complex mixture toxicants. The toxicant used in this investigation is not a single unit toxicant rather a complex heterogeneous chemical, and contains other chemicals along with mercury. Hence, the effect observed in higher concentrations may not only be attributed to mercury, but to other compounds present in the heterogeneous medium of solid waste extract. Stimulation in nitrogen fixing ability at low concentrations of various toxicants have been reported by several authors. Wurtsbaugh and Apperson (1978) reported an increase in nitrogen fixation with some insecticides. Shaw *et al.* (1990) reported a decrease in nitrogen fixation capacity of *Westiellopsis* when exposed to the effluent of a chlor-alkali industry. Rath (1984) described a decrease in the cellular Kjeldhal nitrogen level with an increase in the concentration of $HgCl_2$. Stratton *et al.* (1979) stated that mercuric ion was toxic towards *Anabaena inaqualis's* growth, photosynthesis and nitrogenase activity to be the primary site of toxicity (Karmp-Nielson, 1971).

In the present study, however, mercury with other chemicals does not appear to play any significant role in its highest concentration applied, to inhibit the nitrogen fixation capacity of the BGA, though, in lower concentrations it showed a stimulatory effect which was correlated with the amount of mercury present in the medium. The increase in the amount of extra-cellular and total nitrogen in lower concentrations of the toxicant was also related significantly with the amount of mercury present in the medium, whereas the change in "B" concentration showing a negative significant correlation with residual mercury accumulation, indicated that SWE, at higher concentration is toxic. The solid waste is a heterogeneous compound and its extract containing an objectionable amount of mercury is deadly toxic in nature. This toxicity may be due to the presence of mercury or due to the presence of some other substances; the derogatory effect of which on living organisms in the ecosystem cannot be overruled, because this solid waste extract may show antagonistic or synergistic effect in field conditions rather than in laboratory controlled conditions. *Anabaena cylindrica*, Lemm, the blue-green alga was more sensitive and less tolerant to the solid waste extract (SWE) of the chlor-alkali industry, when compared to *Westiellopsis prolifica*, Janet, which was less sensitive and more tolerant (Sahu and Panigrahi, 2002). On the basis of experimental evidences for the existence of various Hg species in the environment (Nriagu, 1979), elemental Hg and methyl mercury (dimethyl and monomethyl) are the principal candidates for volatilisation (emission and/or re-emission) into the atmosphere. Total vapor-phase Hg fluxes from the agricultural and forest soils were found to be smaller than those from the surface of a lake. Volatilisation and methylation of Hg appear to occur at the highest rates under aerobic conditions (Olson and Cooper, 1976). Under aerobic conditions the rate of volatilisation by Hg reducing strains has been shown to be reduced, but methylation takes place in a natural way as such (Barkey *et al.*, 1979). In the present investigation, it was observed that the, alga *Anabaena cylindrica* can reduce the toxicity of the mercury contained solid waste extract by way of volatilisation of elemental mercury from the medium because the alga did not contain methyl mercury (Sahu and Panigrahi, 2002) even though their total residual mercury levels were high (Holm and Cox, 1974). In the lower concentration of mercury in the medium, higher rate of detoxification was marked which strongly indicated the dilution of the toxicant to a particular level, where volatilisation and absorption become faster (Sahu and Panigrahi, 2002). Sahu (2000) reported a similar finding, while working with the liquid effluent of a chlor-alkali industry containing mercury either in the dissolved form, or in the form microsize elemental mercury. The same author reported that *Anabaena cylindrica* could remove a substantial amount of mercury from the effluent medium. The only question, she raised was the concentration of the effluent. She suggested that the effluent should be diluted sufficiently and a base nutrient medium should be provided, where the BGA can grow. Once the BGA grows, at lower concentration, the alga can remove a lot amount of mercury by way of volatilisation.

We are of the opinion that the solid waste should be diluted proportionately and the extract is also to be diluted to get a value of 0.2 per cent dilution, so that the alga can grow profusely, fix higher amount of atmospheric nitrogen and exude higher amount of extra-cellular nitrogen and simultaneously detoxify the environment by way of removal of mercury from the environment, may be by a way of volatilisation. Zingmark and Miller (1975) reported considerable loss of mercury from culture of dinoflagellates and opined that loss of mercury was probably due to chemical volatilisation, although some might be due to the biological process. We also agree with the remarks of the above authors. The only hint at this point that low doses of mercury in the medium can stimulate growth and nitrogen fixation. We agree with the findings of Rath (1984), Shaw (1987), Sahu (1987), Rath (1991) and strongly disagree with the findings and remarks made by Mohapatra (1992). The residual mercury played the key role for the reduction in growth parameters including nitrogen fixation at higher concentrations in the medium and in the alga. The observed trend and results in concentration A contradicts the trend and results observed in concentration B. In both concentration A and B, the same solid waste extract was used. But the only difference between two sets was the dose/concentration of the SWE and the main factor was mercury. Hence, it can be concluded that mercury at low concentrations acts as a growth stimulant and fixes higher amount of atmospheric nitrogen and the same metal mercury at higher concentration is a growth inhibitor and reduces atmospheric nitrogen fixation. This indicated again that mercury played a dual role depending on the availability and concentration of the chemical.

The situation becomes more difficult because, once they are in the environment, chemicals spread in a very complex way and may be converted into other substances which have different effects. The industrial pollution has spoiled the natural water, air and soil. The effluents and wastes have caused havoc for all living organisms. One such microorganism is blue-green algae inhabiting crop fields whose importance as biofertiliser in rice cultivation is undisputed and well documented. A rice field with a healthy growth of BGA would require no input of nitrogenous fertiliser, India being predominantly an agriculture based country and the agriculture is totally based on the nitrogen economy, any effect on the blue-green algae will affect the nitrogen budget of the paddy fields.

A thorough knowledge of basic mechanisms by which toxic agents impair organisms are fundamental to any eco-toxicological approach. The toxicity of the metal was probably detoxified due to the presence of the masking agents in the SWE, or the nutrients present in the medium or the exudates of the blue-green alga. From the correlation matrix, it can be hinted that the extra-cellular products of the blue-green alga probably helped to change the nature and toxicity of the solid waste extract. Detoxification mechanisms can involve storage of metals at inactive sites within organisms on a temporary or more permanent basis. Temporary storage is generally by binding of metals to proteins. polysaccharides and amino acids in soft tissues or in body fluids. This particular alga detoxified the medium by the process of volatilisation, which showed that like bacteria, this BGA has the capacity to volatilise mercury from the medium, the rate of which was higher in lower concentrations of solid waste extract. So it can be suggested that the waste of this chlor-alkali industry should be diluted sufficiently, before its discharge to outside environment either to be used in the crop fields for irrigation purpose or for any other purpose. In addition, the alga, *Anabaena cylindrica* can be used as an agent, which can remove mercury from a mercury contaminated area as a detoxifier and simultaneously can act as a biofertilizer.

Acknowledgements

The financial support of DOD, Government of India, New-Delhi and OSTC, Berhampur University in the form of a project is sincerely acknowledged.

References

Adhikary, S.P. and Sahu, J., 1985. Effect of distillery effluent on the growth of blue-green alga, *Anabaena* in light and dark. *Environ. and Ecol.*, 3(4): 580–590.

Agarwal, M. and Kumar, H.D., 1978. Physico-chemical and phycological assessment of two mercury polluted effluents. *Ind. J. Environ. Health*, 20, 141–155.

Allen, M.B. and Arnon, D.I., 1955. Studies on nitrogen fixing blue-green algae, growth and nitrogen fixation by *Anabaena cylindrica*. Lemm. *Pl. Physiol.*, Lancaster, 30: 366–372.

Barkay, T., Olson, B.H. and Coldwell, R.R., 1979. In: *Management and Control of Heavy Metals in the Environment*. CEP Consultants Ltd., Edinburgh, U.K., pp. 356–363.

Den Dooren deJong, L.E., 1965. Antonie Vane Leeuwenhoek. *J. Microbiol. Serol.*, 31: 301–313.

Fogg, G.E., 1949. Growth and heterocyst production in *Anabaena cylindrica*. Lemm. II. In relation to carbon and nitrogen metabolism. *Ann. Bot. N.S.*, 13: 241–259.

Herbert, D., Phipps, P.J. and Strange, R.E., 1971. Chemical analysis of microbial cells. In: *Methods of Microbiology*, (Eds.) J.R. Norris and D.W. Ribbons. Academic Press, N.Y., London, pp. 210–344.

Holm, H.W. and Cox, M.F., 1974. In: *Mercury in Aquatic Systems: Methylation, Oxidation-Reduction, and*

Bioaccumulation, (Ecological Research Series) (U.S. Environmental Protection Agency), E.P.A. 660/3-74-021, p. 39.

Jennett, J.C. and Wixson, B.C., 1975. In: Proc. 30th Purdue Ind. Waste Cong., Ann Arbor, Mi: Ann. Arbor Science Publishers, Inc., p. 1173–1180.

Kamp-Nielsen, L., 1971. The effect of deleterious concentrations of mercury on the photosynthesis and growth of *Chlorella pyrenoidose. Plant Physiology*, 24: 556–561.

Mohapatra, A., 1992. Eco-physiology, resistence and ecological implications of a mercurial compound on a blue-green alga. *Ph.D. Thesis*, Berhampur University.

Nriagu, J.O. (Ed.), 1979. *The Biogeochemistry of Mercury in the Environment*. Elsevier/North-Holland Biomedical Press, Amsterdam.

O'Kelley, J.C., 1974. Organic nutrients. In: *Algal Physiology and Biochemistry*, (Ed.) W.D.P. Stewart. Blackwell Scientific Publications, Oxford, p. 610–635.

Olsen, B.H. and Cooper, R.C., 1976. Comparison of aerobic and anaerobic methylation of mercuric Chloride by San Francisco Bay Sediments. *Water Res.*, 10: 113–116.

Pattnaik, H., 1964. Studies on nitrogen fixation by *Westiellopsis prolifica*, Janet, *Ph.D. Thesis*, University of London.

Rai, L.C., Gaur, J.P. and Kumar, H.D., 1981. Protective effects of certain environmental factor on the toxicity of zinc, mercury and methyl mercury to *Chlorella vulgaris. Environ. Res.*, 25: 250–259.

Rath, P., 1984. Toxicological effects of pesticides on a blue-green algae, *Westiellopsis prolifica* Janet. *Ph. D. Thesis*, Berhampur University, India.

Rath, P., Panigrahi, A.K. and Misra, B.N., 1983. Effect of inorganic mercury on growth of *Westiellopsis prolifica* Janet. *J. Environ. Biol.*, 4: 103–109.

Rath, S.C., 1991. Toxicological effects of a mercury contained toxicant on *Anabaena cylindrica*, L. and its ecological implications. *Ph.D. Thesis*, Berhampur University, India.

Sahu, A., 1987. Toxicological effects of a pesticide on a blue-green alga: III. Effect of PMA on a blue-green alga, *Westiellopsis prolifica* Janet, and its ecological implications. *Ph.D. Thesis*, Berhampur University, India.

Sahu, A., 2000. Eco-toxicological effects of mercury contained waste of a chlor-alkali industry on a blue-green alga, *Westiellopsis prolifica*, Janet. *D.Sc. Thesis*, Berhampur University, India.

Sahu, A. and Panigrahi, A.K., 2002. Eco-toxicological effects of a chlor-alkali industry effluent on a cyanobacterium and its possible detoxification. In: *Algological Research in India*, p. 351–430 (Ed.) N. Anand. Published by Bishen Singh Mahendra Pal Singh, Dehra Dun, India.

Shaw, B.P., 1987. Eco-physiological studies of industrial effluent of a chlor-alkali factory on biosystems. *Ph.D. Thesis*, Berhampur University, Orissa, India.

Shaw, B.P., Sahu, A. and Panigrahi, A.K., 1990. Comparative toxicity of an effluent from a chlor-alkali industry and $HgCl_2$. *Bull. Environ. Contam. Toxicol.*, 45: 280–287.

Stratton, G.W., Huber, A.L. and Carke, C.T., 1979. Effect of mercuric ion on the growth, photosynthesis and nitrogenase activity of *Anabaena inaqualis. Appl. Environ. Microbiol.*, 38: 537–543.

Tewari, A., Thampan, S. and Joshi, H.V., 1990. Effect of chlor-alkali industry effluent on the growth and biochemical composition of two marine macroalgae. *Mar. Pollut. Bull.*, 21(1): 33–38.

Venkataraman, G.S., 1972. *Algal Biofertilizers and Rice Cultivation*. Today and Tomorrow's Printers and Publishers, New Delhi.

Venkataraman, G.S. and Rajyalakhmi, B., 1972. Relative tolerance of nitrogen fixing blue-green algae to pesticides. *Ind. J. Agric. Sci.*, 42: 119–121.

Wanntorp, H. and Dyfverman, A., 1955. Identification and determination of mercury in biological materials. *Arkiv for Kemi. Viltry*, 9(2): 7.

Whitton, B.A., 1970. Toxicity of heavy metals to algae: A review. *Phykos*, 9: 116–125.

Wurtsbaugh, W.A. and Apperson, C.S., 1978. Effect of mosquito insecticides on nitrogen fixation and growth on blue-green algae in natural plankton association. *Bull. Environ. Contam. Toxicol.*, 19: 641–646.

Zingmark, R.C. and Miller, T.G., 1975. The effects of mercury on the photosynthesis and growth of estuarine and oceanic phytoplankton. *Belle W. Baruch Libr. Marine Science*, 3: 45–57.

Chapter 3

Comparative Study of Zooplankton Ecology in the Lakes of Mysore, Karnataka

B. Padmanabha and S.L. Belagali
Department of Studies in Environmental Science,
University of Mysore, Mysore – 6, Karnataka

ABSTRACT

Comparative study of Zooplankton Ecology in the four lakes of Mysore city was carried out during July 2004 to June 2005. A total of twenty-nine species of zooplankton were identified (thirteen species of rotifers, six species of cladocerans, five species of copepods and five species of ostracods). Kamana lake recorded lowest water quality index (WQI) and population density but highest species diversity of zooplankton, followed by Karanji lake and Kukkarahally lake. Highest WQI and population density but lowest species diversity was recorded in the Dalvoi lake. Increase in the water quality index means increase in the pollution load. As water quality index increases (WQI) likewise population density of zooplankton increases whereas species diversity decreases.

Keywords: Kamana, Karanji, Kukkarahally, Dalvoi, Zooplankton, Water quality index.

Introduction

Zooplankton constitutes a vital link in the food web of the freshwater ecosystem, which transfers energy from producers to consumers and fish yield is to a greater extent dependent on their abundance. Knowledge of their abundance, composition and seasonal variation, therefore, is an essential pre-requisite for any successful aquaculture programme. Zooplankton is the good indicator of water quality. Among the zooplankton, rotifers are apparently the most sensitive indicators of the water quality. Few workers (Sunkad, 2004; Sheeba and Raamanujan, 2005 and Manzer *et al.*, 2005) have attempted to study the qualitative and quantitative study of zooplankton. Water quality can be assessed either by monitoring physico-chemical parameters or analyzing inhabiting biota. Large amount of water quality data reduced to single numerical value to formulate water quality index (WQI) (Kesharwani *et al.*, 2004), which reflects the composite influence of different water quality parameters on the overall quality of water. In the present context, comparative study of zooplankton may help to assess the environmental degradation in the lakes of Mysore city. These lakes may be polluted not only by the sewage but also by the human activities in and around the lakes. The ecology of zooplankton may be discussed for the first time and there is absolutely no record of such studies from these lakes.

Materials and Methods

Four lakes namely Kamana lake, Kukkarahally lake, Karanji lake and Dalvoi lake were selected, which are located at north, west, east and south ends of Mysore city respectively. The sampling was carried out for a period of one year from June 2004 to May 2005 to calculate Water quality index. The physico-chemical parameters of water samples were estimated according to APHA (1995). The Water Quality Index was computed according to Kesharwani *et al.* (2004). The Zooplankton samples were collected by filtering 70 litres of water through the plankton net and were preserved in 4 per cent formaldehyde. The rotifers were identified with the help of keys by Battish (1992), Dhanapathi (2000) and Altaff (2004). Abundance of zooplankton was estimated by using Sedgwick rafter cell.

Results and Discussion

The water quality index of four lakes during June 2004–July 2005 shown in the Table 3.1. Dalvoi lake has highest WQI (139.81) followed by Kukkarahally lake (106.57), Karanji lake (97.00) and Kamana lake (70.68) has lowest value. Increase in the WQI indicates, increase in the pollution load. So Dalvoi lake was highly polluted than any other lakes, whereas Kamana lake has least pollution load.

Table 3.1: Water Quality Index

Parameter	*Water Quality Rating (q_i)*					*Sub Index (q_iw_i)*			
	Kamana Lake (q_i)	*Karanji Lake (q_i)*	*Kukkara Hally Lake (q_i)*	*Dalvoi Lake (q_i)*	*Unit Weight (W_i)*	*Kamana Lake (q_iw_i)*	*Karanji Lake (q_iw_i)*	*Kukkara Hally (q_iw_i)*	*Dalvoi Lake (q_iw_i)*
pH	119.00	110.00	145.00	78.70	0.20	23.80	22.00	29.00	15.74
Alkalinity	197.70	233.33	250.00	458.33	0.01	1.98	2.33	2.50	4.58
Hardness	93.33	93.33	90.00	156.67	0.005	0.42	0.47	0.45	0.78
Chloride	27.22	56.32	66.93	80.44	0.007	0.19	0.39	0.47	0.56
Calcium	30.00	32.28	25.25	50.52	0.02	0.60	0.65	0.51	1.01
Mg^{2+}	102.48	85.40	95.90	111.76	0.04	1.37	1.28	1.14	1.49
BOD	12.00	72.00	92.00	192.00	0.35	4.20	25.20	32.20	67.20
TDS	280.00	310.00	380.00	300.00	0.004	1.12	1.24	1.52	1.20
DO	106.00	114.00	110.00	135.00	0.35	37.00	39.90	38.50	47.25
				Σq_iw_i	=	70.68	93.18	106.57	139.81

Total twenty-nine species of zooplankton were present in the four lakes of Mysore city during study period, which represented thirteen species of rotifers, six species of cladocerans, five species of copepods and five species of ostracods (Table 3.1). The rotifer species were *Brachionus forficula, Brachionus calyciflorus, Brachionus falcatus, Brachionus quadridentatus, Brachionus caudatus, Brachionus diversicornis,*

Brachionus plicatilis, Plationus patulus, Keratella tropica, Keratella procurva, Keratella cochlearis, Filinia longiseta and Filinia terminalis. The cladocerans were *Moina brachiata, Moina spmosa, Moina macrocopa, Daphnia pulex, Daphnia carinata* and *Ceriodaphnia cornuta.* Copepoda species represented by *Macrocyclops distinctus, Mesocyclops leukarti, Paradiaptomus greeni, Microcyclops minutus* and *Diaphanosoma sarsi.* Represented ostracods were *Cypris protubera, Cyprinotus, Stenocypris major nudus, Strandensia elongata* and *Heterocypris dentata marginatus.*

Table 3.2: Distribution of Zooplankton in the Lakes of Mysore

Sl.No.	Zooplankton	Kamana Lake	Karanji Lake	Kukkarahally Lake	Dalvoi Lake
	Rotifera				
1.	*Brachionus forficula*	++	++	+++	+++
2.	*Brachionus calyciflorus*	++	+++	++++	+++
3.	*Brachionus falcatus*	++	–	–	–
4.	*Brachionus quadridentatus*	–	–	+++	–
5.	*Brachionus caudatus*	+	++	–	–
6.	*Brachionus diversicornis*	++	–	–	–
7.	*Brachionus plicatilis*	+	–	+++	–
8.	*Plationus patulus*	++	+	–	++
9.	*Keratella tropica*	++	++	–	–
10.	*Keratella procurva*	+	–	–	–
11.	*Keratella cochlearis*	+	+	–	–
12.	*Filinia longiseta*	–	+	+	+
13.	*Filinia terminalis*	–	–	+	–
	Cladocera				
1.	*Moina brachiata*	–	++	–	++++
2.	*Moina spinosa*	+	–	–	++++
3.	*Moina macrocopa*	–	–	++	+++
4.	*Daphnia pulex*	+	–	++	–
5.	*Daphnia carinata*	–	–	–	++
6.	*Ceriodaphnia cornuta*	++	–	–	++
	Copepoda				
1.	*Macrocyclops distinctus*	+++	–	++	–
2.	*Mesocyclops leukarti*	+++	–	–	–
3.	*Paradiaptomus greeni*	+++	++	–	–
4.	*Microcyclops minutus*	++++	+	–	+++
5.	*Diaphanosoma sarsi*	–	–	–	–
	Ostracoda				
1.	*Cypris protubera*	+	+++	–	–
2.	*Cyprinotus nudus*	+	++	++	–
3.	*Stenocypris major*	–	++	++	–
4.	*Strandensia elongata*	–	+	++	–
5.	*Heterocypris dentata marginatus*	–	+	–	++

–: Absent, +: Less than 10, ++: 11–20, +++: 21–30, ++++: 31 and above.

Kamana lake recorded lowest water quality index (70.68) and population density (55/L) but highest species diversity (19) of zooplankton (10 species of rotifers, 3 sps of cladocera, 4 sps of copepoda and 2 sps of ostracoda) (Tables 3.3 and 3.4) followed by Karanji lake (WQI-93.18, population density-90/L and species diversity-14sp = 6 sps of rotifers + 1 sps of cladocera + 2 sps of copepoda and 5 sps of ostracoda) and Kukkarahally lake (WQI-106.57, population density-120/L, species diversity-12 sps = 6 sps of rotifers + 2 sps of cladocera + 1 sps of copepoda and 3 sps of ostracoda). Highest WQI (139.81) and population density (165/L) but lowest species diversity was recorded in the Dalvoi lake (11 sps = 4 sps of rotifers + 5 sps of cladocera + 1 sps of copepoda and 1 sps of ostracoda).

Table 3.3: Composition of Zooplankton in the Four Lakes

Sl.No.	*Zooplankton*	*Kamana Lake*	*Karanji Lake*	*Kukkarahally Lake*	*Dalvoi Lake*
1.	Rotifera (13)	10	06	06	04
2.	Cladocera (6)	03	01	02	05
3.	Copepoda (5)	04	02	01	01
4.	Ostracoda (5)	02	05	03	01
	Total 29	**19**	**14**	**12**	**11**

Table 3.4: Water Quality Index (WQI), Abundance and Species Diversity of Zooplankton in the Four Lakes of Mysore City

Sl.No.		*Kamana Lake*	*Karanji Lake*	*Kukkarahally Lake*	*Dalvoi Lake*
1.	Water Quality Index	70.68	93.18	106.57	139.81
2.	Zooplankton (No/L)	55	90	120	165
3.	Species diversity	19	14	12	11

According to data depicted in Table 3.3, Kamana lake represented by ten species of rotifers (77 per cent) out of total thirteen species recorded and four species out of five species of copepods (80 per cent). Karanji lake recorded five species of ostracods out of total five species (100 per cent) and Dalvoi lake was recorded five species of cladocerans (83 per cent) out of total six species reported. This indicates that Kamana lake was dominated by copepoda and rotifers, Karanji lake was dominated by ostracods and Dalvoi lake was dominated by cladocerans. All the four groups (rotifers, copepods, cladocerans and ostracods) more or less evenly distributed in the Kukkarahally lake.

Water quality index (WQI) is positively correlated with zooplankton abundance (r = 0.996) but negatively correlated with species diversity (r = –0.904). This indicates that as water quality index increases, population density also increases but species diversity decreases. In other words, increase in the water quality index means increase in the pollution load, which leads to congenial environment for only to few species with higher abundance.

References

Altaff, K., 2004. Key to common rotifers. In: *A Manual of Zooplankton for the National Workshop on Zooplankton*, Chennai, pp. 19–47.

APHA, 1995. *Standard Methods for the Examination of Water and Wastewater*. Washington D.C., 19th edn.

Battish, S.K., 1992. *Freshwater Zooplankton of India*. Oxford and IBH Publ. Co. Pvt. Ltd., 69–114 pp.

Dhanapathi, M.V.S.S.S., 2000. Taxonomic notes on rotifers from India. *IAAB*, 15–97 pp.

Sadhana Kesharwani, Mandoli A.K. and Dube, K.K., 2004. Determination of water quality index (WQI) of Amkhera pond of Jabalpur City (M.P). *Natl. J. Life Sciences*, 1(1): 61–66.

Manzer, M.B.H., Nehal, M., Rahmatullah, M. and Bazmi, H., 2005. A comparative study of population kinetics and seasonal fluctuation of zooplankton in two diverse ponds of north Bihar. *Nat. Environ. and Poll. Tech.*, 4(1): 23–26.

Sheeba, S. and Ramanujan, N., 2005. Qualitative and quantitative study of zooplankton in Ithikkara river, Kerala. *Poll. Res.*, 24(1): 119–122.

Sunkad, B.N., 2004. Diversity of zooplankton in Rakshakoppa reservoir of Belgum of North Karnataka, India. *Indian J. Environ. and Ecoplan.*, 8(2): 399–404.

Chapter 4

Effect of Nitrogen on Growth, Nitrogen Fixing Activity and Ammonia Excretion of Salt Tolerant Cyanobacteria

P. Amsaveni[1]* *and S. Kannaiyan*[2]

[1]*Dean, Faculty of Biotechnology, International University College of Technology Twintech, Bandar Sri Damansara, 52200, Kula Lumpur, Malaysia,*

[2]*Tamil Nadu Agricultural University, Coimbatore, Tamil Nadu – 641 003, India*

ABSTRACT

The effect of nitrogen on growth, nitrogen fixing activity and ammonia excretion of three salt tolerant cyanobacteria *viz., Anabaena*–TR-52-ST1, *Nostoc*–BS-60-ST2 and *Westiellopsis*–TR-5-ST3 was estimated. Nitrogen sources *viz.*, urea and ammonium sulphate [$(NH_4)_2SO_4$] at 10 and 20 ppm each were employed in this investigation. Among the three salt tolerant cyanobacterial cultures *Westiellopsis*–TR-5-ST-3 has registered higher growth in the presence of ammonium sulphate than in urea and control. Nitrogenase enzyme activity of all the algal cultures were inhibited by the nitrogen sources.

Keywords: Cyanobacteria, saline, sodic, Anabaena, Nostoc, Westiellopsis.

Introduction

Cyanobacteria utilize a number of nutrients *viz.*, nitrogen, phosphorus, potassium, carbon and micronutrients to support the growth and nitrogen fixation. Most of the cyanobacteria are able to

* Corresponding Author: Phone No. 006 036 286 1200; Fax No. 006 036 2742462; HP No. 006 0163 276132; E-mail: amsaashi@yahoo.com.

utilize nitrate or ammonia as the sole source of nitrogen for growth (Fogg *et al.*, 1973). Cyanobacteria whether unicellular or filamentous, nitrogen fixing or not when grown in nitrogen deficient medium resulted in loss of phycobilin protein pigments (Allen and Smith, 1969). The source of nitrogen in the form of nitrate acts as an inducer for nitrate reductase in cyanobacteria (Hattori, 1970 and Herrero *et al.*, 1981). The enzyme nitrate reductase activity in cyanobacteria was reported to vary in response to nitrogen source in growth medium (Bagchi *et al.*, 1985). Fertilizer nitrogen urea was found to stimulate and increase the growth of salt tolerant cyanobacteria. Nitrate uptake by the cyanobacteria has been demonstrated by Singh (1992). The enzyme nitrogenase is suppressed by the nitrogenous compounds such as ammonium, glutamine and nitrite (Reich *et al.*, 1986; Martin-Nieto *et al.*, 1991; Troshina *et al.*, 1992 and Singh and Dash, 1994). This research was conducted with the objective of investigating the effect of different nitrogen sources on growth, ammonia excretion and nitrogen fixing activity of salt tolerant cyanobacteria.

Materials and Methods

Cyanobacterial Cultures

Three number of salt tolerant cyanobacterial cultures *viz.*, *Anabaena*–TR-52-ST1, *Nostoc*–BS-60-ST2 and *Westiellopsis*–TR-5-ST3 isolated from saline and sodic soils of Tamil Nadu were employed in this study.

Nitrogen Sources

Nitrogen sources *viz.*, urea and ammonium sulphate at two different concentration of 10 and 20 ppm each were prepared in 250 ml conical flask containing 100 ml of N-free BG-11 liquid medium (Stanier *et al.*, 1971).

Growth and Ammonia Excretion of Cyanobacteria

Growth of the cyanobacterial cultures were measured spectrophotometrically at 750 nm and expressed as OD values. Ammonia excretion of the cyanobacterial cultures was estimated by the method developed by Solorzano (1969).

Nitrogen Fixing Activity

The acetylene reduction assay (ARA) for nitrogenase activity was carried out as per the method of Hardy *et al.* (1968). Nitrogenase activity (ARA) was expressed as nmole of C_2H_4 produced g^{-1} dry wt h^{-1}.

Results and Discussion

All the three cyanobacterial cultures *viz.*, *Anabaena*–TR-52-ST1, *Nostoc*–BS-60-ST2 and *Westiellopsis*–TR-5-ST3 grown with nitrogen sources have recorded higher growth rate than control (Table 4.1). Among the three salt tolerant algal cultures *Westiellopsis*–TR-5-ST-3 has registered relatively higher growth rate when grown in the presence of ammonium sulphate than in urea and control. But the difference is not significantly high. Fogg *et al.* (1973) have demonstrated that most of the cyanobacteria were able to utilize nitrate and ammonia as the sole source of nitrogen for their growth and biomass production. Ohmori *et al.* (1977) have investigated the role of ammonia and nitrate as nitrogen source in the growth medium and found that cyanobacteria first utilized ammonium and later utilized nitrate. Ammonium and glutamine sources increased the growth and metabolic activity of *Phormidium uncinatum* (Bagchi and Kleiner, 1991). Ammonia excretion was highly inhibited by the presence of nitrogen sources at all concentrations (Table 4.2). The inhibition exhibited by both nitrogen sources at 20 ppm was higher than at 10 ppm.

Table 4.1: Effect of Nitrogen Sources (Urea and Ammonium Sulphate) on Growth of Cyanobacterial Cultures

Cyanobacteria	*Growth (O.D at 750 nm)*			
	7th day	*14th day*	*21st day*	*28th day*
Control				
Anabaena–TR-52-ST1	0.18	0.30	0.36	0.49
Nostoc–BS-60-ST2	0.16	0.26	0.33	0.43
Westiellopsis–TR-5-ST3	0.29	0.34	0.47	0.56
Urea 10 ppm				
Anabaena–TR-52-ST1	0.19	0.32	0.46	0.71
Nostoc–BS-60-ST2	0.18	0.28	0.37	0.61
Westiellopsis–TR-5-ST3	0.30	0.46	0.50	0.72
Urea 20 PPM				
Anabaena–TR-52-ST1	0.19	0.36	0.49	0.72
Nostoc–BS-60-ST2	0.17	0.30	0.39	0.61
Westiellopsis–TR-5-ST3	0.38	0.48	0.52	0.86
$(NH_4)_2$ SO_4 10 PPM				
Anabaena–TR-52-ST1	0.20	0.38	0.47	0.74
Nostoc–BS-60-ST2	0.18	0.26	0.44	0.69
Westiellopsis–TR-5-ST3	0.34	0.51	0.56	0.89
$(NH_4)_2$ SO_4 20 PPM				
Anabaena–TR-52-ST1	0.23	0.41	0.55	0.77
Nostoc–BS-60-ST2	0.22	0.33	0.48	0.76
Westiellopsis–TR-5-ST3	0.38	0.68	0.66	1.08

	SE	*SED*	*CD*
Concentration	0.108	0.153	0.304
Days	0.097	0.137	0.272
Cultures	0.084	0.119	0.235

Both the nitrogen sources (urea and ammonium sulphate) have highly inhibited the nitrogenase enzyme activity in salt tolerant cyanobacterial culture (Table 4.3). Among the three algal culture *Nostoc*–BS-60-ST2 was highly sensitive to nitrogen sources and nitrogenase activity could not be detected at both 10 and 20 ppm of ammonium sulphate. A rapid inactivation of nitrogenase enzyme in *Anabaena variabilis* by incorporation of ammonium into the culture medium was reported by Reich *et al.* (1986). It is interesting to note that *Anabaena*–TR-52-ST1 could tolerate nitrogen sources and produced 190.44 nmoles of ethylene hour^{-1} g dry wt^{-1}. But it is nearly 7 times lesser than that of control which has not received nitrogen source. Algal growth is influenced by the type and mode of application of inorganic fertilizer nitrogen, phosphorus and potassium (Roger *et al.*, 1980). Stewart (1980) has demonstrated that synthesis of nitrogenase enzyme was inhibited when the free living cyanobacteria were exposed to combined nitrogen. Bohme (1986) stated that the addition of nitrite into the culture medium of *Anabaena variablis* have inhibited the nitrogenase enzyme activity. Sarma and Khatta (1994) have reported that the addition of nitrogen into the medium inhibited the heterocyst development and nitrogen fixation in *Anabaena torulosa.* The present results lead to suggest the deleterious effect of fertilizer nitrogen on nitrogen fixing activity in cyanobacteria. The results are in conformity with the above findings.

Table 4.2: Effect of Nitrogen Sources (Urea and Ammonium Sulphate) on Ammonia Excretion by the Salt Tolerant Cyanobacteria

Cyanobacteria	*Ammonia Excretion (nmoles/ml)*			
	7th day	*14th day*	*21st day*	*28th day*
Control				
Anabaena–TR-52-ST1	36.50	108.49	82.31	41.15
Nostoc–BS-60-ST2	24.86	92.25	81.46	40.43
Westiellopsis–TR-5-ST3	39.75	111.34	84.73	39.54
Urea 10 ppm				
Anabaena–TR-52-ST1	14.90	23.31	25.95	19.32
Nostoc–BS-60-ST2	14.93	32.15	20.32	20.95
Westiellopsis–TR-5-ST3	17.88	50.89	28.56	15.30
Urea 20 PPM				
Anabaena–TR-52-ST1	10.51	20.51	10.43	12.12
Nostoc–BS-60-ST2	12.35	20.69	10.75	11.30
Westiellopsis–TR-5-ST3	16.89	21.62	18.13	9.38
$(NH_4)_2SO_4$ 10 PPM				
Anabaena–TR-52-ST1	24.92	14.19	16.12	15.93
Nostoc–BS-60-ST2	14.26	10.03	21.30	12.35
Westiellopsis–TR-5-ST3	26.85	16.30	15.38	12.33
$(NH_4)_2SO_4$ 20 PPM				
Anabaena–TR-52-ST1	17.35	15.59	15.91	13.19
Nostoc–BS-60-ST2	12.21	11.37	12.31	11.32
Westiellopsis–TR-5-ST3	19.63	15.15	14.80	10.30

	SE	*SED*	*CD*
Concentration	0.227	0.331	0.636
Days	0.203	0.287	0.568
Cultures	0.176	0.248	0.492

Table 4.3: Effect of Nitrogen Sources [Urea and $(NH_4) SO_4$] on Nitrogenase Activity of Salt Tolerant Cyanobacteria

Cyanobacteria	*Nitrogenase Activity**				
	Control	*Urea (ppm)*		*$(NH_4) SO_4$ (ppm)*	
		10.0	*20.0*	*10.0*	*20.0*
Anabaena–TR-52-ST1	1273.11	190.44	76.17	110.65	93.87
Nostoc–BS-60-ST2	1286.20	95.28	63.02	0.0	0.0
Westiellopsis–TR-5-ST3	1638.98	72.70	55.25	79.34	84.02

	SE	*SED*	*CD*
Concentration	0.869	1.230	2.519
Cultures	0.673	0.953	1.951
Interaction	1.506	2.130	4.364

*n moles ethylene produced $hour^{-1}$ g dry $weight^{-1}$.

References

Allen, M. and A.J. Smith, 1969. Nitrogen chlorosis in blue-green algae. *Arch. Microbiol.*, 69: 14–120.

Bagchi, S.N., R. Sharma and H.N. Singh, 1985. Inorganic nitrogen control of growth, chlorophyll and protein level in *cyanobacterium, Nostoc Muscorum. J. Plant Physiol.*, 121: 73–81.

Bagchi, S.N. and D. Kleiner, 1991. Interrelations between hydrogen peroxide, ammonia, glutamine, hydroxylamine and nitrite metabolism in the *cyanobacterium, Phormidium unicinatum. Arch. Microbiol.*, 156: 367–369.

Bohme, R., 1986. Inhibition of nitrogen fixation by nitrite in *Anabaene variabilis. Arch. Microbiol.*, 146: 99–101.

Fogg, G.E., W.D.P. Stewart, P. Fay and A.E. Walsby, 1973. *The Blue-green Algae*. Academic Press Ltd., London and New York.

Hardy, R.W.F., R.D. Holsten, E.K. Jackson and R.C. Burns, 1968. The acetylene reduction assays for N_2-fixation: Laboratory and field evaluation. *Plant Physiol.*, 43: 1185–1207.

Hattori, A., 1970. Solubilization of nitrate reductase from the blue-green alga, *Anabaena cylindrica. Pl. Cell. Physiol.*, 11: 975–978.

Herrero, A., E. Flores and M.G. Guerrero, 1981. Regulation of nitrate reductase levels in the *cyanobacteria, Anacystis nidulans, Anabaena* sp. strain 7119 and *Nostoc* sp. strain 6719. *J. Bacteriol.*, 145: 175–180.

Martin-Nieto, J., A. Herrero and E. Flores, 1991. Control of nitrogenase mRNA levels by product of nitrate assimilation in the *cyanobacterium, Anabaena* sp. strain PCC 7120. *Plant Physiol.*, 97: 825–828.

Ohmori, M., K. Ohmori and H. Strotmann, 1977. Inhibition of nitrate uptake by ammonia in a blue-green alga, *Anabarna cylindrica. Arch Microbiol.*, 114: 255–262.

Reich, S., H. Almon and P. Boger, 1986. Short term effect of ammonia on nitrogenase activity of *Anabaena variabilis. FEMS Microbiol., Letts.*, 14: 51–56.

Roger, P.A. and S.A. Kulasooriya, 1980. *Blue-green Algae and Rice*. Intel. Rice Res. Inst., Los Banos, Phillippines, p. 112.

Roger, P.A., S.A. Kulasooriya, A.C. Tird and E.T. Craswell, 1980. Deep placement method of nitrogen fertilizer application, compatible with algal nitrogen fixation in wetland rice soils. *Plant and Soil*, 57: 137–142.

Sarma, T.A. and J.I.S Khattar, 1994. Photoheterotrophic and chemoheterotrophic dinitrogen fixation and nitrate utilization by the cyanobacterium, *Anabaene torulosa. Folia Microbiogica*, 39(5): 404–408.

Singh, P.K. and H.P. Dash, 1994. Blue-green algal growth and N_2-fixation at nitrogen and phosphorus fertilizer in rice field. In: *Recent Advances in Phycology*, (Eds.) K. Kashyap and H.D. Kumar. Rastogi Publications, Varanasi, India, pp. 203–210.

Singh, S., 1992. Regulation of nitrite reductase cellular levels in the cyanobiont, *Nostoc* ANTH. *Biochemie physiologie des dfianzen*, 188(4): 241–246.

Solorzano, L., 1969. Determination of ammonia in natural waters by the phenol hypochlorite method. *Limnol. Oceanogr.*, 14: 799–801.

Stainer, R.Y., R. Kunisawa, M. Mandal and G. Cohen-Bazier, 1971. Purification and properties of unicellular blue green algae Corder (roococcales), *Bacteriol.*, 35: 171–305.

Stewart, W.D.P., 1980. Some aspects of structure and function in N_2-fixing Cyanobacteria. *Annu. Rev. Microbiol.*, 34: 497–536.

Troshina, O.Yu., A.F. Yakunin and J.N. Gogotov, 1992. Growth and activity of nitrogen metabolism enzymes in free living cultures of *Anabaena azollae* and its mutants resistant to ethylene diamine. *Microbiology*, New York, 61(6): 706–710.

Chapter 5

Study of the Effects of Extracts of *Ocimum sanctum* (Basil Herb) on Phlebotomine Sandflies (Diptera : Psychodidae) in Bihar, India

Kundan Lal*, P. Nath and Ragini Mishra

Department of Zoology, BN College, Patna University, Patna – 800 005

ABSTRACT

Phlebotomine species are important members of order Diptera which have very broad effects on public health. Especially in case of Bihar (India) the Kala-azar *i.e.* visceral leishmaniasis is spread by sandflies (*Phlebotomus argentipes*) which is a very dreadful disease affecting very large population. Many steps have been taken by the Govt. and scientists to control this harmful vector by mass spraying synthetic organochlorines like DDT, BHC, Malathion etc. but all such efforts have ended into a big failure. These synthetic insecticides are not very effective as not only these vectors developed a high degree of resistance against them but these are also creating very drastic effects on the ecosystem and environment. Since last few decades scientists are very keen to develop eco-friendly and effective control measures against the insect vectors. Plant derived extracts have shown many beneficial results in this field. Oil extracted from seeds of the *Ocimum sanctum* and other extracts like N-hexane, alcoholic and aqueous extracts of leaves, and fruits of this plant have shown very effective results dealing with control of this insect vector at multiple stages of their life cycle. The Basil plant extracts have shown very good results as larvicide and adult repellant.

Keywords: *Ocimum sanctum, Plant-extract, Phlebotomus argentipes, Herbal control, N-hexane, Glycerin.*

* Corresponding Author: E-mail: kunaatrey@graffiti net.

Introduction

To control the drastic effects of insect vectors in mass spreading of such communicable diseases, Health Ministry of Government of India initiated the large scale DDT spraying program in the affected endemic areas under "National Malaria Control Program". Since 1990–91 special spray program was initiated by Government of India for Kala-azar Control. But the sprays of such organochlorine insecticides are not giving better results now, although in initial periods this was the effective measure. The reason behind it is the resistance power developed in the insect vector against these chemical insecticides. According to the project handled by Environmental Health Project, Office of Health, Infectious Diseases and Nutrition, Bureau of Global Health, Washington DS, in South Asian Countries, it is well proved that almost in all the areas, these insect vectors have developed a good degree of resistance against DDT, BHC, Malathion and other synthetic insecticides. In Bihar the scene is not different and mass spraying of such insecticide may decrease the population density of those vectors temporarily but never eradicate their existence. However, continued spraying of DDT has resulted in the precipitation of DDT resistance in *P. argentipes* in various districts of KA endemic areas in Bihar (Mukhopadhyay *et al.*, 1990; NMEP, WHO, Kaul, 1993; Pillai MKK, 1996). There is some susceptibility of indoor spraying of Malathion and some other insecticides in Bihar, but the environmental hazards are the major problem. But since last few decades the plants extracts have shown very positive results as insecticides. They offer a safer alternative to synthetic chemicals and can be obtained by individuals and communities very easily at very low cost. Basil herb is very frequent in India and it is meant a poisonous plant because of very high concentration of narcotic materials.

Materials and Methods

Sandflies were collected from five districts in North Bihar where Visceral Leishmaniasis is very endemic *viz.* Vaishali, Muzafferpur, Saran, Darbhanga and Patna. Sandflies were caught by using suction-tube, torches and test were from human dwellings, cattle sheds and mixed dwellings (S. Kesari *et al.*, 1999). The collection of Sandflies were repeated in first week of every month except December, January and February when the appearance of adults is very low and not sufficient to maintain the colony in laboratory. The collected sandflies were transferred into the special container made of plastic initially and then transferred into a wooden cage of size 2′ × 2′ × 2 and covered by fine Aurgandy cloth (K. Kishore, 1988). This cage has a cloth bound inlet through which hand may be entered for picking, dropping and feeding the adult sandflies. In every cage both male and female sandflies were put so that random mating may possible. Before egg laying at least three blood meal were essential for *P. argentipes* and this was maintained by entering hand to the cage. After three blood meals the female sandflies were transferred into the special type plastic container which bottom was perforated and covered with Plaster of Paris layering so that moisture may enter through the bottom.

Female sandflies lay eggs on the surface and cracks of Plaster of Paris. After flushing with water, eggs were collected in another container of similar type. As all the stages of sandflies need very high moisture, so to maintain the optimum moisture level all containers were kept on wet cotton towel so moisture easily be transferred into the internal environment of the container. Gradually first instar larvae, second, third and fourth instars of larvae and pupae were obtained respectively and all these stages were transferred separately in different containers. At last adults emerged from pupae and the number of adults were counted and shifted to the cage.

The larvae were fed with special mixture made by mixing sand, rabbit feaces, yeast powder in fixed quantity and some blood was also added to it. The eggs were divided in five groups in which one group were kept in normal environment and other groups were sprayed with oil extract, aqueous extract, benzene extract and alcoholic extract respectively. Similarly all the instars of larvae and pupae were also treated with different extracts exactly in same manner. The temperature was maintained at 30–37° C and humidity was maintained above 80 per cent always.

For preparation of the plant extracts distilled water was taken for preparation of aqueous extract, pure glycerin was taken for preparation of oil extract, N-hexane was taken for preparation of N-hexane extract and alcohol was used for preparation of alcohol extract. Percentage value was maintained as w/W formula *i.e*

$$w/W = \frac{\text{Dry weight of plant leaves and stems}}{\text{Weight of the selected extract medium}}$$

So the 10 per cent extract concentration means 10 gm. of dry plant parts was mixed with 90 gm. of the selected medium (distilled water or glycerin or N-hexane or alcohol). The mixture was then centrifuged for 5 min. and the supernatant was then collected for spraying on the different stages of life-cycles of the species population.

Finally all the observations regarding the stages of sandflies were counted and summarized with statistical analysis and then the final conclusions were plotted and results were analyzed.

Results

Effects of Different Types of Extracts of Plant *Ocimum sanctum* on Sandfly (*P. argentipes*)

Note: For all the following control experiments the number of initial population of all the stages = 500

The Chi-square test of the results obtained independently for different extracts with respect of results obtained without using any extract are showing following trends:

1. For the effect on eggs all the values of X^2 (Chi square) are lying below 3.841 which is the critical value on 5 per cent on degree of independence 1. So we can say that any types of the used extracts are not significant for the effect against eggs of sandflies.
2. Similarly the value of X^2 is below 3.841 for the 10 per cent concentration for the effect on the first larval instars but the value of X^2 is more than 3.841 for the concentrations 20 per cent, 40 per cent and 60 per cent for the first instars *i.e.* these extracts on those concentration show significant effects against the larvae.
3. For second to fourth larval instars, all the critical values of the X^2 lies above 3.841 which means that almost all the extracts of Basil have shown significant effects against the larvae of sandflies.
4. The t-test for significance is showing that the value of significance on d.f. (degree of independence) 4 is maximum for oil extract and minimum for the aqueous extract but all values are above 2.132 which proves that all extracts have their positive significance and the degree of affectivity is oil extract > N-hexane extract > alcohol extract > aqueous extract.

Table 5.1: Effects on Eggs

Types of Extracts	*Conc.**	*F1*	*Conc.**	*F1*	*Conc.**	*F1*	*Conc.**	*F1*	*Conc.**	*F1*
No extract		328		328		328		328		328
Aqueous extract	10%	282	20%	280	40%	259	50%	251	60%	243
Oil extract	10%	98	20%	91	40%	76	50%	68	60%	57
N-hexane extract	10%	171	20%	164	40%	152	50%	150	60%	147
Alcohol extract	10%	206	20%	201	40%	198	50%	201	60%	187

F1: No. of alive 1st larval instars; Conc.*: Concentration of extracts used.

Table 5.2: Effects on 1st Larval Instars

Types of Extracts	*Conc.**	*F2*	*Conc.**	*F2*	*Conc.**	*F2*	*Conc.**	*F2*	*Conc.**	*F2*
No extract		341		341		341		341		341
Aqueous extract	10%	102	20%	84	40%	57	50%	45	60%	42
Oil extract	10%	87	20%	71	40%	41	50%	29	60%	24
N-hexane extract	10%	76	20%	45	40%	35	50%	27	60%	21
Alcohol extract	10%	89	20%	81	40%	59	50%	41	60%	37

F2: No. of alive 2nd larval instars; Conc.*: Concentration of extracts used.

Table 5.3: Effects on 2nd Larval Instars

Types of Extracts	*Conc.**	*F3*	*Conc.**	*F3*	*Conc.**	*F3*	*Conc.**	*F3*	*Conc.**	*F3*
No extract		349		349		349		349		349
Aqueous extract	10%	58	20%	37	40%	13	50%	00	60%	00
Oil extract	10%	41	20%	26	40%	00	50%	00	60%	00
N-hexane extract	10%	46	20%	29	40%	09	50%	00	60%	00
Alcohol extract	10%	52	20%	32	40%	11	50%	02	60%	03

F3: No. of alive 3rd larval instars; Conc.*: Concentration of extracts used.

Table 5.4: Effects on 3rd Larval Instars

Types of Extracts	*Conc.**	*F4*	*Conc.**	*F4*	*Conc.**	*F4*	*Conc.**	*F4*	*Conc.**	*F4*
No extract		342		342		342		342		342
Aqueous extract	10%	38	20%	20	40%	14	50%	09	60%	06
Oil extract	10%	29	20%	00	40%	00	50%	00	60%	00
N-hexane extract	10%	32	20%	18	40%	04	50%	00	60%	00
Alcohol extract	10%	36	20%	21	40%	02	50%	00	60%	00

F4: No. of alive 4th larval instars; Conc.*: Concentration of extracts used.

Table 5.5: Effects on 4th Larval Instars

Type of Extracts	Conc.*	F5	Conc.*	F5	Conc.*	F5	Conc.*	F5	Conc.*	F5
No extract		362		362		362		362		362
Aqueous extract	10%	84	20%	71	40%	55	50%	31	60%	28
Oil extract	10%	51	20%	39	40%	22	50%	15	60%	08
N-hexane extract	10%	56	20%	44	40%	27	50%	22	60%	19
Alcohol extract	10%	69	20%	48	40%	39	50%	34	60%	29

F5: No. of alive pupae; Conc.*: Concentration of extracts used.

Table 5.6: Effects of Different Extracts of Plant *Ocimum sanctum* on Adult Sandflies

Extracts Used	Effects on the Behaviour Observed after Spraying Extracts				
	Hopping	Biting	Egg laying	Mating	Deaths after oviposition
Oil extract	Negative++	Negative+++	Negative+++	No effect	99.76%
Aqueous extract	Negative	Less Negative	No effect	No effect	91.04% (approx.)
N-hexane extract	Negative+++	Negative+++	Negative+	No effect	99.20%
Alcohol extract	Negative+	Negative	Negative+	No effect	92.78% (approx.)

+: Slowed activity.

As per statistical analysis of my experiments and findings shown in the Tables 5.1 to 5.6, I have reached on following conclusions:

1. The extracts of Basil are not very much effective against the eggs. Yet some negative deflections were obtained in these but except the Oil extracts I have not found any noticeable results. Probably any type of Basil extracts is not able to penetrate the covering of the eggs but oil extracts may able to cut the oxygen supply of eggs.
2. These extracts are almost equally affective against the first larval instars and the rate of affectivity was somewhat moderate on the first instars. On 10 per cent and 20 per cent of the concentration the effects were not very much satisfactory but on 40 per cent and 50 per cent of concentration the effects of the extracts were seen extraordinarily well.
3. The effects of the extracts of Basil seemed very much effective against the 1st and 4th larval instars. Even the 10–20 per cent of the extract concentrations was observed well effective against these instars, but the results of the higher concentrations were proved very much effective against these instars. Among all the extracts used, I have found the N-hexane extract and Tulsi (Basil) oil were found extremely effective against the larval instars but the aqueous extract and alcohol extract were also proved very effective.
4. But as in the observation tables, it seemed very clear that the Basil extracts were found most effective against the 2nd and 3rd larval instars. Yet again the N-hexane and oil was observed most effective but all other extracts were also found extremely effective against these instars.
5. The died larvae when examined under the high microscope, it was observed that the larval chitinisation were dissolved and destroyed to some extent due to effect of Basil extracts.
6. After the spray of its extracts on the adult sandflies (especially the N-hexane extract), it was observed that the hopping movement and active biting properties of sandflies were slowed

down too much and these usually try to rest on any corner of the bottom of the cage. In a big cage when at any corner the extracts were sprayed then the adult sandflies were try to avoid that site, so it is well clear that the Basil extracts works as a very effective repellant against the adult sandflies.

7. Almost 80 per cent of female sandflies usually die out after oviposition normally but the Basil herb extracts enhance this rate up to 94–99 per cent.
8. As per the collective results it is well proved that the Basil extracts are extremely effective larvaecides and in some extent work as repellant for the adult sandflies.
9. The average effects of the *Ocimum sanctum* extracts were proved as much as 80 per cent effective against the larvae of sandflies and in higher concentrations, its extracts have been proved almost hundred per cent fatal against few larval instars of sandflies.

Discussion

As per above findings following conclusions can be deducted:

1. All types of extracts of the plant *Ocimum sanctum* have almost no adverse effects on the eggs of the Sandflies (*Phlebotomus argentipes*).
2. These extracts are well effective against 1st instar larvae but this is somewhat lower and the degree is somewhat less than the effects over 2nd and 3rd instars.
3. These extracts are too much effective against the 2nd and 3rd instars of sandfly larvae.
4. These are not effective against the pupae of sandflies and it was observed that in the adverse conditions maximum of the pupae remain in prolonged diapaused state.
5. Among the chosen four types of Basil extracts, all the used types were proved very effective but the N-hexane extracts and oil extracts have been proved more effective.
6. The Basil extracts show high narcotic value, so they have shown to alter the hopping movements of the sandflies. The sandflies are not good fliers and the actual movements of them are only hopping. These extracts make them sluggish and so lowering the speed of hopping movements.
7. The sluggish behaviour of the adult female sandflies have been also observed in biting actions and the vigorously sprayed adults were seen mostly hopping around the mammalian body but the rate of blood-sucking was very low.
8. Even after vigorous spraying no any typical changes were observed in egg laying and mating behaviour. This was due to the fact that the spray only affected temporarily on the adults and as soon as the effects were minimized the sandflies returned to their normal activities. So it was very clear that all the types of extracts of *Ocimum sanctum* are just worked as repellent for the adults and these extracts were note proved fatal for the adult *P. argentipes*.

Final Conclusion

The extracts of the plant *Ocimum sanctum* have very effective larvicidal action and for adult sandflies these work as very good and effective repellant. These extracts are very eco-friendly and will never create any types of environmental hazards. The Basil plants are very frequent in Bihar and almost all over India because its holy values and medicinal use in Ayurveda, and due to its mass availability the estimated cost will be very less and even poor people will also be able to use those measures.

Acknowledgements

We are heartily thankful to Council for Scientific and Industrial Research (CSIR) and UGC for granting us fellowships which helped us financially to work in this field. We are also very much thankful to Rajendra Memorial Research Institute (RMRI), ICMR, Gulzarbagh, Patna for providing us some technical guidance.

References

Campbell, F.L., Sullivan, W.W. and Smith, L.N., 1993. The relative toxicity of nicotine, anabasine, methyl anabasine and lupinine for culicine mosquito larvae. *J. Econ. Entomol.*, 26: 500.

Kaul, S.M., R.K Das, Shivraj, N.B.L. Saxena and M.V.V.L. Narsimham, 1993. Entomological monitoring of Kala-azar control in Bihar state India: Observations in Vaishali and Patna Districts. *J. Commun. Dis.*, 25(3): 101–106.

K. Kishore, 1988. Entomological Studies in Kala-azar. In: *National Seminar on Kala-azar (Report)*. December, 1–2, 1988, pp. 92–99.

Mukhopadhyay, A.K, N.B.L. Saxena and M.V.V.L. Narsimham, 1990. Susceptibility of *Phlebotomus argentipes* to DDT in some Kala-azar endemic areas of Bihar (India). *Indian J. Med. Res.*, 91: 458–460.

Palsson, K. and Jaenson, T.G. 1999. Plant products used as mosquito repellent in Guinea Bissau, West Africa. *Acta. Trop.*, 72: 39.

Pillai, M.K.K., 1996. Vector resistance to insecticides. In: *Proc. Nat. Acad. Sci.*, India, 68(B) Special Issue: 77–97.

S. Kesari, K. Kishore, A. Palit, V. Kumar, M.S. Roy, S. Sivakumar and S.K. Kar, 2000. An entomological field evaluation of larval biology of Sandfly in Kala-azar endemic focus of Bihar: Exploration of larval control tool. *J. Commun. Dis.*, 32(4): 284–288.

Chapter 6

Performance of *Mentha piperita* against *T. castaneum* Herbst (Coleoptera : Tenebrionidae)

Sudhakar Gupta

Department of Animal Science, M.J.P. Rohilkhand University, Bareilly – 243 006, Uttar Pradesh, India

ABSTRACT

The mortality rate of eggs was recorded 22.00 and 22.34 per cent in 1 and 2 per cent diet respectively, as compared to control (15.89 per cent). The rate of larval mortality increased from 6.0 per cent (L_1) to 12.19 per cent (L_3) in 2 per cent diet which was identical to those reared on diet mixed 1 per cent *Mentha piperita* leaf powder. In case of 2 per cent mixture of *Mentha piperita* leaf powder, the highest mortality was recorded in L_3 stage 12.19 per cent. The maximum mortality of pupae was recorded 2.54 per cent in 1 per cent treatment. The fresh biomass of untreated larvae of *T. casteneum* increased rapidly from L_1 = 0.06 to L_6 = 1.9 mg and declined in pupae (1.8 mg). Among the larvae given access to 2 per cent *Mentha piperita* leaf powder, the fresh weights of individuals from L_4 to adult stage (L_4 = 0.91, L_5 = 1.24, L_6 = 1.47, pupae = 1.37 and adult = 1.27 mg) were significantly lower (P/_ 0.01) than those of controls (L_4 = 1.27, L_5 = 1.64, L_6 = 1.91, pupae = 1.80, and adult = 1.70 mg). The dry weights of some of the developmental stages treated with 2 per cent leaf powder of *Mentha piperita* (L_5 = 0.46, pupae = 0.51 and adult = 0.44 mg) were significantly lower (P/_ 0.05 and P/_ 0.01) than those of controls. The absolute growth rates at L_4 stage (control = 0.072, 1 per cent treatment = 0.069, and 2 per cent treatment = 0.053 mg/day) were maximal but afterward it declined negatively is pupae (control = – 0.010, 1 per cent treatment = – 0.011 and 2 per cent treatment = – 0.0216 mg/day) and adults (controls = – 0.011, 1 per cent treatment = – 0.0082 and 2 per cent, treatment = – 0.014 mg/day). The egg production by untreated adults of *T. castaneum* started on the 4th day after emergence of adults and an average of 6.6 eggs were laid on this day. Subsequently, the eggs laid were recorded on the 6th day (19.3), 8th day (23.3), 10th day (24.0), 12th day (26.0), 14th day (26.6), 16th day (27.6), 18th day (26.6) and 20th day (26.0). At 2 per cent treatment, however, the mean daily egg production was lower (4th day = 03.0, 6th day = 8.0, 8th day = 12.0, 10th day = 13.6, 12th day = 14.3, 14th day = 17.0, 16th day = 14.6, 18th day = 15.3 and 20th day = 14.0) than those of controls. These observations indicated that dietary incorporation of *Mentha piperita* leaf powder reduced the number of eggs to 81.09 per cent at 1 per cent treatments and 54.28 per cent at 2 per cent treatment in comparison to those of controls on the 20th day.

Keywords: *T. castaneum, Mentha piperita, Growth-rate, Oviposition.*

Introduction

T. castaneum (Herbst) is the most common pest of stored grain which causes severe losses in almost all parts of the world. The biology and certain aspects of the growth-rates of this red flour beetle on its natural diets have been variously reported by Klekowski *et al.* (1967) and Petrusewicz and Macfadyen, (1970). The effects of dietary incorporation of *Mentha piperita* leaf powder on the absolute daily growth life-table characteristics and oviposition of *T. castaneum* have been investigated in the present study.

Materials and Methods

The stock culture of *T. castaneum* was maintained on refined wheat flour in BOD incubator at 30±2°C and 70±5 per cent RH. In case of the treatments, *Mentha piperita* leaf powder was mixed in wheat flour in two different doses of 1 and 2 per cent (w/w) and the insects were reared on these diets. Desired number of eggs were collected by allowing the adults of similar age-group to lay eggs in refined wheat flour diets and sifting the eggs out with a 70 mesh sieve. Batches of 100 eggs each were sorted out and kept separately in petridishes, (6" diameter) containing these dietary formulations, in BOD incubator for rearing. Five replicates of each experiment were run along with an appropriate control set-up and biomasses (fresh and dry) of the developmental stages were determined gravimetrically (Singh *et al.*, 1976) in order to estimate the growth rates of the immature stages of *T. castaneum*. For this purpose the samples of developmental stages were dried in oven at 60°C for 24 hrs. and the life table characteristics of *T. castaneum* were determined (Odum, 1983 and Krebs, 1985). Accordingly, the following parameters were determined.

Equations

$$I_x = n_x / n_o \qquad ....1$$

$$d_x = n_x - (n_x + 1) \qquad ....2$$

$$q_x = d_x / n_x \qquad ....3$$

$$T_x = \sum_{x}^{\infty} L_x \qquad ....4$$

$$e_x = T_x / n_x \qquad ....5$$

where,

X: Age interval

n_o: Number of individuals at the beginning of the experiment.

n_x: Observed number of alive individuals at the start of age interval x.

I_x: Proportion surviving to start of age interval x.

d_x: Number of organisms dying within age interval x to x +1.

q_x: Rate of mortality during age interval x to x+1.

T_x: Units of individuals times time unit.

e_x: Mean expectation of life for organisms alive at the start of age x.

L_x: Number of individuals alive on the average during the age interval x to x +1.

The absolute growth-rates were estimated using the following equation (Klekowski *et al.*, 1967, Petrusewicz and Macfadyen, 1970 and Campbell *et al.*, 1976).

Absolute daily growth (mg/day) = $W_2 - W_1/t_2 - t_1$

where,

W_1 and W_2: The mean biomasses of the individuals at times t_1 and t_2 respectively.

Results and Discussion

The life-table characteristics of the immature and adults stages of *T. castaneum* reared on refined wheat flour (Control) containing 1 and 2 per cent *Mentha piperita* leaf powder are presented in the (Table 6.1) revealed a total mortality in about 32 days of the life-table of *T. castaneum*. The maximum rate of mortality of eggs was recorded 22.00 per cent in 1 per cent and 22.34 per cent in 2 per cent diet respectively where as control showed only 15.89 per cent. At 1 per cent treatment of *M. piperita* leaf powder the larval mortality was the highest (9.13 per cent). At 2 per cent treatment of *M. piperita* leaf powder the mortality at L_3 stage (12.19 per cent) was highest than that of L_6 (7.96 per cent) stages. Pupae showed a fairly high mortality (2.54 per cent) in 1 per cent diet as compared to that of control (2.04 per cent). The observations indicated that incorporation of *Mentha piperita* leaf powder adversely effected the survival of larvae and promoted survival of pupae of *T. castaneum* contact presumably on account of its repellency or contact toxicity specific to larvae while pupae experience better survival (Tripathi *et al.*, 2000 and Varma and Duby, 2001).

Table 6.1 : Life Table of *T. castaneum* Reared at 30±2°C. (L_1–L_6 = Larval instars)

Life Stage	*Duration (Days)*	n_x	l_x	d_x	q_x	L_x	T_x	e_x
Egg	3.5	100	1.00	20.667	0.2066	89.666	587.307	5.8730
L_1	1.5	79.333	0.7933	03.00	0.0378	77.833	497.641	6.2728
L_2	2.5	76.333	0.7633	03.00	0.0393	74.831	419.808	5.4996
L_3	4.5	73.33	0.7333	4.997	0.0681	70.831	344.977	4.7044
L_4	4.5	68.333	0.6833	5.003	0.0732	65.831	274.146	4.0119
L_5	4.0	63.33	0.6333	02.67	0.0421	61.995	208.315	3.2893
L_6	5.5	60.66	0.6066	2.000	0.0329	59.66	146.32	2.4121
Pupa	6.0	58.66	0.5866	01.33	0.0226	57.995	86.66	1.4773
Adult	5.0	57.33	0.5733	0.00	0.00	28.665	28.665	0.5
1 per cent leaf powder of *M. piperita*								
Egg	3.5	100	1.00	22.00	0.220	89.000	551.980	5.519
L_1	1.5	78.000	0.780	03.670	0.047	76.165	462.980	5.935
L_2	2.5	74.330	0.743	05.000	0.067	71.830	386.810	5.203
L_3	4.5	69.330	0.693	06.330	0.091	66.165	314.980	4.543

Contd...

Table 6.1 –Contd...

Life Stage	Duration (Days)	n_x	l_x	d_x	q_x	L_x	T_x	e_x
L_4	4.5	63.000	0.630	05.000	0.079	60.500	248.820	3.949
L_5	4.0	58.000	0.580	02.000	0.034	57.000	188.320	3.246
L_6	5.5	56.000	0.560	03.670	0.065	54.160	131.320	2.345
Pupa	6.0	52.330	0.523	1.330	0.025	51.660	77.160	1.474
Adult	5.0	51.000	0.510	0.000	0.000	25.500	25.500	0.500
				2 per cent leaf powder of *M. piperita*				
Egg	3.5	100	1.00	22.34	0.223	88.830	539.960	5.399
L_1	1.5	77.660	0.776	04.660	0.060	75.330	451.130	5.800
L_2	2.5	73.000	0.730	04.670	0.0639	70.660	375.800	5.147
L_3	4.5	68.330	0.683	08.330	0.121	64.160	305.14	4.465
L_4	4.5	60.000	0.600	03.000	0.050	58.500	240.980	4.016
L_5	4.0	57.000	0.570	02.670	0.046	55.660	182.480	3.201
L_6	5.5	54.330	0.543	04.330	0.0796	52.160	126.820	2.330
Pupa	6.0	50.000	0.500	00.340	0.006	49.830	74.660	1.490
Adult	5.0	49.660	0.496	0.00	0.00	24.830	24.830	0.500

The mean biomasses (fresh weights) of the untreated individuals of *T. castaneum* increased sharply from: Egg = 0.031, L_1 = 0.06, L_2 = 0.139, L_3 = 0.39, L_4 = 1.27, L_5 = 1.64, L_6 = 1.91 mg followed by a decline in pupae (1.8 mg) and adults (1.7 mg). At 1 per cent treatment with *Mentha piperita* leaf powder, however, the fresh body weight of pupa (1.67 mg) was significantly (P/_ 0.05) lower than that of control although, other stages were not affected (Table 6.2). At 2 per cent treatment, however, the fresh weight all the developmental stages from L_4 stage onwards (L_4 = 0.91, L_5 = 1.24, L_6 = 1.47, pupae = 1.37 and adults = 1.27 mg) were significantly lower (P/_ 0.01) than those of the controls (Table 6.2).

Table 6.2: The Fresh Weights of Immature Adult Stages of *T. castaneum* Reared on 1 and 2 per cent Leaf Powder of *M. piperita* at 30±2°C

Life Stage / Diet Mixture	Egg	L_1	L_2	L_3	L_4	L_5	L_6	Pupa	Adult
Refined wheat flour (mg±SE)	0.0313 ±0.00067	0.0663 ±0.00263	0.1395 ±0.01024	0.397 ±0.02384	1.277 ±0.06092	1.648 ±0.02555	1.919 ±0.06880	1.806 ±0.00702	1.706 ±0.04910
1% *M. piperita* Leaf Powder in wheat flour (mg±SE)	0.03 ±0.00115	0.0645 ±0.00235	0.1332 ±0.00246	0.363 ±0.0470	1.23 ±0.0881	1.57 ±0.03511	1.73 ±0.16623	1.67* ±0.03929	1.63 ±0.0185
2% *M. piperita* Leaf powder in wheat flour (mg±SE)	0.03 ±0.00066	0.0623 ±0.00033	0.1228 ±0.02535	0.328 ±0.1080	0.914** ±0.0062	1.24** ±0.0782	1.47** ±0.1181	1.37** ±0.0504	1.27** ±0.0145
Critical difference at 0.01 level	0.0049	0.01040	0.08310	0.16274	0.32482	0.27097	0.65112	0.19461	0.16477
Critical difference at 0.05 level	0.0032	0.00686	0.05485	0.10742	0.21441	0.17886	0.42981	0.12846	0.10876

Significant at *p/_ 0.05; **p/_ 0.01.

The mean biomasses (dry weights) of the individuals at various developmental stages (Table 6.3) revealed that the standing crop biomass of the earlier larval instars (L_1 to L_3) did not change on account of *Mentha piperita* leaf powder. The biomasses of remaining stages, such as, L_5 (0.46 mg), pupae (0.51 mg) and adult (0.44 mg) reared at 2 per cent treatment of *Mentha piperita* leaf powder, were significantly lower (P/_0.05 and P/_0.01) in comparison to those of control L_5 (0.58 mg), pupae (0.78 mg) and adult (0.72 mg) (Table 6.3).

Table 6.3: The Dry Weights of Immature Adult Stages of *T. castaneum* Reared on 1 and 2 per cent.Leaf Powder of *M. piperita* at 30±2°C

Life Stage / *Diet Mixture*	*Egg*	*L_1*	*L_2*	*L_3*	*L_4*	*L_5*	*L_6*	*Pupa*	*Adult*
Refined wheat flour (mg±SE)	0.0166 ±0.00067	0.042 ±0.00393	0.0606 ±0.01020	0.136 ±0.01850	0.464 ±0.04663	0.581 ±0.01912	0.846 ±0.05174	0.7833 ±0.01202	0.7266 ±0.04055
1 per cent *M. piperita* leaf powder in wheat flour (mg±SE)	0.016 ±0.00033	0.0408 ±0.00475	0.0612 ±0.00553	0.1333 ±0.0225	0.447 ±0.05910	0.527 ±0.02811	0.803 ±0.0964	0.736 ±0.00881	0.695 ±0.03329
2 per cent *M. piperita* leaf powder in wheat flour (mg±SE)	0.0164 ±0.00083	0.040 ±0.00120	.0.059 ±0.0051	0.122 ±0.0018	0.363 ±0.0470	0.464* ±0.0466	0.646 ±0.03711	0.516** ±0.02027	0.445** ±0.01365
Critical difference at 0.01 level	0.00338	0.01902	0.03840	0.08812	0.26854	0.17459	0.34954	0.07613	0.11640
Critical difference at 0.05 level	0.00223	0.01255	0.02535	0.05817	0.17726	0.11525	0.23073	0.05025	0.10826

Significant at *p/_ 0.05; **p/_ 0.01.

These observations revealed that *Mentha piperita* leaf powder affected the biomass accumulation at different developmental stage of *T. castaneum* at 2 per cent dose only perhaps by acting as a powerful repellent adversely affecting the feeding of the larvae as reported by Tripathi *et al.* (2000) and Gupta (2006) and against *Trogoderma granarium* by Gupta and Singh (2005).

The observation on the absolute growth of the various developmental stages of *T. castaneum* revealed an almost identical rate of earlier immature stages (L_1 to L_3) in all the experiments. Beyond L_3 stage, however, the growth rates of developmental stages reared on 1 per cent leaf powder (L_4 = 0.069, L_5 = 0.02, L_6 = 0.050, pupa = – 0.011 and adult = – 0.008 mg/day) and those reared on 2 per cent treatment with *Mentha piperita* leaf powder (L_4 = 0.053, L_5 = 0.025, L_6 = 0.032, pupae = – 0.021 and adult = – 0.0114 mg/day) were lower than those of controls (L_4 = 0.072, L_5 = 0.029, L_6 = 0.048, pupae = – 0.010 and adult = – 0.011 mg/day) (Figure 6.1). Obviously, a negative growth rate in pupae and adults occurred on account of the fact that pupae did not feed and use their body nutrients for metabolism (Campbell *et al.*, 1976 and Singh *et al.*, 1976). The fact that *Mentha piperita* leaf powder brought about a reduction in growth-rate of *T. castaneum* indicated its adverse effects reducing the growth of body tissues in the immature stages of *T. castaneum* as reported by Tripathi *et al.* (2000) and Kokate *et al.* (2000).

The egg production by untreated adults of *T. castaneum* only started on the 4th day (6.66 eggs per pairs) after its emergence. Subsequently, the average number of total eggs laid were recorded on 6th day

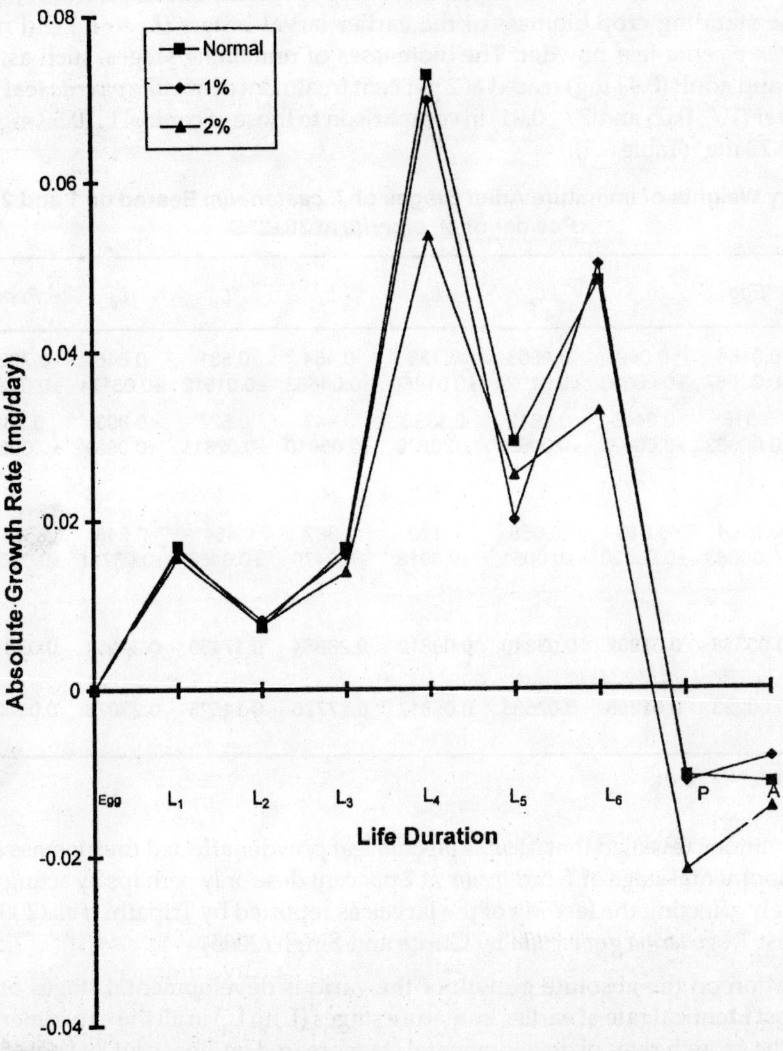

Figure 6.1: Absolute Growth Rates of Immature and Mature Stages of *T. castaneum* Reared on 1 and 2 per cent Leaf Powder of *Mentha piperita* (L_1–L_6 = Larva, P = Pupa, A = Adult)

(19.33) 8th day (23.33), 10th day (24.00), 12th day (26.00), 14th day (26.66), 16th day (27.00), 18th day (26.66) and 20th day (26.00) (Table 6.4). At 1 per cent treatment with *Mentha piperita* leaf powder, however, the mean daily egg laying was lower (4th day = 05.66, 6th day = 12.33, 8th day = 19.00, 10th day = 21.66, 12th day = 22.33, 14th day = 22.66, 16th day = 21.33, 18th day = 22.00 and 20th day = 20.33) than those of controls (Table 6.4). At 2 per cent treatment, however, there were further decline in the number of eggs laid (4th day = 3.00, 6[th] = 8.00, 8th day = 12.00, 10th day = 13.66, 12th day = 14.33, 14th day = 17.00, 16th day = 14.66, 18th day = 15.33 and 20th day = 14.00) in comparison to controls. The egg laying capacity of *T. castaneum* females treated with 1 and 2 per cent *M. piperita* leaf powder was reduced to 81.09 and 54.28 per cent, respectively.

Table 6.4: Mean Daily Egg Production in *Tribolium castaneum* Reared on 1 and 2 per cent Leaf Powder of *Mentha piperita*

Number of Days	*Refined Wheat Flour (Control)*	*1% Mentha piperita Leaf Powder in Wheat Flour*	*2% Mentha piperita Leaf Powder in Wheat Flour*
2	–	–	–
4	06.66	05.66	03.00
6	19.33	12.33	08.00
8	23.33	19.00	12.00
10	24.00	21.66	13.66
12	26.00	22.33	14.33
14	26.66	22.66	17.00
16	27.66	21.33	14.66
18	26.66	22.00	15.33
20	26.00	20.33	14.00
Total	206.30	167.30	111.98
Percentage (%)	100	81.09	54.28

The oviposition of *T. castaneum* given access to *Mentha piperta* leaf powder mixed with wheat flour, (Table 6.4) was adversely affected vis-à-vis control both in case of 1 and 2 per cent treatments lowering the total egg production from 206.3 (control) to 111.98 (2 per cent treatment). The fact that 2 per cent *Mentha piperita* leaf powder brought about a reduction in egg laying capacity of *T. castaneum* to the tune of 54.28 per cent than that of control which indicates its effects either as fumigant or repellent for *T. castaneum*, as reported by Tripathi *et al.* (2000) and Varma and Dubey (2001) in case of other insects, or as a potent inhibitor of egg laying.

Acknowledgement

I express my grateful thanks to Dr. N.B. Singh, Professor and Head, Department of Animal Science, M.J.P. Rohilkhand University, Bareilly (U.P.) for providing worthy guidance and necessary facilities for research work. I also thank Dr. R.K. Agarwal, Deputy Director, Indian Grain Storage Management and Research Institute, Hapur (U.P.) for his valuable help and suggestions in this work.

References

Campbell, A., Singh, N.B. and Sinha, R.N., 1976. Bioenergetics of the granary weevil, *Sitophilus granarius* (L.) (Coleoptera : Curculionidae). *Can. J. Zool.*, 54: 786–798.

Gupta Sudhakar, 2006. Effect of Mint (*Mentha arvensis* Linn.) leaf powder on growth and development of *Tribolium castaneum* (Herbst). *J. Food Sci. Tech.*, 43(1): 22–25.

Gupta, S. and Singh, N.B., 2005. Effect of 'menthol mint' leaf powder on growth rate and life table of *Trogoderma granarium*. *J. Appl. Bio. Sci.*, 31(1): 12–16.

Klekowski, R.Z., Prus, T. and Zyromska-Rudzka, H., 1967. Elements of energy budget of *Tribolium castaneum* (Hbst.) in its developmental cycle, pp. 859–79 In: *Secondary Productivity of Terrestrial Ecosystem*, (Ed.) K. Petrusewicz. PWN, Warszawa, Krakow.

Kokate, C.K., Purohit, A.P. and Gokhale, S.B., 2000. *Pharmacognosy*. Nirali Prakashan, 41, Budhwar Peth, Jogeshwari Mandir Lane, Pune, pp. 314–565.

Krebs, C.J., 1985. *Ecology: The Experimental Analysis of Distribution and Abundance*, 3rd Edn. Harper and Row Publ., New York.

Odum, E.P., 1983. *Basic Ecology*. CBS College Publ., Saunders College Publishing, Philadelphia.

Petrusewicz, K. and Macfadyen, A., 1970. *Productivity of Terrestrial Animals: Principles and Methods*. IBP Handbook No. 13, Blackwell Publ., Oxford.

Singh, N.B., Campbell, A. and Sinha, R.N., 1976. An energy budget of *Sitophilus oryzae* (Coleoptera : Curculionidae). *Ann. Ent. Soc. Am.*, 69 (3): 503–511.

Tripathi, A.K., Prajapati, V., Aggarwal, K.K. and Kumar, S., 2000. Effect of volatile oil constituents of 'Mentha' species against the stored grain pests, *Callosobruchus maculatus* and *Tribolium castaneum*. *J. Med. Aro. Plant Sci.*, 22: 549–556.

Varma, J. and Dubey, N.K., 2001. Efficacy of essential oils of *Caesulia axillaris* and *Mentha arvensis* against some storage pests causing biodeterioration of food commodities. *International J. Food Microbiology*, 68: 207–210.

Chapter 7

An Assessment of Soil Fertility: A Case Study of Varahi River Basin, Udupi District

K.L. Prakash and R.K. Somashekar

Department of Environmental Science, Bangalore University, Bangalore – 560 056

ABSTRACT

The present study aims at a status of fertility of the soils of Varahi river command area. A total of 20 sampling sites spread over different villages were identified and soil samples were collected. Various soil quality parameters *viz.*, pH, Electrical Conductivity, Chlorides, available Calcium and Magnesium, Phosphorus, Nitrate, Fluoride, exchangeable Sodium and Potassium, Ammonia nitrogen were analyzed. The results have been discussed parameter wise. The fertility status have been assessed through nutrient index values.

Keywords: Assessment, Soil fertility, Varahi River Basin, Agricultural crops.

Introduction

In a predominantly agricultural country like India, land is by far the most precious asset. We have within the geographical confines of the country a large landmass, which from historical times to the present day is supporting a large growing population, and the multifarious needs and demands of the populace. Soil being a prime component of the land matrix is getting increasingly lost from a host of land erosion processes, resulting in wide spectra of environmental problems, the principal one amongst them being serious loss of agricultural productivity which needs to be viewed with concern (Trivedi and Gurudepraj, 1992). Though we are becoming slowly aware of this problem, yet we remain silent spectators to the steady loss of destruction of soil, our most valuable resource. Unless this soil erosion or soil loss is checked in time, it would amount to an irreparable loss for the preservation of the entire life system and all living beings. The soils have been degrading at an estimated rate of one million hectares per year due to man made activities mainly agricultural activities; use of fertilizers and Pesticides (Mishra *et al.*, 1989 and Rao *et al.*, 1993) and natural eroding activities. Thus there is an urgent need for developing strategies to slow down the degradation process or reclaim the soils to normalcy and ensure sustainability of our crop production system, which are the major issues confronting us today. It is a great challenge to the scientific community to develop ways and means to increase food production on a sustainable basis. Soil is one of the most important resources of the earth and the science dealing with the soil is known as soil science. Knowledge of soils in respect of their extent, distribution and their characteristics are a pre-requisite for a good understanding of the proper land use and associated problems and potentials and, besides the suitability of soil for various uses has been a matter of great interest throughout the world. Soil also needs protection from being degraded. A systematic and scientific appraisal of soil resource, especially creation of their data base are important parameters which can help in increasing food production, and thus soil resource inventory is the basis for rationalizing land use according to soil types and their capacity.

Study Area

Dakshina Kannada district has three distinct types of soils. These are both *in situ* and transported. Confined to the coast, there is alluvium. All the rivers in the district are west flowing and carry during the monsoon considerable amount of suspended rock debris from the upper reaches of the district. At the river mouth there is a check in the flow velocity and the transported material is thus deposited along the banks.

Varahi river is a major west flowing river in west coast of Karnataka. The river takes its origin near Guddakoppa village in Hosanagar taluk, Shimoga district at an altitude of about 761 mtrs (2500 ft) above MSL. Tributaries like Hungedhole, Kabbenahole, Dasanakatte, Chakranadi etc., join Varahi before it joins the Arabian Sea.

The main objectives aimed at to study the soil physico-chemical characteristics, assess the soil fertility with respect to nutrient index and overview the cropping pattern and land use pattern.

Materials and Methods

A preliminary soil survey was conducted in the study area and villages were identified followed by an extensive sampling programme undertaken in the study area. Soil samples were collected from different agricultural lands in March 2005 as per standard procedure recommended by United States Department of Agriculture (USDA). A total of 20 sampling sites spread over different villages were identified during preliminary survey. Various soil quality parameters *viz.*, pH, Electrical Conductivity, Chlorides, available Calcium and Magnesium, Phosphorus, Nitrate, Fluoride, exchangeable Sodium and Potassium, Ammonia nitrogen etc., were determined employing standard methods of Jackson (1973) and Black (1982).

Results and Discussion

pH of Soil

pH of the soil samples varied between 4.1 to 7.12, *i.e.*, it was observed that a wide variety of soils that are both acidic and alkaline in nature in the study area exists. The lower value of pH was found in Sample No. 17 which belongs to a farmer Chandra Poojari's land of Shiriyur village and the highest pH of 7.12 was observed in Sample No. 5 which belongs to Seetharama's land which lies in Shankaranarayana. All the soil samples have values of below 6.8 are slightly acidic in nature, and this may be due to high amount of leaching which has led to the leach out of exchangeable cations, while pH values above 7 in the soil samples are considered slightly alkaline. Thus the measure of soil pH is an important parameter helps in identification of chemical nature of the soil (Shalini *et al.*, 2003).

Electrical Conductivity

Conductivity, as the measure of current carrying capacity, gives a clear idea of the soluble salts present in the soil. It plays a major role in the salinity of soils. Lesser the EC value low will be the salinity value of soil and vice-versa.

EC Values (μmhos/s)	*No. of Samples*
10 to 500	18
501 to 1000	2
1001 to 1500	0
1501 to 2000 and above	0

The EC values varied from 31.2 and 580.8 μmhos/s. The highest value of EC was observed in Sample No. 18 which belongs to S.R. Acharya's land at Yedthadi village.

EC value of less than 800 μmhos/s are considered as normal nature of soil, while EC values between 800 and 1600 are considered critical for tolerant crops while EC values 1600 and 2500 are considered critical for salt tolerant crops and EC values more than 2500 are not considered safe for most of the crops. In the study area no samples have crossed 1500 hence the soil samples are suitable for agriculture.

Colour of the Soil

Soil colour is one of the visual indicators of the soil type, which varies from region to region. Soil derives its colour from the parent material. However, the colour may also vary due to,

1. Soil forming process
2. Moisture content and drainage
3. Nature and amount of organic matter
4. Mineral content

In the study area villages like Siddapura, Hosangady, Kulanji, Shankaranarayana, Haladi, Hardalli Huvinalli, Ydthadi, Vaderhobli, Koni, Mudalakatte villages had soils with grayish brown to dark grey and at some places dark brown soil were found. Villages like Machotu, Hardalli Mandalli, Molahalli and Basrur villages have red soil.

Exchangeable Calcium

The minimum value of Exchangeable Calcium is found to be 0.0002 while the maximum value is 0.0053 (expressed as Ca, meq/100 g). The minimum value of exchangeable calcium was observed in Sample No. 16 which belongs to Srinivas Kethiah of Haladi Harkadi village of Shirur village and the highest has been observed in Sample No. 8, which belonged to Augumbe road of Haladi-76 village.

Exchangeable Mg

The minimum value of Exchangeable Magnesium is found to be 0.0004 meq/100 g while the maximum value has been found to be 0.0021 meq/100 g. The least value of exchangeable magnesium was found in Sample No. 19 which belonged to Chandrashekar Karant's land of Vaderhobli village and the maximum value of 0.0021 was found in Sample No. 13 which belonged to H. Seetharam Shetty's land of Hardalli Mandalli village.

Per cent Organic Carbon (OC)

Soil forms the base for all life forms, of plants, animals and microorganisms in various stages of decomposition, which constitute the organic matter. The organic substances are a major determinant of soil structure, moisture, pH and the soil nutrient status. It is confined to the top soil. The importance of organic matter in the soil is implied in the definition of soil, which recognizes fertility, as a unique feature distinguishing soil from the parent rock.

In the study area the amount organic carbon showed increasing trend with increase in soil pH. This is because the decomposition rate of organic matter in high pH soils slows down perhaps, due to decrease in microbial activities.

The percentage of organic matter varied from region to region and was generally enhanced in thickly vegetated areas. It is impossible to determine the optimum level of organic matter required by the plants, as it is not a single value required for all the plants, for all the soils. The variation largely depends on soils, climate, plant and animal species (Brady, 1995).

Organic matter content in the soil is the most important parameter to be determined while recommending soil management practices. It increases the soil fertility status and controls erosion and runoff of the soil and water, besides indicating the structure and general nutrient status of the soil.

Depending upon the organic carbon content (per cent), the quality of soil may be graded as per the following practice.

Percentage of Organic Carbon	*Rating*
< 0.40	Low
0.4 to 0.75	Medium
> 0.75	High

In the study area noticed that, per cent organic carbon to be less than 0.4 per cent in fifteen locations were having low rate of organic carbon content. While other five samples have per cent organic content between 0.4 and 1.2 which indicates medium rate. Therefore most of the samples possess medium content of per cent organic carbon content which favour good yield of crops.

Exchangeable Sodium and Potassium

Potassium (K) is the third most required element by the plants. It is not an integral part of any specific compound. But it affects cell division, the formation of carbohydrates, translocation of sugars, activates various enzymatic reactions, cell permeability and improves resistance of some plants to some diseases. It also plays a key role in water balance in plants or regulation of osmosis (Singh and Tripathi, 1993). It is the most abundant metal cation in plant cell (2 to 3 per cent by dry weight).

Concentration of Potassium in Soil

Deficient supply of (K)	Less than 113 kg/ha
Doubtful supply of (K)	113 to 280 kg/ha
Adequate supply of (K)	More than 280 kg/ha

In the study area villages have showed a wide variety of potassium level in the lands. In some villages like Haladi-76 soil series has higher content of available potash, but are medium in exchangeable potassium and sodium, while Hardalli village soil series has medium supply of exchangeable potassium and high in available potash and exchangeable sodium. And rest of the village soil series exhibited low to medium exchangeable potassium, and contained medium to high available potash.

ESP and SAR

Exchangeable Sodium Per cent values for all the soil samples lies between 0.18 to 0.2 while SAR values of the soil samples lie between 0.3 and 18.

In the study area almost all the samples are low in exchangeable sodium per cent, which is good indication of soil fertility, with a potential to get a good yield of crops.

Available Phosphorus

Phosphorus is the second most important macro nutrient in biological systems, which constitutes more than 1 per cent of the dry organic weight. It is a constituent of nucleic acids, phospholipids and many phosphorylated compounds. It is the second most limiting factor often affecting plant growth. Phosphorus exists in the soil in both organic and inorganic forms. Plants take up inorganic phosphorus in the form of phosphate ions. Organic phosphates are important sources of phosphorus in most of the soils. Phosphorus is required in small quantities though; it may be the most likely limiting element in plant productivity and hence is ecologically significant (Bhattacharya and Bhattacharya, 1993). Mineral phosphates are not readily available to the plants. Most of the soluble phosphates get fixed into soluble form complex before the plants can absorb them. Available phosphorus represents a fraction of total phosphorus which is absorbable by plants.

Concentration of Phosphorus in Soil

Low phosphorus	Less than 12.4 kg/ha.
Medium phosphorus	12.4 to 22.4 kg/ha
Adequate phosphorus	More than 22.4 kg/ha
Abundant phosphorus	Still higher

In the study area soils of Machotu, Halady, Koni, and Mudalakatte soil series have medium levels of available phosphorus, while Siddapura and Haladi Harkadi villages have adequate quantity of available phosphorus whereas the rest of the villages like Hosangadi, Kulanji, Shankaranarayana, Haladi-76, Molahalli, Shiriryur, Vaderhobli, Koni and Basrur villages have low content of available phosphorus.

Ammonia Nitrogen

Nitrogen is one among the four primary elements, which make up the plant tissues. It is the major component of proteins, nucleic acids and chlorophyll. Atmospheric nitrogen gets fixed in the soil by electro and photochemical fixation and by the action of microorganisms. Soil nitrogen is made available by mineralization. It is present in the soil in both organic and inorganic forms. However, nitrogen content in organic form is as high as 90 per cent. Soil organic matter decomposed by microbial activity and organic nitrogen gets converted to ammonium, nitrates and nitrites. Nitrogen is responsible for maintaining the soil fertility whilst nitrogen content in most of the soils are very low and is found as nitrates, nitrite and ammonium. Plants take up nitrogen generally as nitrates in aerobic conditions and as ammonium ions in anaerobic conditions. Nitrogen is most often the limiting nutrient for the plant growth.

Quantity of Nitrogen	*Rating*
Less than 272 kg/ha	Low
272 to 554 kg/ha	Medium
More than 554 kg/ha	High

In the study area most of the soil samples are very fertile with low to medium and high quantity of nitrogen. Soil moisture is one of the important factors affecting nitrification. Excess water as found in water logged soils suppress nitrification because of lack of oxygen. Unlike in dry soils as in the case of our study area soils however, does have enough moisture for bacterial metabolism and the moistening of such soils has rapidly increased the rate of biosynthesis of nitrogen which has added to its fertility.

Salinity

Salt affected soils are commonly seen in arid and semi arid regions, in irrigation command areas and in regions with poor drainage and in areas where poor quality water is used for irrigation. Saline soils are those, which contain appreciable quantities of soluble salts to interfere with crop growth. They contain neutral salts such as chlorides and sulphates of sodium, calcium and magnesium (excluding gypsum) in quantities sufficient enough to interfere with growth and crop yield of most crop plants (Lauter and Munns, 1986).

Depending upon the electrical conductivity of the soil, soil salinity can be classified into four classes:

Four salinity classes proposed are:

Water Class	*Electrical Conductivity (μmhos/cm)*	*Approximate Salt Concentration of the Samples (per cent)*
CI–Low salinity	0 to 250	< 0.16
CII–Medium salinity	250 to 750	0.16 to 0.50
CIII–High salinity	750 to 2250	0.50 to 1.50
CIV–Very high salinity	2250 to 5000	1.5 to 3

Salinity Range of Soil Samples

Sample No.	*Village Name*	*E.C. (μmhos/cm)*	*Salt Conc. (%)*
1	Machotu	32.1	0.0193
2	Hosangady	54.9	0.0329
3	Siddapura	57.71	0.0346
4	Kulanji	78.08	0.0468
5	Shankaranarayana	78.25	0.0470
6	Shankaranarayana	103.2	0.0619
7	Haladi-76	136.4	0.0818
8	Haladi-28	95.04	0.0570
9	Hardalli Mandalli (Gude Angadi)	158.2	0.0949
10	Molahalli	54.42	0.0327
11	Huvinalli	440.2	0.2641
12	Haladi Harkadi	243.4	0.1460
13	Shiriyura	516.2	0.3097
14	Yedthadi	580.8	0.3485
15	Vaderhobli	490.57	0.2943
16	Koni	97.2	0.0583
17	Koni (Lake)	256.1	0.1537
18	Koni	31.2	0.0187
19	Mudala Katte	68.65	0.0412
20	Basrur	212.8	0.1277

In the study area none of the soil samples have crossed the EC value above 600 which clearly shows that the soil samples are not very saline, however occasionally pockets of low saline and medium saline type of soils are found in the area. To brief out in figures we can tell that 13 samples out of 20 samples have total salt concentration less than 0.16 per cent and only six samples out of 20 samples have total salt concentration 0.16 to 0.50 per cent. This clearly indicated that all the samples in the command area are having salt concentration of less than 0.5 per cent, which is quite safe for agricultural purposes.

CI water is considered as safe with no likelihood of salinity problems.

CII When used for irrigation, moderate leaching is required.

CIII and CIV cannot be used on soils with inadequate drainage, since salinity develops.

Fertility Levels of the Soils

Based on the results and nutrient indices, one can say whether a particular nutrient is low, medium or high. Rating chart was made use of while rating the soil analysis results.

Rating Chart for Soil Test values and their Nutrient Indices

1. Soil pH			
	Acidity	Neutral	Alkaline
Range	Below 6.0	6.0–8.0	Above 8.0
Soil Reaction Index	I	II	III
2. Electrical Conductivity			
	Normal	Critical	Injurious
Range (μmhos/cm)	Below 1000	1000–2000	Above 2000
Salt index	I	II	III
3. Organic Carbon			
	Low	Medium	High
Range (per cent)	Below 0.5	0.5–0.75	Above 0.75
Nutrient index	I	II	III
4. Available Phosphorus (By Bray's method)			
	Low	Medium	High
Range (kg/ha)	Below 22	22–54	Above 54
Nutrient index	I	II	III
5. Available Potash			
	Low	Medium	High
Range (kg/ha)	Below 123	123–296	Above 296
Nutrient index	I	II	III

Nutrient Index	*Range*	*Remarks (OC, P, K)*
I	Below 1.67	Low
II	1.67–2.33	Medium
III	Above 2.33	High

OC: Organic carbon; P: Available phosphorus; K: Available potash.

The nutrient index values are evaluated for the soil samples analyzed using the following formula:

Nutrient Index

[(1X No. of samples in low category) + (2X No. samples in medium category) + (3X No. of samples in high category)] / Total number of samples

The values are:

Characteristics	*Nutrient Index*	*Remarks*
Organic carbon (OC)	1.35	Low
Available phosphorus (p)	1.3	Low
Available potash (K)	1.0	Low

Crops and Cropping Patterns

There are two distinct cropping season:

1. Kharif
2. Rabi

Kharif and Rabi crops are grown mainly under rain fed conditions. In recent periods bore wells are being drilled in the command area, but their number is less.

Crops grown in Kharif season: Paddy

Crops grown in rabi season: Paddy

One season crop: Sugarcane

Plantation: Arecanut, Coconut garden, Natural vegetation and Cashew garden.

Crop Raised Under Irrigated Condition

1. Sugarcane
2. I crop paddy
3. II crop paddy
4. III crop paddy.

Farmers use fertilizers and pesticides for irrigated crops like paddy and sugarcane. The dozes of fertilizer are usually as per recommendation of package of practice. It is always safe to apply fertilizer as per soil test result. A balance dose of fertilizer boosts the yield especially in case of hybrid crops. Sugarcane is one such crop, which needs higher doses of fertilizer since it is a feeder crop and yield of sugarcane usually depends on the higher balanced doses of fertilizers. There is an increasing tendency to use compound and mixed fertilizers like DAP 17 : 17 : 17 and Ammonium Phosphate 20 : 20 etc., as they provide most of the important nutrients in a balanced proportion. Urea is being recommended and is being used in command area for paddy crop. Almost all the farmers are aware of the benefits of the use of fertilizer in balanced doses.

The popular pesticides and insecticides, which are being used by farmers of the command area, are Endosulphan, Manocrotothas, Quinolphos, Carbondizeem, Parathroides (sinnerin), Malathion dust, Wettable sulphur, Carbaryl and Capton.

Strategy for Cropping Pattern

There are a number of factors that influence the adoption of cropping patterns. These are:

Land and Soil

Suitability of the land for specific crops and type of soil will have to be taken into consideration for deciding the crops to be raised. Low lying areas and upland sandy and clayey soils, etc., are some such considerations.

Irrigation

The availability of irrigation facilities or otherwise will influence the crop as well as the cropping system.

Environmental and Climatic Factors

Precipitation, temperature, soil moisture status, etc., also determine the type and number of crops that can be raised.

Market

The dearth of proper marketing facilities acts as a disincentive for raising two or more crops, particularly if the product is perishable.

Time Gap between two Crops

For raising a crop after harvesting, a minimum time gap is required for processing the first crop and to prepare the land for the second. The time required will depend on the power available with the farmers in terms of tools and equipment and labour force.

Capability of Farmers

The small and marginal farmers in most cases do not have the resources, including capability for investment, to engage in for intensive cropping, whereas tractors and tillers with accessories can be used by well-to-do farmers.

There is an increase in tendency of farmers in the command area to follow improved and scientific methods of agriculture due to a good interphase with departmental staff and university scientists in recent years. As a result farmers as noticed in many villages are obtaining higher yield of product crops.

References

Bhattacharya, B. and Bhattacharya, A.K., 1993. Ecological study of plant species diversity in Delhi region. *J. Ecobiol.*, 5: 207–211.

Black, C.A., 1982. *Method of Soil Analysis*, Part I and II. American Society of Agronomy, Madison, Wisconsin, USA.

Brady, N.C., 1995. *The Nature and Properties of Soils*, 10th Edition, Organisms of Soil. Prentice-Hall of India Pvt. Ltd., New Delhi, pp. 253–277.

Jackson, M.L., 1973. *Soil Chemical Analysis*. Prentice-Hall of India Pvt. Ltd., New Delhi.

Lauter, D.J. and Munns, D.N., 1986. Salt resistance of chick pea geonotype in solution salized with NaCl and Na_2SO_4. *Plant and Soil*, 95: 271–299.

Mishra, S.D., Prasad, D. and Dwivedi, B.K., 1989. Pesticide residue in Soil and soil organisms. In: *Soil Pollution and Soil Organisms*, pp.17–28.

Rao, V.R., Adhya, T.K. and Sethunathan, N., 1993. Effect of Pesticide on soil health. I. In: *Pesticides: Their Ecological impact in Developing Countries*, (Eds.) G.S. Dhaliwal and Balwinder Singh. Commonwealth Publishers, pp. 112–130.

Shalini Kulshreshtha, H.S. Devenda, S.S. Dhindsa and R.V. Singh, 2003. Studies on causes and possible remedies of water and soil pollution in Sanganer town of Pink City. *Ind. J. of Env. Sci.*, 7(1): 47–52.

Singh, K. and Tripathi, D., 1993. Different forms of potassium and their distribution in some representative soil groups of Himachal Pradesh. *J. Potassium Res.*, 9: 196–205.

Trivedi, P.R. and Gurudeep Raj, 1992. *Encyclopedia of Environmental Science*, Vol. 12, Akashdeep Publications, New Delhi.

Chapter 8

Thermal and pH Stability of Dibutyl Phthalate: An Antimetabolite of Proline from *Streptomyces albidoflavus* 321.2

R.N. Roy and S.K. Sen*

School of Life Sciences, Department of Botany, Visva-Bharati, Santiniketan – 731 235

ABSTRACT

Thermal and pH stability of dibutyl phthalate as antimicrobial compound (an antimetabolite of proline) were studied. The compound could be 100 per cent active at 50°C but the activity was only 32 per cent and 37.5 per cent when exposed to 100°C and 121°C temperature, respectively. The compound was found susceptible with the change of pH. At pH 9.5, the loss of activity was 50 per cent whereas in pH 2.0 the loss of activity was only 12.5 per cent.

Keywords: Dibutyl phthalate, Antimicrobial compound, Thermostability, pH stability.

Introduction

Antimetabolites are the compounds, which are able to repress the mechanism for synthesis of cellular metabolites from simple nutrients; this repression can be reversed by concurrent administration of one or more biochemicals (Pruess and Scannell, 1974; Hossain *et al.*, 1987). Several antimetabolites are effective antimicrobials and showed their effectiveness in cancer treatment (Hossain *et al.*, 1987; Rosangkima and Prasad, 2004).

* Corresponding Author: E-mail: sksenvb@rediffmail.com; Tel: 03463-261686, Fax: 03463-261268.

Antimicrobial (El-Naggar, 1997; Lee, 2000) efficacy and antimetabolic (Roy *et al.*, 2005) activity of dibutyl phthalate (DBP) have been studied from *Streptomyces*. This antimetabolic activity of DBP was found to overcome by application of proline. Being an important amino acid, proline plays a notable role in cell metabolism especially under stress *e.g.* water (Martinez *et al.*, 2004), heat (Rivero *et al.*, 2004) and salt (Tonon *et al.*, 2004). DBP is also effective as peroxisome proliferator (O'Brien *et al.*, 2001), controlling demodicidosis (Yuan *et al.*, 2001) and as drug channelling agent (Makhija and Vaiva, 2003). The compound shows no acute toxicity on the plant and animal system (Roy and Sen, 2004). Thus, DBP becomes an important compound for proper understanding of its role in the study of cell physiology.

This article deals with the thermal and pH stability of DBP, an antimetabolite of proline from *Streptomyces albidoflavus* 321.2.

Materials and Methods

Microorganism

The producer, *Streptomyces albidoflavus* 321.2 (Roy and Sen, 2002) a new soil isolate was maintained on Glucose asparagine agar (Krainsky, 1941); the test organism *Escherichia coli* ATCC. 25922 was grown on Nutrient agar medium (Hi Media). Organisms were maintained at 4°C.

Fermentation

Spores (1.75×10^7/ml) of the isolate 321.2 was inoculated into 100 ml of medium (pH 6.75) composed of 0.025 per cent $NaNO_3$, 2.18 per cent Glycerol, 0.05 per cent KH_2PO_4, 0.15 per cent NaCl, 0.03 per cent $MgSO_4$, $7H_2O$, 0.001 per cent $Fe_2(SO_4)_3$ $6H_2O$, 0.0001 per cent $CuSO_4$ $5H_2O$, 0.001 per cent $MnSO_4$ H_2O, 0.0001 per cent $ZnSO_4$ $7H_2O$ in 500 ml of Erlenmeyer flask, sterilized at 120±1°C and cultured in stationary condition at 32°C for 6 days.

Isolation and Purification of Active Compound

The active compound was isolated from fermented broth by solvent extraction method, using ethyl acetate (3 : 1). The crude antimicrobial principal was purified by ascending paper chromatography using a mixture of petroleum ether (60–80°C), ethyl acetate and methanol in the ratio of 10 : 0.5 : 1 as mobile phase. Using end guide strip, the active band was detected by bioassay. The semi purified compound, from the detected band, was eluted in ethanol. Then the compound was further purified by silica gel column of 10 × 2 cm and was eluted with methanol under a flow rate of 3 ml/min fractions (of 1 ml each) were collected separately and the activity was detected by bioassay. The active fractions were pooled together and evaporated. One-dimensional TLC was performed using benzene: ethyl acetate (1 : 1) as mobile phase. The bands were detected by keeping the plate in iodine vapor chamber and were eluted with methanol : ethyl acetate (1 : 1). Confirming homogeneity of compound, the activity was detected by bioassay.

Determination of pH Stability

pH stability of DBP was determined as a measurement of retained activity after incubation at a given pH. The purified antibiotic principal (500 µg/ml) was treated in buffer (Sadasivam and Manickam, 2004) of pH between 2.0–10.0 and was incubated for 24 hr at 35°C. The loss of antibiotic activity was measured by bioassay.

Determination of Thermal Resistance

Thermal resistance was determined by exposing the purified DBP (500 µg/ml) at different temperatures (30–121°C) for 5 min and 10 min period. The loss of antimicrobial activity was detected by bioassay in each case.

Antimicrobial activity was determined by agar cup method (Higashide *et al.*, 1971), using glucose-asparagine agar along with 1 ml cell suspension (1.2 × 10^6 CFU) of the test organism. Each cup was filled with purified principal (0.1 ml) and incubated at 30°C for 24 hr.

Results and Discussion

Commercial feasibility of an active compound depends on maintenance of its self-life. The compound was quite stable in varied pH and temperature. The loss of activity of DBP was more pronounced in alkaline pH (Figure 8.1) and 50 per cent loss was recorded at pH 9.5. Augustine *et al.* (2005) reported about the effect of pH and temperature on the activity and stability of the antibiotics from the *Streptomyces albidoflavus* PU 23 where the active compound was incubated at pH in the range 5.7–8.0, maximum activity was observed at a pH range 6.8–8.0, the activity decreased at pH 6.3 to downward. Thermal loss of activity of DBP was only 32 per cent at 100°C and retained its 100 per cent activity up to 50°C (Table 8.1). The non-polyene antimicrobial principal from isolate PU 23 was stable at different temperature (30–80°C), however the principal compound lost its activity completely when treated in an autoclave at 121°C for 15 minutes (Augustine *et al.*, 2005) but DBP lost only 37.5 per cent of its activity. So, DBP is less susceptible to degradation and or modification in change in pH and temperature.

Table 8.1: Effect of Temperature on Antimicrobial Activity of DBP

Temperature (°C)	*Treatment Period (min)*	*Loss of Activity (%)*
30	05	0
	10	0
40	05	0
	10	0
50	05	0
	10	0
60	05	10
	10	15
70	05	15
	10	18
80	05	20
	10	22
90	05	22
	10	25
100	05	32
	10	32
121	05	37
	10	37.5

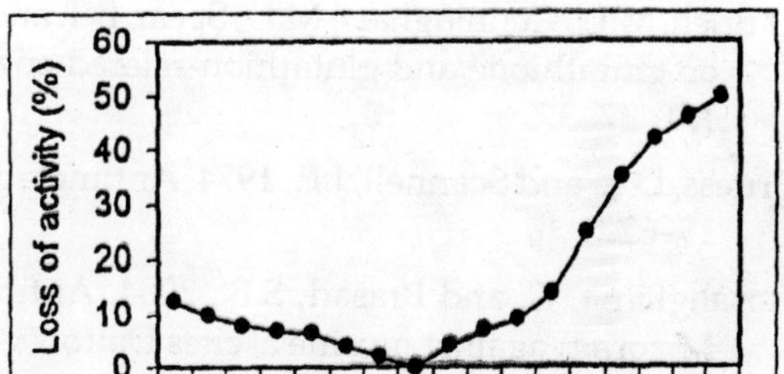

Figure 8.1: Effect of pH on Antimicrobial Activity of DBP

The active principal from the isolate 321.2, defined as dibutyl phthalate ester, was found stable probably due to its aromatic structure with no free reactive group. Further studies regarding the effect of various parameters on the activity of DBP are currently in progress.

References

Augustine, S.K., Bhavsar, S.P. and Kapadnis, B.P., 2005. A non-polyene antifungal antibiotic from *Streptomyces albidoflavus* PU 23. *J. Biosci.*, 30: 201–211.

El-Naggar, M.Y.M., 1997. Dibutyl phthalate and the antitumor agent F5A1, two metabolites produced by *Streptomyces nasri* submutant H35. *Biomed. Lett.*, 55: 125–131.

Higashide, E., Fugono, T., Hatano, K. and Shibata, M., 1971. Studies on T2636 antibiotics. I. Taxonomy of *Streptomyces rochei* var. *volubilis* var nov. and production of the antibiotics and an esterase. *J. Antibiot.* 24: 1–12.

Hossain, M.; Dastidar, S. G. and Chakrabarty, A. N. 1987. Antimetabolic activity of some commonly used drugs. *Indian J. Exp. Biol.*, 25: 866–868.

Krainsky, A., 1941. *Die Aktinomyceten und ihre Bedeutung in der Natur Center* Bakterio. *Parasitenk*, 41: 649–688.

Lee, D.S. 2000. Dibutyl Phthalate: An Glucosidase inhibitor from *Streptomyces melanosporofaciens. J. Biosci. Bioeng.*, 89: 271–273.

Makhija, S.N. and Vavia, P.R., 2003. Controlled porosity osmotic pump-based controlled released systems of pseudoephedrine. I. Cellulose acetate as a semi permeable membrane. *J. Control Release*, 89: 5–18.

Martinez, J.P., Lutts, S., Schanck, A., Bajji, M. and Kinet, J.M., 2004. Is osmotic adjustment required for water stress resistance in the Mediterranean shrub *Atriplex halimus* L? *J. Plant Physiol.*, 161: 1041–1051.

O'Brien, M.L., Cunningham, M.L., Spear, B.T. and Glauert, H.P., 2001. Effects of peroxisome proliferators on glutathione and glutathion-related enzyme in rates and hamsters. *Toxicol. Appl. Pharmacol.*, 171: 27–37.

Pruess, D.L. and Scannell, J.P., 1974. Antimetabolites from microorganisms. *Adv. Appl. Microbiol.*, 17: 19–62.

Rosangkima, G. and Prasad, S.B., 2004. Antitumour activity of some plants from Meghalaya and Mizoram against murine ascites Dalton's lymphoma. *Indian J. Exp. Bicl.*, 42: 981–988.

Rivero, R.M., Ruiz, J.M, and Romero, L.M., 2004. Importance of N source on heat stress tolerance due to the accumulation of proline and quaternary ammonium compounds in tomato plant. *Plant Biol.* (Stutta), 6: 702–707.

Roy, R.N. and Sen, S.K., 2002. Survey of antimicrobial streptomycetes from soils of West Bengal: Characterization and identification of potent broad spectrum antibiotic producing *Streptomyces albiboflavus* 321.2, *Hind. Antibiot. Bull.* 44: 25–33.

Roy, R.N. and Sen, S.K., 2004. Evaluation of biological activity of dibutyl phthalate produced by *Streptomyces albidoflavus* 321.2. *J. Sci. Technol.*, Sambalpur University, 16(A): 125–132.

Roy, R.N., Laskar, S. and Sen, S.K., 2005. Dibutyl phthalate, the bioactive compound produced by *Streptomyces albidoflavus* 321.2. *Microbiol. Res. Germany* (In Press).

Tonon, G., Kevers, C., Faivre-Rampant, O., Grazianil, M. and Gaspar, T., 2004. Effect of NaCl and mannitol iso-osmotic stresses on proline and free polyamine levels in embryogenic *Fraxinus angustifolia* callus. *J. Plant Physiol.* 161: 701–708.

Sadasivam, S. and Manickam, A., 2004. *Biochemical Methods*, 2nd Edition. New Age International Publishers, pp. 246.

Yuan, F.S., Quo, S.L., Qin, Z.X., Deng, S.H. and Huang, G.H., 2001. Effect of dibutyl phthalate on demodicidosis. *Zhongguo Ji Sheng Chong Xue Yu Ji Sheng Chong Bing Za Zhi*, 19: 160–162.

Chapter 9

Biochemical Changes in the Snail *Bellamya bengalensis* (Lamarck) Under Toxic Stress of Sumicidin

P.H. Rohankar and K.M. Kulkarni
Environmental Physiology Laboratory, Department of Zoology, Government Vidarbha Institute of Science and Humanities, Amravati – 444 609

ABSTRACT

The glycogen, protein and lipid levels were studied in body tissues of *B. bengalensis* under pesticidal application of lethal and sublethal doses of synthetic pyrethroidal pesticide-sumicidin. All these constituents decreased significantly suggesting a rapid breakdown of glycogen, protein and lipid in energy metabolism under toxic stress.

Keywords: Pyrethroid, Glycogen, Lipid, Protein, Lethal and Sublethal concentration.

Introduction

The freshwater ecosystem is becoming increasingly polluted due to various pesticides and agrochemicals which also affect non-target organisms like fishes and snails (Holden, 1973). Study of biochemical parameters is helpful in understanding the mechanism of pesticide toxicity to organisms. The metabolites such as carbohydrates, proteins and lipids which play a major role in body construction and energy production, are adversely affected by water pollution which lead to inhibition of important enzymes, retardation of growth, longevity of organisms (Sri Nivasa Murthy, 1986).

As molluscs form an integral part of aquatic ecosystem and are good marker of aquatic pollution and few reports are available on variations in biochemical components in molluscs under pesticide stress (Kulkarni *et al.*, 1985; Rao *et al.*, 1995; Shah *et al.*, 2001; Amanulla *et al.*, 2004) the present work was undertaken.

Materials and Methods

The snails were collected from Wadali tank near Amravati city. Active and healthy snails were used in study after acclimatization. These were divided into two groups as control and experimental. The test solutions prepared for biochemical experiments as LC_{50} for 96 hr (0.15 ml/l which is termed as lethal concentration and 1/3rd concentration of 96 h LC_{50} (0.0475 ml/l) which was taken as sublethal concentration of pyrethroid Sumicidin. At the end of 24, 48, 72 and 96 hrs of lethal and sublethal exposure to pesticide, the required number of treated snails from each group were sacrificed by decapitation, tissues separated, weighed and analysed for biochemical components such as glycogen, proteins and lipids. The glycogen was estimated by Anthrone method (Carrol *et al.*, 1958). Protein was estimated by Biuret method (Lowry *et al.*, 1951) and total lipids, were estimated by chloroform methanol method (Folch *et al.*, 1957) and values were expressed as mg/gm wet wt. tissues.

Results and Discussion

There is maximum reduction in glycogen level of all tissues exposed to lethal concentration as compared to tissues of the snails exposed to sublethal concentration of pyrethroids. There is decreasing trend from 24 hr to 96 hr, the reduction is maximum in glycogen level of hepatopancreas of snail. This might be due to high metabolic potency and efficiency of the gland as compared to other tissues. In general, there is significant and gradual decrease in the levels of glycogen in all tissues tested.

From the Table 9.2 it is evident that there is total decrease of lipid content of all tissues continuously at all hours, hepatopancreatic tissue exposed to lethal and sublethal concentrations show maximum decrease of lipid level than mantle, foot, gills and gonad tissue. The decrease is more significant at lethal concentration of pesticide.

Changes in protein level of body tissue are recorded in Table 9.3. Protein also show decreasing trend in body tissue with increase in concentration of pesticides and time exposure. There is maximum decrease at 96 hr exposure in all tissues of the snail. The hepatopancreas tissue show maximum decrease in trend than other tissues exposed to sumicidin.

Changes in macromolecular components like glycogen, protein and lipids are considered to be the indicators of pollution stress. In present study, marked decrease in glycogen content suggesting the possibility glycogenolysis, probably caused by a stress induced by pollutant mainly due to hypoxia. Many investigators observed similar changes (Sheela *et al.*, 1992; Vasanthi and Ramaswamy, 1987; Patil and Kulkarni, 1993; Gupta and Agrawal, 1995; Muley and Mane, 1995; Shaikh and Yeragi, 2004) decrease in glycogen content might be for high rate of energy production, at onset of various enzymatic blockage due to pesticidal toxicity (Kulkarni and Chaudhari, 1995).

Lipid is also a major source of energy after carbohydrates as it is reserved stock of food for animals. Since more energy is needed to minimize any stress condition, lipid source may be used after depletion in glycogen as immediate source (Gupta, 1987; Padmaja and Rao, 1994; Muley and Mane, 1995; Bhavan and Geraldine, 1997).

Protein is a complex organic substance of dual importance for organism as a building material and as a source of energy. Several investigators reported that pesticide interferes with protein synthesis

Table 9.1: Changes in Glycogen Content of Different Tissues of Snail *B. bengalensis* Exposed to Lethal and Sublethal Concentration of Sumicidin (mg/100 mg wet wt. tissue)

Time Interval	*Hepatopancreas*		*Foot*		*Mantle*		*Gill*		*Gonad*	
	Lethal	*Sublethal*	*Lethal*	*Sublethal*	*Lethal*	*Sublethal*	*Lethal*	*Sublethal*	*Lethal*	*Sublethal*
Control	16.11±1.70	17.18±1.67	10.93±1.74	11.00±1.91	16.25±2.13	16.43±1.26	10.11±1.01	19.13±0.93	9.11±1.19	9.14±1.12
24	13.24*±2.00	15.14±1.61	10.66±1.96	10.86±1.91	15.37±1.97	16.56±1.45	9.15**±0.87	9.01±0.99	8.62*±1.41	9.08±1.00
	(–25.62)	(–3.98)	(–2.89)	(–4.14)	(–2.26)	(–0.70)	(–2.53)	(–1.84)	(–9.41)	(–10.86)
48	12.48**±2.26	14.37±1.76	8.64*±2.11	9.64**±1.53	14.70±2.11	15.56±1.45	8.12±1.172	8.38**±1.37	8.25*±1.45	8.69*±1.32
	(–25.62)	(–16.35)	(–21.38)	(–12.36)	(–9.53)	(–5.29)	(–19.68)	(–8.21)	(–9.44)	(–4.92)
72	10.58±1.53	12.59*±2.05	7.47**±1.25	9.38±2.01	13.42±1.72	14.62***±1.71	7.51±1.34	8.19***±1.32	7.54**±1.51	8.24±1.13
	(34.32)	(26.71)	(–32.02)	(–14.72)	(–17.41)	(–11.01)	(–25.71)	(–10.29)	(–17.23)	(–9.84)
96	8.49±1.71	10.72±.1.70	6.71**±1.37	8.68**±1.49	11.79*±1.78	13.46±1.40	6.39±1.28	7.47±1.05	6.84*±1.42	8.01±1.09
	(–47.29)	(–37.60)	(–39.23)	(–21.09)	(–22.44)	(–18.07)	(–36.79)	(–18.18)	(–24.91)	(–12.36)

* $P < 0.05$, ** $P < 0.01$, *** $P < 0.001$.

Figures in parenthesis indicate percent change value.

Table 9.2: Changes in Total Lipid Content of Different Tissues of Snail *B. bengalensis* Exposed to Lethal and Sublethal Concentration of Sumicidin (mg/100 mg wet wt. tissue)

Time Interval	*Hepatopancreas*		*Foot*		*Mantle*		*Gill*		*Gonad*	
	Lethal	*Sublethal*	*Lethal*	*Sublethal*	*Lethal*	*Sublethal*	*Lethal*	*Sublethal*	*Lethal*	*Sublethal*
Control	9.06±0.97	8.65±1.29	9.18±1.17	8.86±1.10	6.15±1.14	5.91±1.17	5.58±1.28	5.52±1.35	9.10±1.07	9.07±0.99
24	8.11**±0.95	8.93±1.03	8.77**±1.61	8.66**±1.54	5.86*±0.93	5.71*±1.40	5.15±1.19	5.49±1.24	8.62±1.31	8.94±1.05
	(–6.24)	(–1.43)	(–7.40)	(–2.59)	(–4.71)	(–3.38)	(–6.53)	(–0.54)	(–5.27)	(–1.43)
48	7.70±1.42	8.70***±1.31	8.06±1.25	8.36±1.31	5.40**±0.90	5.50±1.35	4.71±1.23	5.25±1.18	8.08±1.13	8.62±1.04
	(–10.98)	(–3.97)	(–18.19)	(–7.67)	(–12.19)	(–6.93)	(–10.88)	(–4.89)	(–11.20)	(–4.96)
72	7.17±1.11	8.03±1.14	7.28±1.19	8.15±1.04	4.63±1.08	5.21±1.04	4.15***±0.79	4.85*±1.16	7.57±1.38	8.35±1.33,
	(–17.68)	(–11.36)	(–28.97)	(–12.64)	(–24.71)	(–11.84)	(–24.68)	(–12.13)	(–16.81)	(–7.93)
96	6.60±1.07	7.26***±1.33	5.59±1.34	7.71±1.12	4.37**±0.86	4.90***±0.84	3.86***±0.86	4.53±1.16	7.28±1.33	8.01±0.96
	(–26.01)	(–19.36)	(–39.76)	(–23.25)	(–28.94)	(–17.08)	(–29.94)	(–17.93)	(–20.00)	(–11.68)

* $P < 0.05$, ** $P < 0.01$, *** $P < 0.001$.

Figures in parenthesis indicate percent change value.

Table 9.3: Changes in Total Protein Content of Different Tissues of Snail *B. bengalensis* Exposed to Lethal and Sublethal Concentration of Sumicidin (mg/100 mg wet wt. Tissue)

Time Interval	Hepatopancreas		Foot		Mantle		Gill		Gonad	
	Lethal	*Sublethal*	*Lethal*	*Sublethal*	*Lethal*	*Sublethal*	*Lethal*	*Sublethal*	*Lethal*	*Sublethal*
Control	11.28±1.49	11.79±1.18	7.26±1.56	8.45±1.31	7.50±2.02	7.09±1.21	2.52±1.28	5.43±1.07	9.24±1.30	9.57±1.24
24	8.39*±1.71	11.32±1.35	7.05**±1.27	8.10**±1.43	7.33±2.25	7.04±1.07	2.38±1.15	5.33±1.02	8.37±1.05	8.53±0.83
	(–25.62)	(–3.98)	(–2.89)	(–4.14)	(–2.26)	(–0.70)	(–2.53)	(–1.84)	(–9.41)	(–10.86)
48	8.49**±2.01	10.92±1.06	6.57**±1.25	7.60±1.01	6.56±1.33	6.32***±0.92	4.89***±0.77	5.20±0.91	7.49±1.35	7.98±1.25
	(–25.62)	(–7.38)	(–9.50)	(–10.05)	(–12.53)	(–10.86)	(–11.41)	(–4.23)	(–18.93)	(–16.61)
72	7.24±1.58	10.58±1.10	6.35**±1.15	7.39±0.96	6.24±1.15	6.13***±0.79	4.72***±0.73	5.03±0.81	7.26±1.44	7.90±1.20
	(–35.81)	(–10.26)	(–12.53)	(–12.54)	(–16.80)	(–13.54)	(–24.49)	(–7.36)	(–21.42)	(–17.45)
96	6.51±1.34	9.91±1.06	4.43***±1.01	6.90***±0.83	5.60±1.05	5.44***±0.82	4.55***±0.67	4.83±0.78	7.09±1.38	7.72±1.17
	(–42.28)	(–15.95)	(–25.20)	(–18.34)	(–25.33)	(–23.27)	(–17.57)	(–11.04)	(–23.26)	(–19.33)

* $P < 0.05$, ** $P < 0.01$, *** $P < 0.001$.

Figures in parenthesis indicate percent change value.

and degradation resulting in alteration of equilibrium (Sivakami *et al.*, 1994; Singh and Bhati, 1994; Sinha, 1997; Rajamannar and Manohar, 1998) depletion in tissue protein may show a prior increased energy cost of homeostasis, tissue repair and detoxification during stress. The present work is in agreement with these workers.

References

Amanulla Hameed, S.V.S., Nazir Ahmed T.A. and Shah D.S.M. 2004. Effect of Butyltin toxicity on lipid metabolism in an estuarine Mussel *Sunetta scripta. J. Ecotoxicol. Environ. Monit.*, 14(3): 185–190.

Bhavan, P.S. and Geraldine, P. 1997. Alteration in concentration of protein, carbohydrate, glycogen, free sugar and lipid in the Prawn *Macrobrachium malcomsonii* on exposure to sublethal concentration of endosulfan. *Pestic. Biochem. Physiol.*, 58: 89–101.

Carrol, N.V., Longley, R.W. and Roe, J.H. 1950. Glycogen determination in liver and muscle by the use of anthrone reagent. *J. Biol.*, 22: 583.

Folch, J., Less, M. and Sloane-Stanley, G.H. 1957. A simple method for the isolation and purification of total lipids from animal tissues. *J. Biol. Chem.*, 26: 499–509.

Gupta, S. 1987. Effect of vegetable oil factory effluent on the lipid content in the liver of *Channa punctatus* (Bloch) studies *in vitro* incorporation of radio labelled substances. *J. Environ. Biol.*, 8: 353–357.

Gupta, R.C. and Agrawal, Sapna. 1995. Alteration in glycogen level under the stress of Azodyes in three freshwater teleost, *Channa punctatus, Channa striatus* and *Catla catla. Ad. Bios.*, 14(1): 97–104.

Holden, A.V. 1973. Effects of pesticides on fish. In: *Environmental Pollution by Pesticides,* (Ed.) C.A. Edwards. Plenum Press, London.

Kulkarni, A.N., Kamble, S.M. and Keshvan, R. 1985. Metabolic depression in the freshwater bivalve *Lamellidens corrianus* exposed to insecticide Hildane. *Proc. Symp. Asserr. Environ. Pollut.*, p. 225–227.

Lowery, O.H., Rosenbrough, N.T., Farr, L.R. and Randall, J. 1951. Protein measurement with Folin-phenol reagent. *J. Biol. Chem.*, 193: 265–275.

Muley, D.V. and Mane, U.H. 1995. Endosulfan toxicity to freshwater mussel *Lamellidens marginalis* and pH induced changes, a biochemical approach. *Indian J. Comp. Anim. Physiol.*, 13(1): 21–26.

Padmaja, R.J. Balaparameswara Rao M. 1994. Effect of an organochlorine and three organophosphate pesticides on glucose, glycogen, lipid and protein content in tissue of the snail *Bellamya dissimilis* (Muller). *Bull. Environ. Contam. Toxicol.*, 53: 142–148.

Patil, P.S. and Kulkarni, K.M. 1993. Biochemical response of cythion exposed frog *Rana cyanophlyctis. Ad. Bio. Sci.*, 12: 101–108.

Rajamannar, K. and Manohar, L. 1998. Sublethal toxicity of certain pesticides on carbohydrate, protein and aminoacids in *Labeo rohita. J. Ecobiol.*, 10(3): 185–191.

Rao, K.R., Shejule, K.B., Sasane, S.R., Subhas, N., Patil, P.N. and Choudhari, T.R. 1995. Impact of decis on carbohydrate metabolism of the Gastropod *Indoplanorbis exustus* from Panzara river, Dhule. *Proc. Acad. Environ. Biol.*, 19(3): 215–220.

Sheela, M., Mathivanan, R. and Muniandy, S. 1992. Impact of fenvalerate on biochemical status of different tissues in the fish *Channa striatus* (Bloch). *Environ. and Ecol.*, 10(3): 547–549.

Sinha, G.M. 1997. Effect of cadmium on protein content in plasma, body muscle and liver in an Indian major carp *Labeo rohita* (Hamilton). *Environ and Ecol.*, 15(2): 275–282.

Singh, S. and Bhati, D.P.S. 1994. Evaluation of liver proteins due to stress under 2,4-D-intoxication in *Channa punctatus* (Bloch). *Bull. Environ. Contam. Toxicol.*, 53: 149–152.

Shah, P.M., Rojaramani, V., Sathik, O. and Sathiya Priya, R.C. 2001. Effect of tributyltin oxide in lipid metabolism in earthworm edible blood clam *Anadara rhombea* (Born). *J. Environ. Pollut.*, 8(1): 7–11.

Shaikh, Nisar and Veragi, S.G. 2004. Effect of Rogor 30E (organophosphate) on muscle: protein in the freshwater fish *Lepidocephalecthyes thermalis. J. Ecotoxicol. Environ. Monit.*, 14(3): 233–235.

Sivakami, R., Premkishore, G. and Chandran, M.R. 1994. Effect of chromium on the metabolism and the biochemical composition of selected tissues in the freshwater catfish *Myctus vittatus. Environ. Ecol.*, 12(2): 259–266.

Srinivasa Moorthy, K., Kasi Reddy, B., Swami, K.S. and Chetty, S.R. 1986. Dichlorovas, induced metabolic changes in tissues of freshwater mussel *Lamellidens marginalis. J. Environ. Biol.*, 7(2): 101–106.

Chapter 10

Influence of Load Carrying in Cross Country Mode on Physiological Parameters of Yak (*Poephagus grunniens* L.) in Mountainous Terrain of Arunachal Pradesh

B.C. Das[1], M. Sarkar[2], D.N. Das[3], D. Gogoi[2], A. Basu[2], D.B. Mondal[4], M. Mazumder[2], P. Bora[2] and M. Ahmed[2]*

[1]*Department of Veterinary Physiology, WBUA&FS, Kolkata – 37*

[2]*National Research Centre on Yak (ICAR), Dirang, West Kameng, Arunachal Pradesh – 790 101*

[3]*Scientist, NDRI, SRS, Adugodi, Bangalore*

[4]*Senior Scientist, Veterinary Medicine Division, IVRI, Izanagar, Bareilly – 243 122, U.P.*

ABSTRACT

Yaks were subjected to load 15 per cent of their body weight and allowed to walk in cross country mode on ups and downs of the hilly terrain almost at around 10,000 ft altitude. Pulse rate, Respiration and body temperature differed significantly ($p < 0.01$) in different treatment group like before, between and after load carrying.

Keywords: Yak, Physiological parameters, Load.

Introduction

Yak is a multipurpose bovid of high altitude and mostly found at 6000–10,000 ft and even as high as upto 30,000 ft above msl. Surprisingly at such altitude apart from producing meat and milk, it is

* Corresponding Author.

used as pack animals by yak herders, where partial pressure of oxygen is very less. It is a sure footed animal and carry the load in hilly terrain without much difficulty. It is the only means of transportation for hilly people as because of difficult topography and landscape. By rearing yak at high altitude not only a balance in nature's ecosystem maintained but also through them man has successfully managed a system of resources of food and energy. Like other animals, work leads to adjustments in the various organ systems involved in work process. Sustained muscular activity over prolonged periods depends mainly on aerobic process and oxygen supply may be a limiting factor in case of yak as its habitat is high altitude various organ systems involved in work process. So far no systematic investigations have been carried out for assessing the work performance of yak. In the context of overall energy strategy this should include, identifying better quality of yak for draught breeding, better feed availability and improved health care, so that it could act in a better way of means of transportation, could pay a noble service to army placed at remote high hills border areas as well as to local yak herders for their day to day need. Thus a right approach in this direction would be not only measuring draught power but also testing the physiological work performance. In the present investigation efforts have been made to study physiological responses.

Materials and Methods

The present study has been conducted at Nuykmadung yak farm of National Research Centre on yak. The altitude of farm is around 8500 ft from msl. Meteorological attributes during experimental period was moderately cold humid. Temperature in morning and afternoon was 18±0.4 and 18.2±0.3°C respectively. Relative humidity morning to evening ranges from 74.2±5.2 to 73.7±2.6 per cent.

Selection of Animals

Apparently healthy and free from respiratory diseases, eight adult male yak, above 2–3 years of age selected from the herd of NRC Yak, Dirang, Arunachal Pradesh. These animals were of similar age and body weight(s).

Management of Experimental Animals

All the experimental animals were maintained under identical conditions of feeding and management. Animals were fed as per schedules adopted at NRC Yak and feeds and fodder were fed depending upon the availability from time to time.

Training of Animals for Walking

A total of ten male yak selected from farm taken daily in the morning for 30 min walk and this training continued for a period of 20 days. Yak in the farm is maintained through semi-range system of management and stall-feeding is practiced. Yaks are not habituated for any kind of exercise and carrying load in the farm premises. Initial 7 days there was a simple walk without any load and afterwards they were subjected for load with gradual increment for another 13 days or so before taking up the actual exercise. During training period no recording have been taken for experiment purpose. When trained sufficiently, actual experiment started on day 21st. All the animals were subjected to weight 15 per cent of their body weight and allowed to move for about 11 km in downslide through approachable kachha road. Loads were placed on the back of yak with the help of a wooden shaddle locally made. Before starting the load carrying all the physiological responses have been recorded. Animals were made to walk around 22 km up and down. After reaching farm all observations were made. So the to total journey made is 22 km out of which 11 km downhill and 11 km upward hilly tract.

A total of six observations were taken, maximum of twice in a week. The rectal temperature was recorded by a clinical thermometer, which was inserted about 5cm deep in the rectum, and it remained in contact with mucus membrane for at least 1–2 minute. The observations was recorded in °F.

The respiration rate was observed by Flanks methods in which inward and outward movements were counted as one complete respiration. The 'respiration rate (Rt) was also recorded by counting expiration of a air over the fingers per unit time (breaths/min). For pulse rate (bpm), coccygeal arteries is monitored and counted.

Results and Discussion

For all 7 different yak individual variations of different parameters were considered before and after load carrying.

Pulse Rate

The pulse rate before load carrying found to be 56.305±1.95 and after walking around 11 km *i.e.* between load carrying it was observed 67.68±1.95 and after load carrying, when travelled around 22 km, it reached to 73.093±1.95 (Tables 10.1 and 10.2). Pulse rate increased significantly ($p < 0.01$) between all the treatment group, which is much similar with the findings of Mondal, 1998; where in yak have been subjected to different weight load irrespective of their body weight and pulse rate in every occasion increased significantly. The same rising trends were also observed in different other animals like cattle and buffalo after work performance (Upadhyay, 1982; Parveen *et al.* 1987; Upadhyay and Madan, 1985 and Zerbini *et al.*, 1992). Heart rate regulation during exercise is effected locally for the most part by level of metabolism in the muscle cell.

Table 10.1: Physiological Parameters of Yak While Carrying Load (15 per cent) of Body Weight in Cross Country Road

Treatment Group	*Observation*	*Pulse/min*	*Respiration/min*	*Temperature (°F)*
Before load carrying	7	56.305±1.95[a]	19.831±2.11[a]	99.986±0.41[a]
Between load carrying	7	67.686±1.95[b]	99.496±2.11[b]	102.189±0.41[b]
After load carrying	7	73.093±1.95[c]	112.991±2.11[c]	103.346±0.41[c]

Different superscript indicates significant difference ($p < 0.01$).

Table 10.2: ANOVA of Different Physiological Parameter Cross Country Mode

Source	*df*	*Pulse*		*Respiration*		*Temperature*	
		MSS	*F*	*MSS*	*F*	*MSS*	*F*
Between Treatment	2	513.98	19.37**	17741.872	568.630**	20.395	17.522**
Within Treatment	18	26.56		31.201		1.164	

**Indicates significance at 1 per cent level.

This local control of blood flow is the most important factor in securing an efficient blood supply to the working muscle (Rowell, 1974). For the present investigation heart rate increase upto a maximum of 62 bpm, which was optimum and further increase was not necessitated or the level in heart rate stabilized at around 62 for these levels of activity and to perform these levels of activity probably further increase was not required. Oxygen transporting capacity assessments are based on linear increase in heart rate with increasing O_2 uptake or work load (Ramesch and Becker, 1991). On the other hand, that heart rate changes quite accurately reflect the physiological state during work in animals (Upadhyay, 1982, Romesch and Becker, 1991). The increase in heart rate as influenced by pulling loads was reported unto 131 from resting value of 76 bpm in yak males by Mondal, 1998.

Respiratory Rate

The respiratory rate before load carrying found to be 19.831±2.11 and after walking around 11 km *i.e.* between load carrying it was observed 99.496±2.11 and after load carrying, when travelled around 22 km, it reached to 112.991±2.11 (Tables 10.1 and 10.2). Respiratory rate increased significantly ($p < 0.01$) between all the treatment group, which is much similar with the findings of Mondal, 1998; where in yak have been subjected to different weight load irrespective of their body weight and respiratory rate in every occasion increased significantly. The resting pre and post work values of respiratory frequency (Rf) varied from 17 to 26, 90 to 107 and 112 to 120 per min respectively in yak. These values are in general agreement with the reported findings by Mondal, 1998. It becomes evident on analysis that in terms of changes in Rf during work may not increase to far greater levels, mainly due to the reason that inspiratory and expiratory centres have to co-ordinate to accommodate more number of Rf in a stipulated time and requirements of oxygen to sustain work has to be achieved precisely (Astrand and Rodahl, 1977). Increase Rf due to load carrying has been reported in other animals like cattle and buffalo (Upadhyay, 1982, Parveen *et al.*, 1987, Upadhyay and Madan, 1985 and Zerbini *et al.*, 1992).

Rectal Temperature

The respiratory rate before load carrying found to be 99.986±0.41 and after walking around 11 km *i.e.* between load carrying it was observed 102.189±0.41 and after load carrying, when traveled around 22 km, it reached to 103.346±0.41 (Tables 10.1 and 10.2). Respiratory rate increased significantly ($p < 0.01$) between all the treatment group, which is much similar with the findings of Mondal, 1998; where in yak have been subjected to different weight load irrespective of their body weight and body temperature in every occasion increased significantly. The resting values of respiratory frequency (Rf) varied from 98 to 101, 101 to 103 and 101 to 104 °F respectively in yak. These values are in general agreement with the reported findings by Mondal, 1998. For yak the rise of temperature is very high which is significant ($p < 0.01$) may be due to strenuous exercise against gradient indicating high metabolic activity and heat of work accumulation was higher. The resultant rise in rectal temperature of yak due to effects of work increased in relation to magnitude of work. The results have similar trends as reported by Mondal *et al.*, 1998, but rising magnitude was not like present investigation. This could be attributed to the minimum number of sweat glands and heat of work was found more in yak who were probably able to increase their heat loss by increased sweating and were able to maintain equilibrium in the heat produced and heat loss. Increase in rectal temperature in relation to type of work and duration have been reported by several workers in different species (Upadhyay, 1982; Parveen *et al.*, 1987; Upadhyay and Madan, 1985 and Zerbini *et al.*, 1992). Deep body temperature is mainly a function of the energy output and is with a wide range independent of ambient air temperature. Actually total aerobic energy production seems to be more decisive for the final body temperature than heat production and to sustain physiological adjustments (Astrand and Rodahl, 1977).

References

Astrand, P.O. and Rodahl, K., 1977. *Textbook of work physiology*, 2nd Edn. McGraw Hill, New York.

Bhattacharya, S., Acharya, A.K., Chowdhary, T.M. and N.C. 1965. Seasonal variation in body temperature, pulse rate, respiration rate and Hb concentration of blood in different breeds of Indian heifers and growing bulls. *Indian J. Vet. Sci.*, 3: 38–49.

Gaalaas, R.F., 1945. Effects of atmospheric temperature on body temperature and respiration rate of joursey cattle. *J. Dairy Sci.*, 28: 555–563

Goswami, S.B. and Premnarain, 1962. The effect of air temperature and relative humidity on some physiological indices of buffalo bulls. *Indian J. Vet. Sci. and A.H.*, 32: 112–118.

Mondal, D.B., 1998. Effect on normal haemato-physiology due to pack on yaks: A preliminary study. In: *Proceedings of the Second International Congress on Yak*, Beijing, pp. 63–65.

Mullick, D.N., 1960. Effect of humidity and exposure of sun on pulse rate, rectal temperature and haemoglobin level in different sexes of cattle and buffaloes. *J. Agric. Sci.*, 54: 391–394.

Parveen Kumar, Rao, M.V.N. and Upadhyay, R.C., 1987. Physiological responses of young crossbred bullocks during load pulling operations. *J. Vet. Phyiol. and Allied Sci.*, 6(1): 1–5.

Rowell, L.W., 1974. Human cardiovascular adjustments to exercise and stress. *Physiol. Rev.*, 54: 75–158.

Romesch, M. and Becker, K., 1991. Methodological experiments on suitability of heart rate as a parameter for characterising the draft potential of oxen. *Draught Animal News*, 14: 2–3.

Upadhyay, R.C., 1982. Work efficiency and associated physiological changes in crossbred and Haryana bullock. *Ph.D. Thesis*, Kurushetra University, Kurushetra.

Upadhyay, R.C. and Madan, M.L., 1985. Studies on blood acid base status and muscle metabolism in working bullock. *Anim. Prod.*, 40: 11–16.

Zerbini, E., Gemeda, T., Oneil, O.H., Howell, P.J. and Schroter, R.C., 1992. Relationship between cardiorespiratory parameters and draught work output in F1 crossbred dairy cows under field condition. *Anim. Prod.*, 55(1): 1–10.

Chapter 11

Seasonal Impact on Per Ovarian Oocyte Retrieval Rate in Buffalo

B.C. Das[1], M.L. Madan[2], R.S. Manik[2] and M. Sarkar[3]*

[1]Deptt of Veterinary Physiology, WBUA&FS, 37, K B Sarani, Kolkata – 37

[2]Embryo Biotechnology Centre, National Dairy Research Institute, Karnal – 132 001, Haryana

[3]Scientist, National Research Centre on Yak, Dirang, West Kameng, Arunachal Pradesh – 790 101

ABSTRACT

Oocytes recovery rate per ovary has been studied in the experiment to assess the oocytes potential in buffalo of ovaries using aspiration techniques. A total of 1137 ovaries collected during 13 trial and 923 oocytes were aspirated from surface follicles and further distributed for the A and B type. The overall oocyte recovery, type A and type B were 0.81,0.43 and 0.37 per ovary respectively. The maximum recovery rate was in the winter month and low in summer period due to seasonal impact on buffalo physiology.

Introduction

In the past few decades, rapid progress has been observed in the development of new technology of reproduction for the genetic improvement in the farm animals, especially cattle. Application of technology to buffalo had limited success due to the fact that this species has some inherited problem of low reproductive efficiency. There are less number of primordial follicles and primordial oocytes in the ovary of buffalo as compared to cattle, which is directly related to low reproductive potential. In cattle, an average of four to six embryos have been successfully collected through super ovulation using exogenous hormones. Among buffaloes, the above response is up to two embryos, only fifty per cent of these are of normal quality and transferable. An attempt has been made to study the per ovarian oocyte recovery potential in buffalo using different aspiration techniques, so that these oocytes could

* Corresponding Author.

be used for further genetic improvement through IVM. IVF and IVC. Primordial follicles in the ovaries of buffalo (average number 11,384) are less than that in cattle (average number 50,000). It appears to be related to its low reproductive potential (Danell, 1987). Many factors have been assumed to be holding the oocytes at dictyate stage of arrest. The factors include gonadotrophins (Tsafriri *et al.*, 1982; Fulka *et al.*, 1985) and interaction between follicular cells and fluid (Guraya, 1985).

Successful in vitro fertilization in most species is dependent on many factors including availability of good number of superior quality oocytes. This is more important in buffalo where yield of follicular oocytes is poor (Madan *et al.*, 1992; Majumder *et al.*, 1988; Singh *et al.*, 1989; Sharma, 1990 and Totey *et al.*, 1992) and all possibilities of low yield during summer months. To see the seasonal impact this study was taken up. There are three techniques commonly used to collect follicular oocytes *in vitro i.e.* dissection, aspiration and slicing.

Materials and Methods

Buffalo ovaries collected (January–June) from Delhi slaughterhouse, which is 130 km away from laboratory. The ovaries collected from matured female buffalo irrespective of their body condition. The ovaries were removed within 2h after animals had been stunned. The ovaries were washed and brought to the laboratory in sterilized saline at 37°C in a thermos flask. The collection and transport of ovaries to laboratory took about 6 hrs.

The ovaries were washed 2–3 times with sterilized saline (37°C) and then lightly dried on sterilized blotting paper. A 20-21-gauge needle attached to 5ml sterilized glass syringe containing 1 ml oocyte collecting media was inserted into the ovary to enter into the follicle of 2–5 mm size from the backside rather than from the surface. The aspirated fluid along with oocyte collection media was collected in a sterilized oocyte searching dish (90 × 15 mm, Nunc, USA) containing 2–3 ml collection media.

A stereozoom microscope (Nikon) was used to locate the oocyte in the media at 20–25 X magnification. The oocytes were removed with glass Pasteur pipette attached to a rubber tubing for mouth control. The collected oocyte were washed three or four times in oocyte culture media and quality of oocytes was judged based upon their morphology and integrity of cumulus cells. For present investigation, the oocytes were grouped into two types. Type A: Oocytes with 3–4, 2 or 1 layers of cumulus cells were taken into consideration. Type B: Completely denuded oocytes were taken into consideration. Partially denuded oocytes were made denuded by repeated pipetting.

It is opined that the oocytes having layers of compact granulosa cells around the cumulus oophorus cells of secondary or tartiary follicles are good for maturation than denuded (Totey *et al.*, 1992). The oocytes here collected from the ovaries using aspiration techniques.

Results and Discussion

A total of 1137 ovaries were collected during 13 visits to slaughterhouse. A total of 923 oocytes were aspirated from the surface follicles of 2–5 mm size from the ovaries which were further distributed to Type A (497) and Type B (429) depending upon the granulosa cells surrounding the oocyte. The data was further subjected to calculate per ovarian availability of various types of oocytes. The overall oocyte recovery, oocyte with cumulus mass (A) and denuded oocytes (B) were 0.81, 0.43 and 0.37 per ovary respectively with the range of 0.63 to 1.21 for total oocyte, 0.26 to 0.74 for A type and 0.27 to 0.47 for B type. The critical evaluation of data reveals that maximum (1.21/ovary) recovery of the oocytes was in the early months of the experimental trial, which incidentally coincides, with the winter months. As depicted in Table 11.1, oocytes recovery rate reduces as the trial progresses from winter to summer. Therefore, differences may be attributed to the well-known seasonal effect on buffalo physiology.

Totey *et al.* (1992) reported low yield of oocyte recovery by aspiration of buffalo ovaries. They aspirated the oocyte from 1226 ovaries and per ovarian availability was low (0.73) in comparison to present results. However, usable oocytes (with cumulus complex) per ovary reported by them were 0.43, which was almost same as in the present study. Madan (1992) also reported similar results with the recovery rate of 0.42 (OCC) per ovary.

Table 11.1: Per Ovarian Oocyte Recovery Rate During Different Trial

Sl.No	No. of Ovaries	Oocytes Aspirated	Type of Oocytes		Per Ovarian Recovery
			OCC (A)	DO (B)	
1.	100	121	74	47	1.21
2.	65	50	32	18	0.76
3.	75	65	32	33	0.86
4.	102	86	51	35	0.84
5.	87	71	46	25	0.81
6.	102	88	48	40	0.86
7.	106	87	46	41	0.82
8.	105	77	39	38	0.73
9.	103	70	31	39	0.67
10.	68	43	18	25	0.63
11.	67	47	19	28	0.70
12.	85	61	32	29	0.71
13.	72	57	29	28	0.79

The overall yield of total and OCC per ovary was low in buffalo compared with cattle in which the number of oocytes per ovary has been reported to be up to 10 (Gordon and Lu, 1990). The low recovery of oocytes per ovary in buffalo can be due to the low average number of primordial follicles in the ovaries (12636 in cycling buffalo and 10132 in non-pregnant cycling buffalo). Danell (1987), Samad and Nasseri (1979) have also reported low number of follicles in the Surti breed (12000 primordial follicles) and Nilli Ravi breed of buffalo (19000 primordial follicles) in contrast to cattle (10,000 primordial follicles).

In conclusion, oocytes retrieval rate is much higher during winter month and reduces gradually when summer approaches in buffalo due to its physiological and endocrinological system unlike cattle and other domestic animals.

References

Danell, B., 1987. Oestrus behavior, ovarian morphology and cyclical variation in follicular system and endocrine pattern in water buffalo heifers. Sveriges Lantbruk suniver sitet. Markentil-Tryckeriet AB, Uppsala, Sweden.

Fulka, J. Jr., J. Motlik, J. Fulka and N. Crozet, 1985. Inhibition of nuclear maturation in fully-grown porcine and mouse oocytes after their fusion with growing porcine oocytes. *J. Exp. Zool.*, 235: 255–259.

Guraya, S.S., 1985. *Biology of Ovarian Follicles in Mammals*. Springer-verlag.

Gordon, I. and Lu, K.H., 1990. Production of embryo *in vitro* and its impact on livestock production. *Theriogenology*, 33: 77–87.

Madan, M.L., 1992. A project report on cattle herd improvement for increased productivity using embryo transfer technology. March, 1992.

Majumdar, A.C., Katiyar, P.K., Taneja, V.K. and Bhat, P.N., 1988. Maturation of slaughterhouse ovarian follicular oocytes of buffaloes in culture and subsequent IVF. In: *Proceedings 2nd World Buffalo Congress*, Delhi, 1(54).

Samad, H.A. and Nasseri, A.A., 1979. A quantitative study of primordial follicles in buffaloes heifer ovaries. In: *Compendium 13th FAO/SIDA International Course on Animal Reproduction*, Uppasala, Sweden.

Sharma, D., 1990. *MVSc Thesis*, IVRI, Izatnagar.

Singh, S., Totey, S.M and Talwar, G.P., 1989. *In vitro* fertilization of buffalo (*Bubalus Buubalis*) oocytes matured *in vitro*. *Theriogenology*, 31(1): 255.

Totey, S.M., Singh, G., Taneja, M., Pawshe, C.H. and Talwar, G.P., 1992. *In vitro* maturation, fertilization and development of follicular oocytes from buffalo. *J. Reprod. Fert.*, 95: 597–607.

Tsafrini, A., Pomerantz, S.H. and C.P. Channing, 1982. Inhibition of oocyte maturation inhibitor in follicular regulation of oocyte maturation. *J. Reprod. Fertil.*, 64(2): 541–551.

Chapter 12

Genetic Diversity Studies in Introgressed Lines of *Gossypium hirsutum* Cotton Using Cluster Analysis

J.S.V. Samba Murthy and N. Chamundeswari

Regional Agricultural Research Station,
Acharya N.G. Ranga Agricultural University, Lam, Guntur

ABSTRACT

Eighty two lines of *Gossypium hirsutum* along with two local checks developed out of introgression involving different wild species for tolerance to biotic and abiotic stresses and improvement in fibre quality at different Cotton Research Centres across the country were evaluated at Regional Agricultural Research Station, Lam during Kharif 2003-'04 to elucidate genetic divergence using a non-hierarchical Euclidean cluster analysis for yield and its components. The genotypes were grouped into ten clusters irrespective of geographic and genetic diversity. Cluster IX contained the largest number of sixteen genotypes and had the genotypes of heterogeneous origin, which showed that there was no parallelism between genetic and geographic diversity. The maximum genetic distance occurred between cluster III and VI. Cluster VI is monogenotypic and had high mean values for seed cotton yield, number of sympodia, boll weight, seed index, lint index, ginning out turn, 2.5 per cent span length and maturity coefficient. It is further suggested to go for a series of diallel analysis with the genotypes grouped in the above clusters for identifying superior heterotic combinations and isolating desirable recombinants in segregating generations. Genotypes were also identified which may serve as potent genetic donors for some metric and quality traits.

Introduction

Reduction in genetic variability of cultivated cotton makes it highly vulnerable to biotic as well as abiotic stresses. Besides above, changing national and international textile scenario, availability of quality cotton with better fibre qualities plays a major role. So genetic variability was created by introgression breeding utilizing wild sources of cotton. Mahalanobis D_2 statistic designed by Rao (1952) has been utilized by a number of workers for estimating genetic divergence. Although D_2 statistic has been widely used as a quantitative measure of genetic divergence, yet the clustering pattern of the genotypes is orbitrary (Spark, 1973). In the present study, eighty-two introgressed lines of *Gossypium hirsutum* cotton were subjected to non-hierarchical Euclidean cluster analysis to over come the limitations of D_2 statistics. Information on the nature and magnitude of genetic diversity present in superior parents for developing cotton varieties with multiple pest and disease tolerance besides arriving for fibre quality improvement.

Materials and Methods

The experimental material comprised of eighty-two variable introgressed lines along with two local checks was laid out in a randomized block design with two replications during Kharif 2003-04 at Regional Agricultural Research Station, Lamfarm, Guntur. Each entry was sown in two rows of six meters length. The spacing adopted between rows was 120cm while within the row it was 60cm. Normal agronomic practices recommended to the region were followed. Observations were recorded on five randomly selected plants in each entry for seventeen characters *viz.*, days to 50 per cent flowering, number of monopodia, number of sympodia, number of bolls/ plant, boll weight (g), number of seeds/boll, seed index (g), lint index (g), ginning out turn (per cent), 2.5 per cent span length (mm), maturity coefficient, uniformity ratio (per cent), micronaire (10^{-6} g/inch), bundle strength (g/tex), fibre elongation (per cent), fibre quality index, seed cotton yield/plant (g). The genetic divergence was carried out using non-hierarchal Euclidean cluster analysis as described by Beale (1969) and further elaborated by Spark (1973).

Results and Discussion

The analysis of variance revealed highly significant differences among the genotypes for all the seventeen characters under study. On the basis of Non-Hierarchical Euclidean cluster analysis, eighty-four genotypes were grouped into ten clusters (Table 12.1), which revealed existence of genetic diversity. The cluster IX had the highest sixteen genotypes followed by fourteen in cluster IV, thirteen in cluster III, seven in cluster V and VII, four in cluster VIII, three in cluster X and cluster VI was a single genotypic cluster. The clustering pattern of genotypes revealed no parallelism between genetic and geographic diversity. These findings are in agreement with Murthy *et al.* (1995b), Gururajan and Manickam (2002) and Altaher and Singh (2003).

Cluster means (Table 12.2) showed appreciable differences for all the characters, particularly in cluster VI which had high mean values for days to 50 per cent flowering, number of sympodia, boll weight, seed index, lint index, ginning out turn, 2.5 per cent span length, maturity coefficient and seed cotton yield per plant. Cluster I had high mean value for fibre elongation. Cluster II had the highest mean value for number of seeds per boll. Cluster V had the highest mean value for number of bolls per plant and bundle strength. Cluster VII had high mean value for fibre quality index. Highest mean value of uniformity ratio was observed in cluster VIII. Cluster X had high mean value for micronaire and number of monopodia per plant.

Inter and intra cluster distances are presented in Table 12.3. The magnitude of intra cluster distance measures the extent of genetic diversity between the genotypes of same cluster. Inter cluster distance is measuring of genetic distance between two clusters. The intra cluster distance was maximum in cluster IV (238.94) followed by cluster VIII (236.81) and minimum of zero in cluster VI. The maximum inter cluster distance of 1115.59 was observed between cluster III and VI followed by cluster IV and VI (961.84).

Table 12.1: Distribution of Genotypes in Different Clusters by Tocher's Method

Sl.No.	Cluster No.	No. of Genotypes	Genotypes
1.	I	11	IH 63, TCH 1649, IS 376/412/57, IS 376/411/31/27, GISV 25/1699XGCot 10, NA1325 X Palmeri, IS 376/411/20/25, IS 376/411/35-II, IS 376/411/34-I, IS376/411/37-I
2	II	14	GCot16XGISV 16, AKH 2053, AKH8828, IS376/411/13-II, GCot16XGISV61, TCH 1696, Abadita X Surat 5, RAI 9, MSH Sp91, AKH 2053-I, Hisutum X Barbadense X Wild-I, Hisutum X Barbadense X Wild-II, Abhadita X Surat 2, Hisutum X Barbadense X Wild-III
3	III	13	GISV 23XGCot16, 8-6PROG (Self) X AKH 2053, Abadita X TCH-1, Abadita X Surat 9, CWROK 165 X Palmeri, L 603 X Palmeri, TCH 1599, IS376/411/34-II, IS376/411/14, Abhadita X AKH 2053, Abadita X TCH-2
4	IV	8	GISV185 X Surat dwarf, GISV 61, MSM, 321-2PI (Self) X AKH2053, Hirusutm X Raimondii, Hirsutum X Anamolum, Rai 1, TCH 1654
5	V	7	Gcot 10 X GISV162, 8-6 PDx55 High fibre strength, IS376/411/33/21, GISV 154, TCH 1692, DSFH-I, IS376/411/1/87
6	VI	1	IH 35
7	VII	7	GCotl0X GISV79, TCH 1648, MCU 5 X Palmeri, GISV 2, Rai 7B-I, TCH 1650, TCH 1651
8	VIII	4	MAH 345, IS376/411/39, Abadita X Suart 2, TCH 1695
9	IX	16	GISV 197 X GISV61, IS376/411/7/64-II, TCH1652, GISV142, Abadita X Surat IV-3, TCH1691, MSH Sp53, IS376/411/68, Abadita X Surat IV-5, Abadita X TCH 6, Surabhi (Check 1), L 604 (Check 2), TCH 1653, TCH 1693, Abadita X Surat 5, GISV 201
10	X	3	Rai 7B-2, AKH 2031, GCot 10X GISV 61

Table 12.2: Intra and Inter Cluster Distances Values for 84 Genotypes

Cluster	I	II	III	IV	V	VI	VII	VIII	IX	X
I	193.16	245.38	236.30	336.66	431.63	780.59	404.63	520.88	325.47	378.38
II		121.72	212.83	258.20	516.05	816.49	298.16	317.54	230.81	359.32
III			141.95	267.47	593.42	1115.59	422.46	607.61	332.80	512.39
IV				239.94	574.16	961.84	371.09	552.55	384.16	658.50
V					188.06	422.00	535.90	552.93	362.52	492.01
VI						0.00	658.71	544.62	635.47	738.10
VII							178.01	338.50	314.74	612.99
VIII								236.81	329.03	433.16
IX									195.20	365.72
X										183.00

The clustering pattern of genotypes in the present study has not been distinctly influence by the source and origin. It should be logical to attempt the crosses between the genotypes of clusters separated by large estimated inter cluster distances to obtain high heterotic combinations and recombinants in segregating generations. It is further suggested that the genotypes which fall under cluster III and VI may be subjected to a series of diallel analysis to identify superior parents for future breeding programmes based on their per se performance together with general and specific combining ability effects.

Table 12.3: Cluster Means for Eleven Characters of Cotton Based on D_2 Values

Sl.No.	Source	I	II	III	IV	V	VI	VII	VIII	IX	X
1.	Days to 50% flowering	56.73	57.50	55.54	56.50	57.71	60.00	56.71	58.50	56.56	56.83
2.	No. of monopodia	1.81	1.88	1.71	1.84	1.64	2.00	1.96	1.30	1.82	2.03
3.	No. of sympodia	15.58	14.19	16.60	12.83	16.63	18.10	15.49	13.98	15.13	14.93
4.	No. of bolls/plant	32.85	18.90	24.97	27.14	40.31	34.10	21.83	14.98	24.07	18.27
5.	Boll weight (g)	3.15	3.78	3.07	3.46	3.48	4.57	3.48	4.40	3.55	3.86
6.	No. of seeds/boll	27.56	28.45	27.09	26.84	26.99	28.20	27.33	27.45	27.06	25.97
7.	Seed index (g)	8.07	7.75	6.76	7.84	8.18	10.60	9.93	9.90	7.68	7.83
8.	Lint index (g)	4.63	5.18	4.46	4.46	6.03	6.41	5.62	6.09	6.08	5.85
9.	Ginning out turn (%)	34.79	35.82	35.67	35.33	34.64	37.60	33.41	36.10	37.10	36.00
10.	2.5% span length (mm)	25.69	25.58	24.97	24.60	25.31	29.03	27.63	27.42	25.79	24.10
11.	Maturity coefficient	0.63	0.62	0.63	0.64	0.62	0.69	0.62	0.63	0.63	0.62
12.	Uniformity ratio	47.60	48.75	49.34	49.20	50.13	45.10	45.91	50.63	49.27	50.13
13.	Micronaire (10^{-6} g/inch)	4.11	3.88	3.65	3.18	3.69	4.02	3.08	3.82	3.64	4.75
14.	Bundle strength (g/tex)	23.49	24.06	23.95	23.95	26.36	18.20	22.80	23.93	23.62	21.83
15.	Fibre elongation (%)	6.57	6.47	6.48	6.45	6.50	6.30	6.45	6.50	6.48	6.43
16.	Fibre quality index	298.85	312.64	313.28	330.59	349.10	263.67	360.63	336.38	318.97	241.34
17.	Seed cotton yield/plant (g)	86.88	58.99	63.57	79.50	131.30	145.44	66.01	59.74	73.91	70.76

Acknowledgement

The authors are highly thankful to the authorities of Cotton Technology Mission and Acharya N.G Ranga Agricultural University for providing necessary financial help and infra structure facilities to undertake these studies.

References

Altaher, A.F. and Singh, R.P. 2003. Yield component analysis in upland cotton (*Gossypium-hirsutum* L. *J. Ind. Soc. Cotton Improv.*, 28: 151–157.

Beale, E.M.L. 1969. Euclidean cluster analysis. *Bull. Int. Stat. Inst.*, 43: 92–94.

Gururajan, K.N. and Manickam. 2002. Genetic divergence in Egyptian cotton (*Gossypium barbadense* L.). *J. Ind. Soc. Cotton Improv.*, 27: 77–83.

Murthy, J.S.V.S., Reddy, D.M. and Reddy, K.H.G. 1995. Genetic divergence for lint characters in upland cotton (*Gossypium hirsutum* L.). *Ann. Agrl. Res.*, 16: 357–359.

Rao, C.R. 1952. *Advance Statistical Methods in Biometrical Research.* John Wiley and Sons, Inc., New York.

Spark, D.N. 1973. Euclidean cluster analysis algorithm. *Appl. Stat.*, 22: 126–130.

Chapter 13

Present Pollution Level in Kolkata and its Abatement

Debojyoti Mitra
Lecturer, Department of Mechanical Engineering, Jadavpur University, Kolkata – 700 032

ABSTRACT

The present article senses the urgeness in evaluating the present situation of air pollution. The causes of air pollution in industrial and metropolitan areas are ventured with special attention to the grave situation at Kolkata and its suburbs. The various preventive measures, already taken or to be taken, to abate air pollution in Kolkata, are put forward for the general environmental awareness.

Keywords: *Air pollution, Kolkata, SPM, RPM, SO_x, NO_x, Abatement.*

Introduction

The environment is the totality of the physical conditions on earth or part of it, especially as affected by human activities. An environmentalist is concerned with the biosphere which extends from the deep sea to flying heights in the sky. The man is the best and the most intelligent creature in the biosphere. He continually changes his environment to meet his biological and social needs. He extracts the materials from the environment to sustain his life or for his ease and comfort. After utilization of the materials, the end products are wastes of all description some of which are harmful for the biosphere. The wastes that are harmful are called pollutants. They contaminate or defile the environment.

The byproducts of the industrial activities of human beings are mostly harmful which cause pollution of air, water and the land. These are known accordingly as atmospheric, hydrospheric and lithospheric pollution. Amongst them, the most detrimental is the air or atmospheric pollution. The human being inhales about 22000 times a day and takes about 16 kg of air per day. Air inhaled is far greater than the amount of food or water taken by a human being per day. Polluted air seriously affects the health and depending on the degree of pollution it can become life taking! Therefore, air pollution is the greatest environmental evil towards mankind.

The man cannot prosper and be comfortable without industrial activity, consider the situation if we stop power producing industries to reduce pollution. So, its growth must continue. But the industrial activities must be regulated and controlled in a manner such that the pollution caused by them is within statutory limits.

Another potential source for air pollution are the automobiles. The numbers of automobiles are larger in industrial areas. Therefore the pollution loads of the automobiles are superimposed on the existing pollution load of the industries, which make the situation grave. Congested cities like Kolkata, Delhi and Mumbai are most adversely affected by vehicular pollution.

The situation is becoming alarming day by day in metropolitan and industrial areas. Hence, proper measures must be taken to abate pollution to protect the environment from harmful effects arising out of all kinds of socio-economic activities and impact of growing population. The abatement must be in a planned way–it must be effective as well as economic. The present article concentrates on the present situation of pollution in and around the industrial belt of Kolkata and discusses various measures to abate such a menace.

Air Pollution in and around Kolkata

By most accounts, Kolkata is a prime candidate for any award for the most polluted city in the world. Long on culture and political consciousness, the city is, paradoxically, an environmentalist's nightmare because of the poor quality of life it offers. The near-poisonous air, deafening noise levels and degraded water reserves have contributed to low quality environment. In conventional wisdom, Kolkata's woes arise mostly from an unsuitable geographical location and other hydro-geological factors. But the city's environment appears to have worsened over the years, primarily because of haphazard land use, inadequate civic facilities and poor public response to the need for environmental protection.

Figure 13.1 sums up the condition of air pollution in Kolkata as found out by the West Bengal Pollution Control Board during the last one year. It indicates alarming conditions so far as SPM and RPM concentrations are concerned, especially during the winter season.

Abatement of Air Pollution in Kolkata

Abatement of air pollution calls for proper monitoring and control. Monitoring of pollutants in the environment may either be made continuously or intermittently. West Bengal Pollution Control Board has been working quite efficiently in monitoring the pollution level in Kolkata and its suburbs.

Pollution control is made either by (*i*) diluting and dispersing or (*ii*) by concentration and containing. A third method may be recycling. While speaking of dilution and dispersion, one must also plan to disperse the industries. The places, which are already very polluted *e.g.* Kolkata or Durgapur should not be made available for further growth of industries. If by dilution and dispersion the pollution density can be kept below the statutory limit it will not be very harmful for the biosphere. For this reason, the chimney heights of the boilers are increased so that the exhaust gases and the SPM may get dispersed.

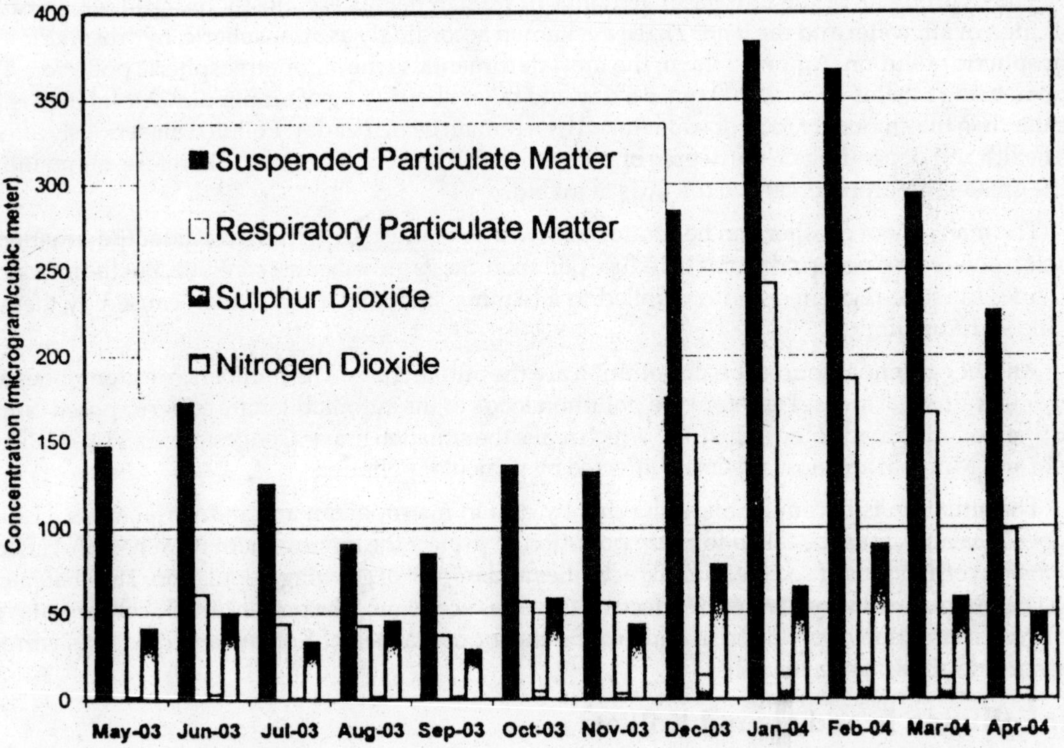

Figure 13.1: Comparative Study of Variation of Pollutants in Air of Kolkata during 2003–04

The problem of pollution control has become difficult to solve due to indiscriminate growth of industries in and around some cities like Kolkata, Mumbai etc. and also because most part of such industries had their birth in an age when people were not environmentally conscious and hence proper protective measures were not taken during their birth. But fortunately now the people have become aware of the environment and they have been able to appreciate that a corporate ethics must be developed in order to protect it.

The control of pollution caused by the industries should be given top priority. Standards of air pollution control have been prescribed for different areas and are given in Table 13.1.

Table 13.1

Area Category	*Concentration in Micro-gm/m³*			
	SPM	*SO_2*	*CO*	*NO_x*
Industrial and mixed area	500	120	5000	120
Residential and rural area	200	80	2000	80
Sensitive area	100	30	1000	30

The automobile exhausts contain many harmful gases and the pollution caused by them is significant in the ground level. Pollutants are exhausted from three points in an automobile–the exhaust point, crankcase blow by and the evaporative losses. Only hydrocarbons are emitted from the last two sources as a result of fuel evaporation.

The Government of West Bengal has now become quite aware of the devastation the menace of air pollution can produce in the near future. Some strong measures have already being planned by the Government to abate pollution in Kolkata and its suburbs. The then Hon'ble Minister-in-Charge Environment, Tourism and Youth Affairs, West Bengal, Sri Manabendra Mukhopadhyay, in a Press Conference held at Paribesh Bhawan on 27th January, 2001 announced the actions which are under consideration of West Bengal Pollution Control Board and Department of Environment, Government of West Bengal for improving air quality of the city. The proposed steps include conversion of all the coal fired boilers operating within the city limit to oil fired ones and exploring the option of closing the operation of New Cossipore Thermal Power Station of CESC during the winter months when the air pollution level of the city is at its worst.

In terms of resolutions adopted by the Board in its 117th meeting held on 19.12.2000 and in exercise of the powers conferred upon the Board under section 15 of the Air (Prevention and Control of Pollution) Act, 1981 and sub-section (3B) of section 12 of the Water (Prevention and Control of Pollution) Act, 1974, the West Bengal Pollution Control Board is pleased to empower in addition to the Member Secretary of the State Board under section 25 of the Water (Prevention and Control of Pollution) Act, 1974 and under sections 21, 24, 25, 26 of the Air {Prevention and Control of Pollution) Act, 1981 as amended from time to time the in-charge of Regional Offices of the State Board within their respective region to grant or to refuse 'Consent to Establish' for the Small Scale Special Red category (excepting those attracting EIA notification of Ministry of Environment and Forests, Government of India) and other Small Scale ordinary Red/Orange/Green category of industries which are not looked after by the General Manager, District Industries Centres.

West Bengal Pollution Control Board and the Department of Environment, Government of West Bengal in exercise of the powers conferred under section 17(1)(g) and section 20 of the Air (Prevention and Control of Pollution) Act, 1981 issued the following order: "All four wheeled vehicles (other than passenger vehicles) with GVW equal to or less than 3500 Kg and vehicles with GVW exceeding 3500 Kg brought for registration before Registering Authorities in Kolkata Metropolitan Area shall on and from 23rd October, 2001 conform to the respective norms specified for these categories of vehicles in the Central Motor Vehicles (2nd Amendment) Rules, 2001 published vide Government of India Notification No. G.S.R. 286(E) dated 24.04.2001."

Conclusion

All the socio-economic activities aiming at betterment of life causes environmental degradation - the degradation of air in particular. So, time is ripe to take strong measures for the abatement of pollution all over the world. New laws are being introduced and standards are being set on pollution concentration. Industries have been enforced to accept the legislation and to keep their discharges within the statutory limits. The environmental standards have been fixed up with a look to the trend lines of industrial growth, population growth and increasing need for food and the changes in social patterns and economic activities. If by proper regulatory measures the pollution is kept within the specified standards, it will not be very much harmful for the biosphere.

References

West Bengal Pollution Control Board Website.

The Telegraph, various issues.

Chapter 14

Analysis of Physico-chemical Characteristics to Study the Water Quality Index, Algal Blooms and Eutrophic Conditions of Lakes of Udaipur City, Rajasthan

Dilip K. Rathore[1], *P. Sharma*[2], *G. Barupal*[3], *S. Tyagi*[1], *and Krishna Chandra Sonie*[4]

[1]*Laboratory of Anatomy and Ecophysiology of Medicinal Plants and Cyanobacterial Research,*
[2]*Microbial Research Laboratory,* [4]*Laboratory of Biomolecular Technology*
Department of Botany, University College of Science, Mohanlal Sukhadia University, Udaipur – 313 001, Rajasthan
[3]*Government Dungar College, Bikaner, Rajasthan*

ABSTRACT

A total of 16 physico-chemical parameters for important drinking water bodies (Lake Pichhola, Lake Fateh Sagar, Swaroop Sagar and Doodh Talai) of Udaipur city in Rajasthan State were monitored bimonthly for one year (2004–05). The Water Quality Index (WQI) was calculated for all the 4 lakes using 12 parameters only. It has been found that eutrophic conditions prevail which supports the growth of bloom-forming algae due to higher concentration of phosphate and nitrate. This makes water of these lakes unsuitable for human consumption unless it is treated and disinfected properly. So, a few remedial measures have been also proposed.

Keywords: Lakes, Eutrophication, Algal Blooms, Water Quality Index.

Introduction

Udaipur is popularly known as the "City of Lakes" and there are as many as 160 lakes situated within the district of Udaipur. The bigger lakes situated within the Udaipur city are Lake Pichhola, Fateh Sagar, Swaroop Sagar and Doodh Talai. Except for some studies made by Vyas and Kumar (1968) and Vyas (1968) on phytoplanktons, little work has been so far done on the hydrobiology of these lakes. Udaipur is an important tourist centre mainly due to enchanting scenes, calm, glittering, blue waters, boating and tranquil atmosphere of these lakes. The increasing human activities have slowly made these lakes dumping grounds for city wastes and even part of the sewage. The result is that these lakes have come under severe pollution stress and have already shown strong eutrophic tendencies.

Eutrophication can be defined as a process of continuous enrichment of waters by the addition of substances. Stumm and Stumm-Zollinger (1972) who reviewed the role of phosphorous concluded that enrichment of phosphorous in lake waters is one of the most potent cause which brings about eutrophication. Nitrogen enrichment of lake water has been considered a less potent factor in eutrophication than phosphorous (Goering, 1972). The major consequences of eutrophication are water blooms. Blue-green algal blooms are the most unmistakable indicators of eutrophication (Lung and Hans, 1988).

Also, these lakes are an important source of municipal water supply for the city. So the aim of water quality study is to minimize the role of water in transmission of diseases. According to WHO (1999) data, contaminated drinking water has contributed directly to the annual 2.2 million diarrhoea related deaths. Basic hydrobiological data which is useful in the proper management of these lakes especially to check eutrophication are not sufficient to plan measures for reversing the trends of eutrophication.

Lake Pichhola is situated between the parallels 24°33′15″ and 24°34′59″ North latitudes and 73°39′32″ and 73°41′08″ East longitudes. This legendary lakh is 7 km long and 2 km broad. The present water-holding capacity of is 3,55,10,699 cubic meters. Doodh Talai is situated near to Lake Pichhola. This is the deepest spot (maximum depth 3.35m).

Fateh Sagar is situated to the north of Lake Pichhola and connected to it by a canal. This lovely lake is with an embankment 861 metres long and 15 metres broad.

The status of water quality of these lakes has been assessed by calculating Water Quality Index (WQI) using the water quality parameters (APHA, 1989). The present study has been conducted over a period of eight months (August'04 to March'05) to examine the variations in the physico-chemical characteristics and growth of algal flora.

Materials and Methods

Sampling Site

A single site has been selected in each of the 4 lakes for physico-chemical analysis. These are:

Lake Pichhola: Near water purify pump

Doodh Talai: Near musical fountain road

Fateh Sagar: Fateh Sagar gate

Swaroop Sagar: Near Dr. Ambedkar park

Collection of Algal and Water Sample

Surface water sample has been collected bimonthly from these sampling sites for the period of eight months. The sampling was done early in the morning and the water samples were collected in plastic bottles of two litre capacity. Separate samples for D.O. and B.O.D. were collected in well stoppered pyrex glass bottles of the 300 ml capacity. Initial fixation was done in the field itself.

Both floating and attached algal samples were collected. Microalgae were collected with the help of planktonic net and the samples were identified in the laboratory using authentic algal identification Keys (Desikachary, 1958; Bold and Wynne, 1978). Some dominant forms were isolated, grown in culture conditions and were maintained as pure cultures by subculture methods. These include *Chlorococcum infusionum*, *Nostoc muscorum*, *Anabaena doliolum*, *Scytonema* sp. and *Scenedesmus bijugatus*.

Physico-chemical parameters were analysed to determine temperature, pH, Free CO_2, Dissolved Oxygen, Biochemical Oxygen Demand, Chemical Oxygen Demand, Total alkalinity, Electrical conductivity, Total Dissolved Solutes, Hardness and concentrations of Nitrate, Phosphate, Sulphate, Chloride and Linear Alkyl Benzene Sulphonate (LABS) as per the methods given in APHA 1989; Trivedi and Goel, 1984; Manivasakam, 1996.

Results and Discussion

The physico-chemical characteristics of the mentioned lakes during the study period are presented in Tables 14.1 and 14.2.

With the season, a change in water temperature ranging between 14.5° to 26°C has been recorded for all the four lakes. The pH was found above 7.0 in all the lakes making them slightly alkaline that may be due to the presence of alkaline substances, detergents as a component of industrial effluents and household sewage. Dissolved oxygen (D.O.) has proved to be an important parameter in studying the quality of water. The maximum amount of dissolved oxygen was observed during December'04 to January'05 (13.70 ppm) for Lake Swaroop Sagar while minimum during February'05 to March'05 (4.2 ppm) for Doodh Talai (Figure 14.1). Another important parameter is phosphate concentration creating

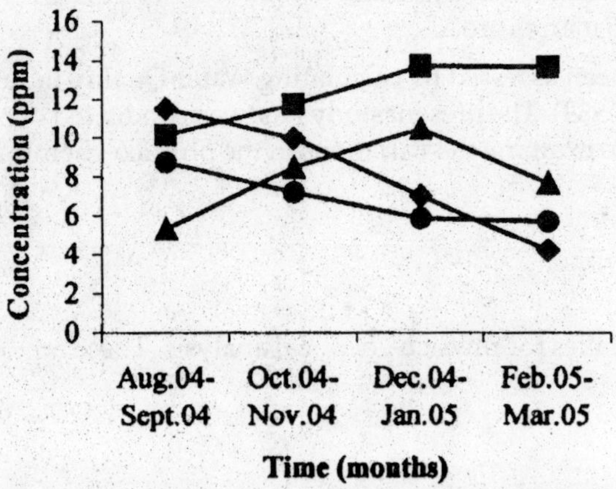

Figure 14.1: Dissolved Oxygen of Studied Lakes of Udaipur

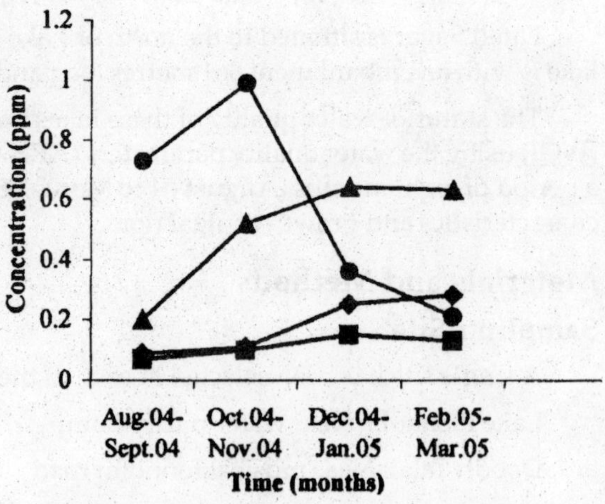

Figure 14.2: Phosphate Concentration of Studied Lakes of Udaipur

Table 14.1: Analysed Physico-chemical Parameters of Water Sample of Lakes

Sl.No.	Parameters	Pichhola Lake				Doodh Talai			
		Aug.'04–Sept.'05	Oct.'04–Nov.'05	Dec.'04–Jan.'05	Feb.'05–March'05	Aug.'04–Sept.'05	Oct.'04–Nov.'05	Dec.'04–Jan.'05	Feb.'05–March'05
(A)	Physical Characteristics								
1.	Water temperature	26	20	15	22	25.5	18	14.5	23
2.	Colour	Yellowish	Yellowish	Yellow green	Brownish yellow	Yellow green	Light green	Brownish green	Muddy green
3.	pH	7.35	7.46	7.84	7.66	7.23	7.57	7.64	7.73
(B)	Chemical Characteristics								
4.	Free CO_2	57	72	64.5	77.1	28.7	23.2	36.3	31.25
5.	Total alkalinity	133	204	156	223	163	198	209	236
6.	Chloride	63.55	67.51	71.47	85.91	49.71	55.12	56.29	67.5
7.	Hardness	180.1	245.8	240.2	210.3	245.1	238.33	240.2	211.89
8.	Dissolved oxygen	5.32	8.42	10.40	7.63	11.51	9.92	6.98	4.2
9.	B.O.D.	4.623	–	–	1.69	–	1.48	5.395	6.102
10.	C.O.D.	33.042	26.33	69.93	104.59	26.13	22.03	34.08	43.44
11.	Nitrate	1.143	1.980	5.43	7.309	9.061	11.171	14.176	3.15
12.	Phosphate	0.203	0.517	0.639	0.629	0.087	0.109	0.245	0.275
13.	Sulphate	2580	3913.42	6500.1	6783	2610	–	3875	4150
14.	T.D.S.	1532	1709	1764	1623	512	681	564	476
15.	Electrical conductivity	2.92	2.31	2.58	3.12	1.21	1.15	1.34	0.930
16.	LABS	0.013	0.019	0.053	0.024	0.009	0.016	0.08	0.04

Except colour, Temperature (°C), pH, electrical conductivity (mS/cm), all other parameters are expressed in ppm.

Table 14.2: Analysed Physico-chemical Parameters of Water Sample of Lakes

Sl.No.	Parameters	Fateh Sagar				Swaroop Sagar			
		Aug.'04–Sept.'05	*Oct.'04–Nov.'05*	*Dec.'04–Jan.'05*	*Feb.'05–March'05*	*Aug.'04–Sept.'05*	*Oct.'04–Nov.'05*	*Dec.'04–Jan.'05*	*Feb.'05–March'05*
(A)	Physical Characteristics								
1.	Water temperature	26	26	19	14.5	21	19.5	15	21.5
2.	Colour	Yellowish	Slightly green	Yellowish green	Brownish green	Muddy	Yellow green	Greenish	Dark green
3.	pH	7.61	7.88	8.15	8.42	"8.48	7.99	8.64	8.68
(B)	Chemical Characteristics								
4.	Free CO_2	23	37	23	31	30.5	22	24	26
5.	Total alkalinity	142	366	343	409	415	161	172	170
6.	Chloride	56.36	98.012	115.32	155.84	161.23	72.01	81.39	86.90
7.	Hardness	142.1	259.68	285.89	390	441.31	166.5	164.5	170.01
8.	Dissolved oxygen	10.11	8.71	7.17	5.84	5.69	11.71	13.70	13.62
9.	B.O.D.	1.13	1.03	–	1.98	2.18	1.56	1.82	1.74
10.	C.O.D.	–	64.39	–	139.84	123.44	39.06	–	72.91
11.	Nitrate	1.012	1.13	1.54	1.33	1.61	3.61	4.43	4.21
12.	Phosphate	1.012	3.61	4.43	4.21	1.13	1.54	1.33	1.61
13.	Sulphate	0.067	0.101	0.147	0.129	0.732	0.99	0.358	0.210
14.	T.D.S.	4150	9780	9810	1030	1350	323	417	421
15.	Electrical conductivity	0.705	1.87	1.83	1.03	1.35	0.769	0.829	0.699
16.	LABS	0.0391	0.034	0.21	0.045	0.019	0.035	0.0295	0.032

Except colour, Temperature (°C), pH, electrical conductivity (mS/cm), all other parameters are expressed in ppm.

nutrient-rich conditions in particular water body. Its value was found to be very high (203.01 to 638.77 ppm) in all the four lakes during late winter season (Figure 14.2). Nitrate concentration was found to be maximum (14.176 ppm) for Doodh Talai during December'04 to January'05 and minimum (1.012 ppm) for Fateh Sagar during August'04 to September'05. Nitrate concentration was found to be high during all the study period ranging from 3.15 to 14.176 ppm for Doodh Talai in comparison to other three lakes (Figure 14.3). These conditions favour the growth of many bloom-forming alga.

Hardness value was found to be very high (245.8 ppm) for Lake Pichhola during October'04–November'04; 259.68 ppm to 441.31 ppm for Lake Fateh Sagar during February'05 to March'05; and 211.89 ppm to 245.1 ppm for Doodh Talai during all the studied time period. Some bloom-forming dominant algal forms were found to flourish therein due to eutrophic conditions of these lakes (Table 14.3). These species mainly belong to class Chlorophyceae, Cyanophyceae and Bacillariophyceae.

Table 14.3: Some Bloom-forming Dominant Algal Species Identified During the Study Period

Lakes	*Dominant Species*
Pichhola	*Chlorococcum infusionum, Cladophora glomerata, Spirogyra* sp., *Fragillaria* sp., *Navicula* sp., *Cyclotella* sp.
Doodh Talai	*Cladophora glomerata, Spirogyra* sp.
Fateh Sagar	*Microcystis aeruginosa, Chara brachypus, Fragillaria capucina, Navicula* sp., *Nitzschia gracilis*
Swaroop Sagar	*Microcystis aeruginosa, Microcystis lamillaris, Hydrodictyon indicum, Cladophora glomerata*

Calculation of WQI

WQI is defined as the composite influence of different water quality parameters in the quality of water. Total 12 parameters were chosen to calculate Water Quality Index (WQI) of the above mentioned lakes. The parameters are enumerated in Tables 14.1 and 14.2. The Water Quality Index of Swaroop Sagar during August'04–September'05 as calculated is given as an example in Table 14.4.

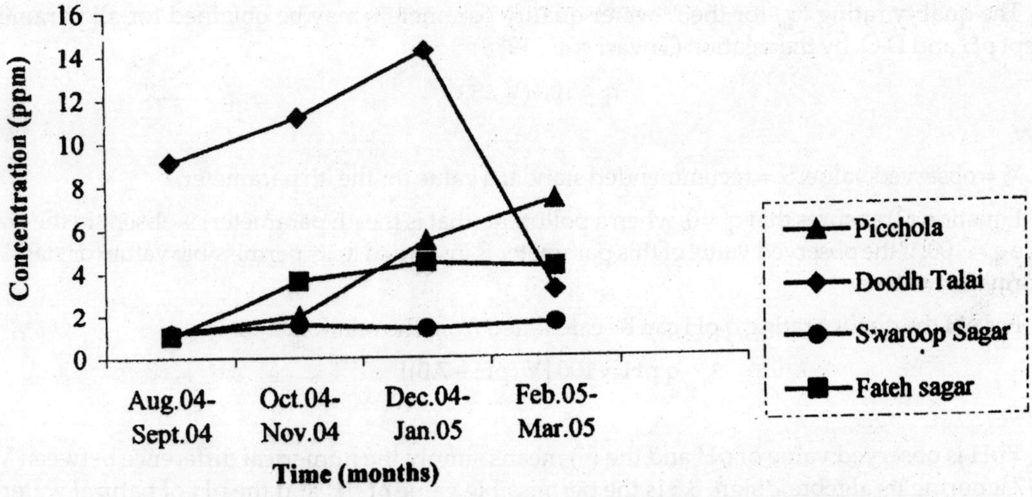

Figure 14.3: Nitrate Concentration of Studied Lakes of Udaipur

Table 14.4: Water Quality Index of Swaroop Sagar During August'04–September'05

Sl.No.	Parameters	Observed Vn	Standard	Wn = k/Sn	q_n	$Log.q_n$	$Wn\ log\ q_n$
1.	pH	7.88	8.5	0.0235	58.66	1.7683	0.0416
2.	T.D.S.	978	500	0.0004	195.6	2.2913	0.0009
3.	Electrical Conductivity	187	300	0.0006	62.33	1.7947	0.0011
4.	Total alkalinity	366	120	0.0016	305	2.4843	0.0040
5.	Chloride	98.012	250	0.0008	39.20	1.5933	0.0013
6.	Hardness	259.68	150	0.0013	173.12	2.2383	0.0029
7.	D.O.	8.71	6	0.033	145.17	2.1617	0.0713
8.	BOD	1.03	5	0.04	20.6	1.3139	0.0526
9.	COD	64.39	10	0.02	643.9	2.8088	0.0562
10.	Nitrate	1.13	50	0.004	2.26	0.3541	0.0014
11.	Phosphate	0.732	0.11	1.818	665.45	2.8231	5.1324
12.	Sulphate	2.18	200	0.001	1.09	0.0374	0.0003
							$\Sigma Wn\ log\ q_n$ 5.3657

Except colour, Temperature (°C), pH, electrical conductivity (mS/cm), all other parameters are expressed in ppm.

$$WQI = \text{anti}\log \Sigma_{n=1} W_n \log q_n \qquad ......1$$

where,

Wn = K/Sn; Wn = unit weight for the *n*th parameters; Sn = (n = 1, 2, 3........12) refers to water quality parameters; K = constant of proportionality for the sake of simplicity we assume K = 1; q_n = quality rating of *n*th water quality parameter.

The quality rating (q_i) for the ith water quality parameters may be obtained for all parameters except pH and D.O. by the relation (Tiwari *et al.*, 1988).

$$q_i = 100\,(V_i/S_i) \qquad ......2$$

where,

V_i = observed value; S_i = recommended standard value for the *i*th parameter.

Equation (2) ensures that $q_i = 0$, when a pollutant (that is the *i*th parameter) is absent in the water, while $q_i = 100$, if the observed value of this parameter is just equal to its permissible value (or standard) for drinking water.

For pH the quality rating q pH can be calculated from the relation.

$$q\,pH = 100\,[V\,(pH - 7.0)] \qquad ......3$$

where,

VpH is observed value of pH and the (–) means simply the numerical difference between VpH and 7 ignoring its algebraic sign. 8.5 is the permissible value of pH and the pH of natural water is 7 which is ideal value.

For D.O., the quality rating $q_{DO} = 100\,[(V_{DO} - 14.6)/(6.0 - 14.6)]$

where,

V_{DO} is the observed value of D.O. and the symbol (–) means simply the numerical difference between V_{DO} and 14.6 ignoring its algebraic sign.

The ideal value of D.O. may be taken as 14.6 ppm, which is solubility of O_2 in pure water at 0°C. The standard value for drinking water is 6 ppm.

The Water Quality Indices as calculated are given in Table 14.5 whereas the standard WQI and status are given in Table 14.6.

Table 14.5: Water Quality Indices of Lakes of Udaipur City

Lakes	*Water Quality Index (WQI)*			
	August'04–September'04	*October'04–November'04*	*December'04–January'05*	*February'05–March'05*
Pichhola	231.1	1045	4979	1795
Doodh Talai	39.96	69.87	382	416.7
Fateh Sagar	25.62	61.27	105.2	95.08
Swaroop Sagar	5630	7155	913.8	251.8

Table 14.6: Standard WQI and Status of the Water Body

Water Quality Index	*Status*
0–25	Excellent
26–50	Good
51–75	Poor
76–100	Very poor
100 and Above	Unsuitable for drinking

According to WQI of Fateh Sagar and Doodh Talai as calculated during August'04 to November'04, it shows that water is suitable for human consumption but for the next four months, it was found to be in very poor conditions rendering it unsuitable for drinking. The WQI of Pichhola and Swaroop Sagar during the whole study period indicates that the water is completely unfit for human consumption because (*i*) it is highly turbid, (*ii*) high content of organic matter which is mostly sewage, (*iii*) high phosphate content, (*iv*) very often obnoxious odour and greenish colour.

The data shows that Fateh Sagar is much less polluted in comparison to the other three lakes and can be used as water reservoir for municipal water supply for a long time.

A higher concentration of phosphate and nitrate led to nutrient rich conditions, called eutrophic condition which favours luxuriant growth of bloom-forming algal forms (Vollenwider, 1997; McCaull and Crossland, 1974; Michaud *et al.*, 1979; Forsberg and Ryding, 1980; Zdanowski, 1982). There is a fast growth of one or other alga, so much so that the entire surface becomes quickly covered with it. Such a condition is known as water bloom. In these water blooms, free-floating angiospermic plants such as *Eichhornia crassipes*, spp. of *Wolffia* and *Lemna* were found to be the notorious agents.

Growing urbanization around these lakes is likely to degrade water quality. It could be a threat to public health in future years since these water bodies makes an important source of drinking water for township of Udaipur city. So on the basis of present study, we propose few suggestions with the aim to monitor pollution and to reduce Water Quality Index (WQI) of these lakes to a much safer level, *i.e.*, below 100.

1. Proper treatment of water (biological/chemical) is required to be done before human consumption.
2. Those rivers which had been dammed to create these artificial lakes pass through several villages bringing with them all the garbage, pesticidal residues, industrial effluents, even rotting dead cattle and other animals etc. Reversal of these processes would definitely bring about ameliorated conditions in lakes.
3. It must be ensured that no sewage or garbage of any kind is discharged into these lakes by hotels situated in and around.

Acknowledgement

We are grateful to CSIR for their financial help. The authors are also thankful to Mr. Lakhpat Meena and Mr. Praveen Kumar Chandel, Department of Botany, University College of Science, Udaipur for their valuable suggestions and encouragement.

References

APHA, 1989. *Standard Method for Examination of Water and Wastewater*, 17th edn. American Public Health Association, Washington, D.C.

Bold, H.C. and Wynne, M.J., 1978. *Introduction to the Algae*. Prentice-Hall of India Private Limited, New Delhi.

Desikachary, T.V., 1958. *Cyanophyta*. Indian Council of Agricultural Research, New Delhi.

Forsberg, Curt. and Ryding, Sven, Olof, 1980. Eutrophication parameters and trophic state, indices in 30 Swedish waste receiving lakes. *Arch. Hydrobiol.*, 89: 189–207.

Goering, C.R., 1972. The role of minor nutrients in limiting the productivity aquatic ecosystem. *Am. Soc. Limnol. Oceanogr*. (Spl. Symposium), 1: 21–38.

Lung, Wu-Seng and Hans W. Paerl, 1988. Modeling blue-green algal blooms in the lower Neuse River (North Carolina U.S.A.). *Water Res.*, 22: 895–906.

Manivasakam, N., 1996. *Physico-chemical Examination of Water, Sewage and Industrial Effluents*. Pragati Prakashan, Meerut, India, pp. 201–203.

McCaull, J. and Crossland, J., 1974. *Water Pollution*. Harcourt Brace Javanovich Publisher, Inc., New York, pp. 206.

Michaud, M.T. *et al.*, 1979. Changes in phosphorus concentrations in a eutrophic lake as a result of macrophyte kill following herbicide application. *Hydrobiologia*, 66: 105–111.

Naik, S. and Purohit, K.M., 1997. Status of water quality a Bondomanda of Rourkela Industrial Complex, Part-I: Physico-chemical parameters. *I.J.E.P.*, 18: 346–353.

Naik, S. and Purohit, K.M., 1996. Physico-chemical analysis of some community ponds of Rourekela. *Indian Journal of Env. Protection*, 16: 679–684.

Stumm, W. and E. Stumm-Zollinger, 1972. The role of phosphorus in eutrophication. In: *Water Pollution Microbiology*, (Ed.) R. Mitchell, pp. 11–42.

Sudan Madhu, Sharma, L.L. and Durve, V.S., 1984. Eutrophication of the lake Pichhola in Udaipur, Rajasthan. *Poll. Res.*, pp. 39–44.

Tiwari, T.N. and Manzoor, Ali, 1988. *Water Quality Index of Indian River,* (Ed.) R.K. Trivedi. Ashish Publishing House, New Delhi, pp. 771–786.

Trivedi, R.K. and Goel, P.K., 1984. *Chemical and Biological Methods for Water Pollution Studies.* Environmental Publications, Karad, India.

Vollenweider, R.A., 1997. Scientific fundamentals of the eutrophication of lakes and flowing waters with particular reference to nitrogen and phosphorus as factors in eutrophication. OECD Report, Paris 30 September, 1970.

Vyas, L.N., 1968. Studies in phytoplankton ecology of Pichhola lake, Udaipur. In: *Proc. Symp. Recent. Adv. Trop. Ecol.,* pp. 334–347.

Vyas, L.N. and Kumar, H.D., 1968. Studies on the phytoplankton and other Algae of Indrasagar Tank, Udaipur, India. *Hydrobiologia,* 31: 421–434.

WHO, 1999. *The World Health Report.* World Health Organisation, Geneva.

Zdanowski, Boguslaw, 1982. Variability of nitrogen and phosphorus contents and lake eutrophication. *Pol. Arch. Hydrobiol.,* 29: 541–598.

Chapter 15

Larvicidal Effect of Quinalphos Against Three Clinically Important Mosquito Species

N. Arun Nagendran

P.G. and Research Department of Zoology and Microbiology, Thiagarajar College, Madurai – 625 009

ABSTRACT

The efficacy of quinalphos, an organophosphate. against larvae of *Anopheles stephensi, Aedes aegypti* and *Culex quinquefasciatus* was assessed. The fourth instar larvae of all the three species were treated with quinalphos and the LC_{50} values determined by probit analysis indicate that, *Culex quinquefasciatus* was more susceptible with high larval mortality than the other two species.

Keywords: *Anopheles stephensi, Aedes aegypti, Culex quinquefasciatus, Quinalphos, Susceptibility.*

Introduction

In India, mosquito fauna include 255 species grouped under 16 genera. Fifty eight species belonging to the genus Anopheles, 57 species to Culex, 111 species to Aedes and 7 species to Mansonia (Nagpal and Sharma, 1995). It has been reported that in Madurai there are 27 species of mosquitoes grouped under the genera Aedes, Anopheles, Armigeres, Culex and Mansonia (Pandian, 1998). Of which, mosquitoes belonging to four genera *viz.*, Anopheles, Aedes, Culex and Mansonia are considered to be most important because of their potential ability in transmitting fatal diseases like malaria, dengue, filariasis and Japanese encephalitis respectively. One of the methods available for the control of mosquitoes is the use of insecticides. Chemical control using synthetic insecticides had been favoured so far, because of their speedy action and ease application. The relative toxicity of insecticides to various mosquito species has been studied by entomologists in detail (Hagstrum and Mulla, 1968; Holex and Meek, 1987; Rajavel *et al.*, 1987; Saxena and Kaushik, 1988).

A major drawback of using chemical insecticides is that they are non selective and develop resistance. So, it is essential to know the insecticide susceptibility status of the vector mosquitoes for developing appropriate strategy for their control (Mahanta *et al.*, 1999). Sublethal dosage of insecticides is therefore required and enough to restrict the mosquito population (Moriarty, 1969). Considering the development of resistance to most insecticides, the present work has been designed to study the larvicidal effect of an organophosphorus insecticide, Quinalphos, and to determine the minimum dosage requirement to control the important vector mosquitoes, *Anopheles stephensi*, *Aedes aegypti* and *Culex quinquefasciatus*.

Materials and Methods

The larvae of *Anopheles stephensi, Aedes aegypti* and *Culex quinquefasciatus* were obtained from Centre for Research in Medical Entomology (CRME), Madurai as egg/egg rafts. The egg/egg rafts were inoculated in sterilized tap water. The temperature was maintained as 29–31°C, which is optimal for normal ground. The larvae hatch with overnight incubation. The larvae were fed with yeast tablets and dog biscuits in the ratio of 3 : 2. Water is changed for once in two days. The early fourth instar larvae were selected for the study. The insecticide quinalphos was procured from commercial sellers.

Larval bioassay was performed on early fourth instar larvae according to the standard procedure of WHO (1992). The tests were carried out by exposing 20 early fourth instar larvae in glass jars containing 249 ml of tap water and 1 ml of different concentrations of the insecticide.

Results and Discussion

Since the discovery of involvement of mosquitoes in the transmission of many diseases, considerable data have been collected in India on the distribution, systematics and bionomics of mosquitoes (Reuben, 1978; Rajput and Singh, 1988; Pandian, 1990). Subsequently, along with the development of series of pesticides, many scientists investigated the particular dose required to combat target organisms and the relative toxicity of various insecticides against mosquitoes had been studied by entomologists (Helson and Surgeoner, 1983; Holck and Meek, 1987; Saxena and Kaushik, 1988).

The results of the above mentioned studies indicate the fact that the susceptibility vary according to the nature of the insecticides. The dose requirement also varies with the type of insecticide and the insect species. Hence, the efficacy of quinalphos under laboratory conditions against fourth instar larvae of *Anopheles stephensi, Aedes aegypti* and *Culex quinquefasciatus* was tested in the present study. The concentration of pesticide, mortality of larvae, LC_{50} value, regression equation, 95 per cent confidential limits, standard error and slope for the effect of quinalphos against *Anopheles stephensi, Aedes aegypti* and *Culex quinquefasciatus* are given in Table 15.1. The toxicity based on LC_{50} values has proved that quinalphos, a systemic insecticide, is found toxic to the three mosquito species selected for the study.

In agreement to the effect of insecticides on mosquitoes observed in the present study, Saxena and Kaushik (1988) reported total developmental arrest of fourth instar arrest of fourth instar larvae of *Culex quinquefasciatus* treated with penfluron. Effrid *et al.* (1991) found that resemthrin continues to offer excellent control of *Anopheles quadrinaculatus.* Chockalingam *et al.* (1991), Venkatesan *et al.* (1991), Floore *et al.* (1992), Saxena *et al.* (1992), Mahanta *et al.* (1999) and Dhanraj and Thirunavukarasu (2005) have also reported the efficacy of various chemical pesticides against various species of mosquitoes.

Table 15.1: Toxicity of Quinalphos to the Fourth Instar Larvae of *Anopheles stephensi*, *Aedes aegypti* and *Culex quinquefasciatus*

Species	*Concentration of Monocrotophos (in ppm)*	*% of Mortality*	*LC_{50} (in ppm)*	*Regression Equation*	*S.E.*	*95% Confidential Limit*		*Slope*
						Upper Limit	*Lower Limit*	
	0.05	10.00						
Anopheles	0.10	31.66	0.1700	Y = 6.800421 +	0.64	0.199	0.145	2.66
stephensi	0.15	43.33		2.340208 X				
	0.20	56.66						
	0.30	46.66						
	0.35	53.33						
Aedes aegypti	0.40	63.33	0.3135	Y = 6.08834 +	1.306	0.376	0.260	2:886
	0.55	68.33		2.160784 X				
	0.03	43.33						
Culex	0.05	53.33	0.040	Y = 7.663176 +	0.517	0.490	0.329	3.323
quinquefasciatus	0.10	76.66		1.90736 X				
	0.15	86.66						

Though the use of insecticides is hazardous, chemical the insecticides are preferred because of their effectiveness, speedy action and easy application. The optimum use of chemicals will reduce their ill effects. Hence, the determination of minimum dose requirement is essential. The LC_{50} value of quinalphos against the larvae of *Anopheles stephensi, Aedes aegypti* and *Culex quinque fasciatus* were found to be 0.17 ppm, 0.31 ppm and 0.04 ppm respectively. The toxicity studies of quinalphos against *Anopheles stephensi, Aedes aegypti* and *Culex quinquefasciatus* based on probit analysis reveal that quinalphos was more toxic to *Culex quinquefasciatus* (LC_{50} value is 0.04 ppm) when compared to the other two species and *Aedes aegypti* was found to be relatively resistant to quinalphos with LC_{50} value of 4.01 ppm. Similar to this result, Baktharatchagan and David (1991) have also reported that *Culex quinquefasciatus* was more susceptible than *Anopheles stephensi* and *Aedes aegypti* when tested against cypermethrin.

Moriarty (1969) had reported three possible means of control of insect population by the sublethal dosage of insecticides *viz.*, (*i*) affecting survival, (*ii*) affecting reproductive and (*iii*) affecting the genetic make up of future generations. Of the three suggested means, in the present investigation. control of the three selected mosquito species by affecting their survival by quinalphos was found evident.

Acknowledgement

The author acknowledges the financial assistance rendered by University Grants Commission, New Delhi under UGC-MRP 219/04.

References

Baktahratchagan, R. and David, B.V., 1991. Evaluation of Trebon (ethofenprox) against *Cx. quinquefasciatus An. stephensi* and *Ae. aegypti* and on non target organisms. *Ind. J. Malario.* 28: 249–253.

Chockalingarn, S., Kuppusamy, A., Punithavathy, G. and Manoharan, T., 1991. Synergistic effect of insecticide with plant extracts against mosquito vectors. *J. Appl. Zool. Res.*, 2(2): 92–95.

Dhanraj, B. and Thirunavakkarasu, B., 2005. Evaluation of sumilarv 0.5 per cent GC (pyripyroxifen) against *Cx. quinquefasciatus* mosquito larvae. *Pestology.*

Effrid, P.K., Inman, A.D., Dame, D.A. and Meisch, M.V., 1991. Efficacy of various ground applied cold aerosol adulticides against *Anopheles quadrimaculatus*. *J. Am. Mosq. Cont. Ass.*, 7: 207–209.

Floore, T.G., Dukes, J.C., Boike, A.H., Greer, M.J. and Coughlin, J.S., 1992 Evaluation of three cypermethrin piperonyl butoxide formulation compared with scourage (R) against adult *Cx. quinquefasciatus. Indian J. Med. Res.*, 8: 97–98.

Hagstrum, D.W. and Mulla, S.M., 1968. Petroleum oils as mosquito larvicides and pupicides. *The Entomological Society of Africa*, 61: 220–222.

Helson, B.V. and Surgeoner, G.A., 1986. Efficacy of cypermethrin for control of mosquito larvae and pupae and impact of non-target organisms including fish. *J. Am. Mosq. Cont. Ass.*, 2: 269–275.

Holck, A.R. and Meek, C.L., 1987. Dose mortality responses of crawfish and mosquitoes to selected pesticides. *J. Am. Mosq. Cont. Ass.* 3: 407–410.

Mahanta, B., Handique, R., Dutt, P. and Mahanta, J., 1999. Susceptibility status of adult and larval *Culex quinquefasciatus* collected from tea gardens of Assam to different insecticides. *Geobios*, 26: 195–198.

Moriarty, F., 1969. The sublethal effects of synthetic insecticides on insects. *Bio. Rev.*, 44: 321–357.

Nagpal, B.N. and Sharma, V.P., 1995. *Indian Anophelenes.* Oxford and IBH publishing Co. Pvt. Ltd., Delhi, India.

Pandian, R.S., 1990. The seasonal prevalence of adults of *Armigeres subalbatus* (Diptera : Culicidae) in Madurai, Tamil Nadu. *J. Ecobiol.*, 2: 172–174.

Pandian, R.S., 1998. Biodiversity of mosquito fauna and efficacy of biopesticides against mosquitoes in an urban area in Tamil Nadu. *Indian J. Env. Sci.*, 2(1): 159–162.

Rajavel, A.R., Vasuki, V., Paily, K.P., Ramaiah, K., Mariappan, T., Kaldaram, M., Tyagi, B.K. and Das, P.K., 1987. Evaluation of synthetic pyrethriod cyfluthrin for insecticidal activity against different mosquito species. *Indian J. Med. Res.*, 85: 168–175.

Rajput, K.B. and Singh, T.K., 1998. Dusk biting mosquitoes of Manipur. *Entomon.*, 13: 173–178.

Reuben, R., 1978. A report on mosquitoes collected in the Krishna–Godhavari delta, Andhra Pradesh. *Indian J. Med. Res.*, 68: 603–609.

Saxena, S.C. and Kaushik, R.K., 1988. Total developmental arrest of fourth instar larvae of *CuJex quinquefasciatus* treated with penfluron. *Curr. Sci.*, 57: 1196–1199.

Venkatesan, P., Baghyalakshmi, M. and Tena D'Sylva, 1991. Effect of decamethrin on bioagents in mosquito control. *J. Ecobiol.*, 3: 261–266.

World Health Organization, 1992. Instruction for determining the susceptibility of resistance of mosquito larvae of insecticides. In: 5th *Report of WHO Expert Committee on Vector Biology and Control.* WHO. Tech. Rep. Ser. 818.

Chapter 16

Dry Matter, Leaf Area Index, Root Mass Density and Yield of Bed Planted Wheat Under Irrigation and Different Plant Population

Sukhvinder Singh, H.S. Uppal, S.S. Mahal, Avtar Singh and R.K. Mahey

Department of Agronomy and Agrometeorology, Punjab Agricultural University, Ludhiana – 141 004

ABSTRACT

The grain yield of wheat increased significantly with the application of irrigations under IW : CPE ratio 0.95 over irrigation level IW : CPE ratio 0.65 during both the years. The magnitude of increase in the grain yield with IW : CPE ratio 0.95 over irrigation level IW : CPE ratio 0.65 was 7.9 and 7.1 per cent during 1997–98 and 1998–99, respectively. During 1997–98, dry matter production was not affected by treatments IW : CPE ratio 0.65 and 0.95 at 60 DAS. At 75 DAS, irrigation was applied in IW : CPE ratio 0.95 and this resulted in higher dry matter production by 7.9 per cent over IW : CPE ratio 0.65 at 90 DAS. Then after 90 DAS, similar trend continued and at harvest 7.7 per cent higher dry matter production was observed under IW : CPE ratio 0.95 over the treatment of IW : CPE ratio 0.65. But during 1998–99, dry matter production increased under IW : CPE ratio 0.95 from 120 DAS on ward and the final dry matter increase at harvest under the said treatment was 7.9 per cent. The leaf area index (5.53) recorded under IW : CPE ratio 0.95 was significantly higher than under IW : CPE ratio 0.65 during 1997–98. The root mass density decreased with increased profile depth from 0–180 cm during both the years in all the treatments. In 1997–98, the root mass density under IW : CPE ratio 0.95 was more than that of IW : CPE ratio 0.65. The roots 73.1 and 72.9 per cent were confined to surface 0–30 cm layer under IW : CPE ratio 0.95 and 0.65, respectively.

Introduction

In light soils, water standing after irrigation application is not a problem due to deep percolation but water standing or excessive wetting of soil profile caused damage to the crop under heavy soils.

The resource conservation technology, the growing of wheat on raised beds and irrigation in furrows to overcome this problem and are also expected to reduce the water losses and consequently increased the water use efficiency with or without increase in yield of wheat. Therefore, keeping in view the above facts the investigations were conducted to know the effect of irrigation and seed rate cum spacing on the various parameters of wheat.

Materials and Methods

Field studies were conducted to see the effect of irrigation and seed rate cum spacing on bed planted wheat on loamy sand soil during *Rabi*, 1997–98 and 1998–99 at Ludhiana. In 1997–98, heavy rainfall (10.71 cm) was received during second week of December at CRI stage and 10.37 cm rainfall was received from mid February to mid March, whereas in 1998–99, 4.17 cm rainfall was during January out of total 5.79 cm rainfall. The field capacity and 15 bar tension storages 0 to 180 cm soil profile were 33.6 and 11.11 cm. The soil of the experimental site was low in available N, P and K which was 113.8 and 120.7 N, 10.1 and 10.6 P_2O_5 and 173.0 and 177.6 K_2O kg/ha during 1997–98 and 1998–99, respectively. The experiments were laid out in Randomized block design with 4 replications keeping gross plot size 2.70 × 10 m. The treatments comprised of 4 levels of seed rate cum spacing of 100 kg seed/ha in 3 rows/bed, 66.7 kg seed/ha in 2 rows/bed, 66.7 kg seed/ha in 3 rows/bed and 44.5 kg seed/ha in 2 rows/bed were tested under two levels of irrigation application *i.e.* irrigation scheduling at IW: CPE ratio of 0.65 and 0.95. The width of seeding bed was 37.5 cm and distance between center to center of furrow 67.5 cm. The crop was sown with wheat variety PBW 343 on 14 November and 15 November during 1997–98 and 1998–99, respectively. Seeding was done as per treatments with bed maker cum planter seed drill. Seed rate and number of rows were kept as per treatments. Recommended dose of 125 kg N, 62.5 kg P_2O_5 and 30 kg K_2O/ha was given to the crop. Half nitrogen, full of P and K were broadcasted just before planting of wheat. Remaining half dose of nitrogen was broadcasted on beds after rainfall at CRI stage during first year and during second year nitrogen was drilled one day before first irrigation at CRI stage. Source of N, P and K was urei, single superphosphate and muriate of potash, respectively. Irrigation was applied to the experimental plots as per requirement of treatments after application of common irrigation at CRI during 2nd year of study whereas, during 1st year of study, no irrigation at CRI was given as there was 10.71 cm rainfall at this Stage. The depth of irrigation water was measured with partial flume. Irrigation were applied 1 (11/2 at 89 DAS), 2 (28/1 at 75 DAS, 25/3 at 131) and 3 (5/12,7/3 at 112 DAS, 26/3 at 132 DAS) and 4 (5/12, 23/2 at 100 DAS, 13/3 at 118 DAS, 24/3 at 130 DAS) under IW: CPE ratio 0.65 and 0.95 during 1997–98 and 1998–99, respectively. Isoproturon @ 1.25 kg (75 WP) + 2, 4-D ester (36 per cent) @ 625 ml/ha (Tank mix) was sprayed at 40 days after sowing using 500 liters water with a knapsack sprayer to control grass and broad leaf weeds. Metasystox (25 EC) @ 375 ml/ha was sprayed in second week of March, using 250 liters of water with a knapsack sprayer to control the aphid. Soil moisture content at different intervals during crop growth period was calculated thermo-gravimetrically by drying the samples in the oven at 105°C till constant weight. A core sampler from 0–7.5, 7.5–15, 15–30, 30–45, 45–60, 60–90, 90–120, 120–150, 150–180 cm soil depth at 103 DAS took root samples during both the years. The samples were taken from 1 site of representative plants, selected in each experimental unit at the base of plant. The samples containing roots from different soil depths were thoroughly washed by putting on a 1 mm sieve kept on moving water. The roots left on the sieves were collected with tweezers and dried at 65°C and weighed and density was expressed as root weight in g m^{-3}. The crop was harvested on April 18, 1998 (156 DAS) and April 9, 1999 (146 DAS) during first and second year, respectively from net plot size of 10 m^2 area (1.35 × 7.41 m). The threshing was done on last week of April with a thresher and grain and straw yield was recorded from each plot and reported as q/ha.

Results and Discussion

Dry Matter Production

During 1997–98, dry matter production was not affected by treatments IW : CPE ratio 0.65 and 0.95 at 60 DAS as the differential irrigations were not started by that time (Table 16.4). At 75 DAS, irrigation was applied in IW : CPE ratio 0.95 and this resulted in higher dry matter production by 7.9 per cent over IW : CPE ratio 0.65 at 90 DAS. Then after 90 DAS, similar trend continued and at harvest 7.7 per cent higher dry matter production was observed under IW : CPE ratio 0.95 over the treatment of IW : CPE ratio 0.65. But during 1998–99, dry matter production increased under IW : CPE ratio 0.95 from 120 DAS on ward and the final dry matter increase at harvest under the said treatment was 7.9 per cent. The increase in dry matter production under IW : CPE ratio 0.95 could be ascribed to adequate water availability which had enhanced the root growth (Table 16.3) and this might have increased the absorption of nutrients, which led to increased dry matter production of IW : CPE ratio 0.95 treatment. Dubey and Sharma (1996) observed irrigation application at IW : CPE ratio 0.9 (IW = 5.0 cm) produced significantly higher dry matter production over IW : CPE ratio 0.6 in flat planted wheat. Dry matter production at 60 DAS during both the years of study was significantly affected by the seed rate cum spacing treatments. Dry matter under 100 kg/ha seed rate with 3 row/bed was significantly higher (284 and 300 gm^{-2} in 1997–98 and 1998–99, respectively) than that of produced under 44.5 kg/ha seed rate with 2 row/bed. Treatments of 66.7 kg/ha (3 row/bed) or 66.7 kg/ha (2 row/bed) also tended to produce higher dry matter production than under 44.5 kg/ha (2 row/bed). Under higher seed rate, since the plant population was more so the dry matter was also more. During later stages of growth, dry matter production was not affected by the seed rate cum spacing treatments. Visual observation showed that stem diameter was considerably higher under lower seed rate of 44.5 kg/ha (2 row/bed) than that of higher seed rate but it could not compensate the effect of higher plant population on dry matter production. Randhawa *et al.* (1997) also reported that increased seed rate from 50 to 100 kg/ha failed to increase the total biomass (grain and straw) significantly in flat planted wheat.

Table 16.1: Effect of Levels of Irrigation and Seed Rate-cum-Spacing on Yield of Wheat

Treatment	*Grain Yield (q/ha)*			*Straw Yield (q/ha)*			*Biological Yield (q/ha)*		
	I	*II*	*Mean*	*I*	*II*	*Mean*	*I*	*II*	*Mean*
Irrigation level IW : CPE ratio									
0.65	56.0	50.8	53.4	74.4	68.2	71.3	130.3	119.0	124.7
0.95	60.4	54.4	57.4	80.2	73.7	77.0	140.5	128.1	134.3
CD (P = 0.05)	3.9	3.6	2.7	4.9	4.8	3.5	8.7	8.3	6.2
Seed rate-cum-Spacing									
100–3R	58.6	54.7	56.7	78.5	74.9	76.7	137.1	129.6	133.4
66.7–2R	57.4	52.6	55.0	77.0	71.3	74.2	134.3	123.9	129.1
66.7–3R	59.7	51.8	55.8	78.9	70.5	74.7	138.6	122.3	130.5
44.5–2R	57.0	51.4	54.2	74.7	67.1	70.9	131.7	118.5	125.1
CD (P = 0.05)	NS	NS	NS	NS	NS	NS	NS	NS	NS

I: 1997–98; II: 1998–99, respectively.

Table 16.2: Effect of Levels of Irrigation and Seed Rate-cum-Spacing on Yield Attributes of Wheat

Treatment	*Effective Tiller (m^{-2})*		*Spike Length (cm)*		*No. of Grains per Spike*		*Test Weight (g)*	
	I	*II*	*I*	*II*	*I*	*II*	*I*	*II*
Irrigation level IW : CPE ratio								
0.65	330	327	10.5	10.6	49.5	48.8	42.0	40.4
0.95	343	347	11.1	10.8	51.8	50.9	43.3	42.1
CD (P = 0.05)	12	14	0.1	NS	1.7	1.8	0.4	0.6
Seed rate-cum-Spacing								
100–3R	350	354	10.6	10.4	48.6	48.2	42.2	40.5
66.7–2R	337	339	10.8	10.6	50.8	49.6	42.6	41.4
66.7–3R	338	336	10.7	10.5	50.8	49.7	42.7	41.2
44.5–2R	322	320	11.2	11.3	52.5	52.0	43.0	42.0
CD (P = 0.05)	16	20	0.2	0.3	2.3	2.6	0.6	0.9

I: 1997–98; II: 1998–99, respectively.

Table 16.3: Effect of Levels of Irrigation and Seed Rate-cum-Spacing on Root Mass Density (g m^{-3}) at 103 Days After Sowing

Soil Profile Depth (cm)	*Irrigation Levels (IW : CPE Ratio)*		*Seed Rate-cum-Spacing*			
	0.65	*0.95*	*100–3R*	*66.7–2R*	*66.7–3R*	*44.5–2R*
			1997–98			
0–7.5	492.2	515.4	507.1	505.1	506.5	496.6
7.5–15	339.8	351.2	347.7	346.7	348.5	339.4
15–30	248.2	258.4	256.9	253.3	254.6	248.4
30–45	146.7	154.2	153.7	151.1	153.1	144.1
45–60	109.9	119.9	115.8	115.5	116.0	112.0
60–90	76.8	80.0	79.7	79.3	80.4	74.3
90–120	33.1	33.8	34.7	33.5	33.9	31.7
120–150	20.2	18.1	19.0	19.5	19.9	17.6
150–180	11.8	10.4	11.8	11.4	11.0	10.3
			1998–99			
0–7.5	460.5	461.9	470.1	468.3	462.2	444.2
7.5--15	313.7	314.2	319.8	319.4	316.4	300.3
15–30	231.4	233.7	239.1	237.1	235.0	219.2
30–45	131.3	131.1	135.5	133.6	130.7	124.9
45–60	96.1	97.1	101.4	98.9	96.5	89.4
60–90	67.6	69.5	70.9	69.5	69.5	64.3
90–120	30.3	30.7	32.3	30.6	29.9	29.4
120–150	17.3	17.2	17.9	17.9	17.4	15.5
150–180	7.6	8.2	8.3	8.3	7.6	7.2

Table 16.4: Effect of Levels of Irrigation and Seed Rate-cum-Spacing on Periodic Dry Matter Production (g m^{-2})

Treatment	60 DAS		90 DAS		120 DAS		At Harvest	
	I	II	I	II	I	II	I	II
Irrigation level IW : CPE ratio								
0.65	272	281	657	666	1107	1068	1279	1171
0.95	270	282	709	671	1215	1200	1378	1263
CD (P = 0.05)	NS	NS	41.1	NS	100	100	81	90
Seed rate-cum-spacing								
100–3R	284	300	707	705	1194	1180	1351	1281
66.7–2R	272	280	678	664	1166	1154	1326	1218
66.7–3R	273	281	683	666	1188	1166	1356	1202
44.5–2R	255	264	663	640	1096	1035	1281	1167
CD (P = 0.05)	19	24	NS	NS	NS	NS	NS	NS

I: 1997–98; II: 1998–99, respectively.

Leaf Area Index

The leaf area index (5.53) recorded under IW : CPE ratio 0.95 was significantly higher than under IW : CPE ratio 0.65 during 1997–98 (Table 16.6). The higher water availability under IW : CPE ratio 0.95 led to increase in growth of plant, whereas a 14 days spell of water stress in IW : CPE ratio 0.65 might have caused reduction in leaf area. It was also observed visually that lower leaves of plants under IW : CPE ratio 0.65 were severely wilted due to water stress in bed planted wheat. Ali *et al.* (1999) observed that green leaf area index was significantly lower in drought plants (10 days after initiation of drought from flag leaf started to emerge) than in well watered plants due to wilting of lower leaves and this period coincided with the growth period of crop in the present study.

Table 16.5: Effect of Levels of Irrigation and Seed Rate-cum-Spacing on Periodic Plant Height (cm)

Treatment	60 DAS		90 DAS		120 DAS		At Harvest	
	I	II	I	II	I	II	I	II
Irrigation level IW : CPE ratio								
0.65	12.1	13.1	46.1	47.4	62.7	63.8	89.1	90.0
0.95	12.0	13.3	47.9	47.5	64.1	65.1	90.1	91.5
CD (P = 0.05)	NS	NS	NS	NS	NS	NS	NS	NS
Seed rate-cum-spacing								
100–3R	12.3	13.4	47.9	48.0	64.3	65.7	90.6	91.6
66.7–2R	12.0	13.3	47.1	47.9	63.6	64.7	89.9	91.5
66.7–3R	12.0	13.2	46.7	47.6	63.0	64.5	89.6	90.7
44.5–2R	11.9	12.8	46.2	46.5	62.6	62.8	88.2	89.3
CD (P = 0.05)	NS	NS	NS	NS	NS	NS	NS	NS

Table 16.6: Effect of Levels of Irrigation and Seed Rate-cum-Spacing on Leaf Area Index at 90 DAS

Irrigation Level	*Leaf Area Index*		*Seed Rate-cum-Spacing*	*Leaf Area Index*	
IW : CPE Ratio	*1997–98*	*1998–99*		*1997–98*	*1998–99*
0.65	5.28	5.23	100–3R	5.45	5.28
0.95	5.53	5.26	66.7–2R	5.43	5.27
–	–	–	66.7–3R	5.45	5.27
–	–	–	44.5–2R	5.30	5.15
CD (P = 0.05)	0.17	NS	CD (P = 0.05)	NS	NS

Root Mass Density

The root mass density decreased with increased profile depth from 0–180 cm during both the years in all the treatments (Table 16.3). In 1997–98, the root mass density under IW : CPE ratio 0.95 was more than that of IW : CPE ratio 0.65. The roots 73.1 and 72.9 per cent were confined to surface 0–30 cm layer under IW : CPE ratio 0.95 and 0.65, respectively. Chaudhary and Bhatnagar (1980) obtained maximum root weight from 0–30 cm layer and it decreased with increasing soil depth. The roots were 4.7, 3.4, 4.2, 5.1, 9.1 and 4.2 per cent higher in 0–7, 5–7, 7–15, 15–30, 30–45, 45–60 and 60–90 cm soil layers under IW : CPE ratio 0.95 than IW : CPE ratio 0.65, respectively. This was due to more of easy availability of water in surface layers under IW : CPE ratio 0.95. In 1998–99, irrigation level IW : CPE ratio 0.95 produced approximately equal root mass density than that of under IW : CPE ratio 0.65, because both the treatments of irrigation received same amount of water upto 100 DAS. Second irrigation was applied just 3 days before sampling of roots in IW : CPE ratio 0.95, which failed to influence the root mass density under IW: CPE ratio 0.95. Sharma *et al.* (1990) found that application of irrigation at IW : CPE ratio 0.9 (IW = 7.5 cm) gave higher root density than under IW : CPE ratio 0.6 in 0–20 cm soil layer, at flag leaf initiation. Seed rate of 100 kg/ha (3 row/bed) during first year of study produced higher root mass density by 2.1, 2.4, 3.4, 6.7, 3.4, 7.3, 9.5, 8.0 and 14.6 per cent over the lower seed rate of 44.5 kg/ha (2 row/bed) in 0–7, 5–7, 7–15, 15–30, 30–45, 45–60, 60–90, 90–120, 120–150 and 150–180 cm, respectively. The corresponding per cent increase in root mass density was 5.8, 6.5, 9.1, 8.5, 13.4, 10.3, 9.9, 15.5 and 15.2 in 1998–99. However, 100 kg/ha seed rate (3 row/bed) gave almost similar root mass density than that of 66.7 kg/ha seed rate (2 or 3 row/bed) during both years. Particularly in the surface 0–15 cm layer treatments of 100 kg/ha (3 row/bed), 66.7 kg/ha (2 row/bed) and 66.7 kg/ha (3 row/bed) produced 2.3, 1.9 and 2.3 per cent more root weight density as compared to 44.5 kg/ha (2 row/bed), respectively during 1997–98 and former set of treatments also produced 6.1, 5.8 and 4.6 per cent more root density over that in 44.5 kg/ha (2 row/bed), respectively during 1998–99.

Grain Yield

The grain yield of wheat increased significantly with the application of irrigations under IW : CPE ratio 0.95 over irrigation level IW : CPE ratio 0.65 during both the years (Table 16.1). The magnitude of increase in the grain yield with IW : CPE ratio 0.95 over irrigation level IW : CPE ratio 0.65 was 7.9 and 7.1 per cent during 1997–98 and 1998–99, respectively. The increase in grain yield at higher IW : CPE ratio 0.95 was due to the significant increased values in plant height (Table 16.5), straw yield (Table 16.1), biological yield (Table 16.1) and yield attributes (Table 16.2) *viz.* effective tillers, spike length, grains per spike and 1000 grain weight. Earlier, Dubey and Sharma (1996) reported that application of irrigation at IW : CPE ratio 0.9 (IW = 5.0 cm) gave significantly higher grain yield over IW : CPE ratio 0.6. From the data, it was observed that various levels of seed rate *viz.*, 100 (3 row/bed), 66.7 (2 rows/bed), 66.7 (3 rows/bed) and 44.5 (2 rows/bed) kg/ha did not exhibit any significant variation on grain yield during both the years. Increased values of spike length, number of grains per spike and thousand grain weight under low seed rate could nullify the effect of higher number of tillers in higher seed rate and resulted in statistically equal yields. However, mean yield tended to increase with the increased seed rate. Hobbs *et al.* (1998) and Dhillon *et al.* (2000) revealed that planting of wheat on beds with 50 kg/ha seed rate gave similar grain yield as compared to 100 kg/ha seed rate. Randhawa *et al.* (1977) also reported similar results in flat sown wheat.

References

Ali, M., Jensen, C.R., Mogensen, V.O., Andersen, M.N. and Henson, I.E., 1999. Root signaling and osmotic adjustment during intermittent soil drying sustain grain yield of field grown wheat. *Field Crops Res.*, 62: 35–52.

Chaudhary, T.M. and Bhatnagar, V.K., 1980. Wheat root distribution, water extraction pattern and grain yield as influenced by time and rate of irrigation. *Agric. Water Mgmt.*, 3: 115–124.

Dhillon, S.S., Hobbs, P.R. and Samra J.S., 2000. Investigation on bed planting system as an alternative tillage and crop establishment practice for improving wheat yields sustainably. In: *Proceedings of the 15th Conference of International Soil Tillage Research Organization*, held at Fortworth Texas, U.S.A. from July 2–7, 2000.

Dubey, Y.P. and Sharma, S.K., 1996. Effect of irrigation and fertilizer on growth, yield and nutrient uptake by wheat (*Triticum aestivum* L.). *Indian J. Agron.*, 41(1): 48–51.

Hobbs, P.R., Sayre, K.D. and Ortiz-Monsteno, J.I., 1980. Increasing wheat yields sustainability through agronomic means. *Natural Resources Group Paper*, Mexico, DF: Mexico, pp. 98–101.

Randhawa, A.S., Jolly, R.S. and Dhillon, S.S., 1977. Effect of seed rate and row spacing on the yield of dwarf wheat (Kalyansona) under different sowing dates. *J. Res. Punjab Agric. Univ.*, 14: 5–8.

Sharma, B.D., Cheema, S.S. and Kar, S., 1990. Water and nitrogen uptake of wheat as related to nitrogen application rate and irrigation water regime. *Fertil. News*, 35(9): 31–35.

Chapter 17

Allelopathic Effect of *Amaranthus sp.* on Growth of *Oryza sativa*

R. Antony Pathrose[1], X. Rosary Mary[1] and P. Dhasarathan[2]*

[1]*Research Centre for Plant Sciences, St. Mary's College, Tuticorin*

[2]*Department of Biotechnology, Sri Kaliswari College, Sivakasi*

ABSTRACT

Allelopathic plays a significant role in both natural and manipulated ecosystem. Allelopathic substances are not only significant in the function of plant communities. Allelopathic plants *Amaranthus viridis* and *Amaranthus caudatus* also inhibit the growth parameters of plant *Oryza sativa.* In the present study selected plant extracts exposed to *Oryza sativa* significantly affect the seed germination in different concentration. In low concentration of plant extract influence less inhibition, 5 per cent of extract inhibit 10 per cent of seed germination. On the other hand the high concentration of plant extract inhibit maximum level, 25 per cent of extract inhibits 33 per cent of seed germination. The different concentration (5, 10 and 25 per cent) of selected plant extracts exposed to *Oryza sativa* in different duration (10, 20 and 30 days), during experimental period the growth parameters and seedling growth (shoot length, root length, number of rootlets and dry weight of seed) was reduced in *Oryza sativa.* The growth parameters were observed that there is a gradual decrease in their 25 per cent concentration and increase in 5 per cent concentration.

Keywords: Allelopathy, Seed germination Amaranthus viridis and A. caudatus.

* Corresponding Author.

Introduction

Allelopathic plays a significant role in both natural and manipulated ecosystem (Rice, 1984). It plays an active role in terrestrial as well as phytoplankton succession, imparting dormancy to the seeds and preventing decay, plant succession and vegetative patterning. In manipulated ecosystems it results in a number of direct and indirect effects. Allelopathic phenomena become, extremely important because of their extracts on the seed germination, crop sequence and inhibition of biological nitrogen fixation. In such a situation, allelopathy can play a key role in future, to control the weeds. Allelopathic substances art not only significant in the function of plant communities, they are part of extensive traffic in chemical influences relating organisms of all major groups to one another in natural communities.

The germination process involves the inception of rapid metabolic activity with in the seed. Result in perceptible growth of the embryo, first the radicle and then the aerial parts. Radicle emergence has been one of the best visible criteria for identifying the initial stage of germination. The growth and productivity of the grains are depending upon the nature of the seeds. The percentage of germination is decreasing with storage time, seedling vigour declined at a similar rate following storage (Ghosh *et al.*, 1978). Hence, in the present study we are analysed growth parameters of *Oryza sativa* with exposed different concentration of *Amaranthus* sp. extraction..

Materials and Methods

Test Plant

Amaranthus viridis L. grow on all plain districts as a weed of waste and cultivated land. An erect, glabrous, annual, the leaves much used as spinach. *Amaranthus caudatus* L. present in the hills of Deccan and Western Ghats. This plants cultivated chiefly in gardens, perhaps also sometimes found run wild and also tall herb with long heavy drooping thyrses of crimson flowers.

Experimental Plant

Oryza sativa is an annual grass with round hollow jointed culms, rather flat, leaves directly attached to the nodes of the stem and a terminal panicle. It is adapted to grow in flooded soil. It is an important cereal for life.

Preparation of Extract

In the present investigation *Amaranthus viridis* and *Amaranthus caudatus* were chosen and the fresh leaves of selected plants collected in early morning and used to prepare extraction. The leaf extract were prepared with the help of Mortar and Pestle and make up it 5, 10 and 25 per cent concentration using sterile distilled water.

Estimation of Growth Parameters

The various growth parameters of *Oryza sativa* were assessed after exposure of test plant extracts.

Percentage of Germination

Healthy 25 seeds were placed in sterile pertiplates, then different concentration of test plant extracts were over layered to pertiplates separately. The percentage of germinate is calculated after 3 days by seeing the presence of the radicle.

Shoot Length

Shoot length was measured by choosing five seedlings at random from each of the control and treated petridish. The average of the seedlings were taken as measure of shoot length in cm p^{-1}.

Root Length

Root length was measured by choosing five seedlings at random from each of the control and treated petridish and the average of the seedlings were taken as measure of root length in cm p^{-1}.

Root Lets

Number of lateral roots were counted by choosing five seedlings at random from each control and treated and the average number was taken per plant.

Fresh Weight

Five seedlings were taken from the control and treated petriplates. It was weighed for their fresh weight and represented in mg p^{-1}.

Dry Weight

The seedlings after weighing their fresh weight dried on 60°C for 2 days and weighed after drying to get a constant weight, it represented as mg p^{-1}.

Estimation of Protein

The protein content was determined according to the method of Lowry *et al.* (1951). The leaf sample, 100 mg was ground in 20 ml of 80 per cent acetone and the homogenate was centrifuged at 5000 rpm for 5 minutes. The resulting while pellet was dissolved in 0.1N NaOH. After complete solubilisation 0.1 ml of the extract was added to 3 ml of alkaline copper reagent (2 per cent Na_2CO_3) in 0.1N NaOH, 0.5 per cent $CuSO_4$ and 1 per cent Sodium potassium tartarate solution). To this 0.5 ml folin ciocalteu reagent (1 : 1 diluted) was added and vortexed. After a brief incubation absorbance was read at 750 nm, Bovine serum albumin was used as the standard. The content of protein was expressed as mg g^{-1} fresh weight of the leaf tissue.

Estimation of Starch

The fresh leaves about 500 mg ground in distilled water; the homogenate was centrifuged at 4500 rpm for 5 minutes. He supernatant was collected and made up to a known volume with distilled water to 2.0 ml of this extract, 0.1 ml lugol's iodine was added and make up to 5.0 ml distilled water. The optical density at 640 nm was read and the amount of starch was calculated from the standard graph.

Result and Discussion

Allelopathic effect of two leafy medicinal plants such as *Amaranthus caudatus* and *Amaranthus viridis* significantly affected the seed germination percentage, seedling growth, shoot length, root length, root lets, fresh weight and dry weight over the control. The degree of inhibition based on the concentration of extract and also suggested the presence of inhibitory substance or chemicals. The germination decreased with increasing concentration of the extract over the control in selected plant extract. The germination rate observed in the control was 100 per cent. The leaf extract of *A. caudatus* also reduce the rate of germination of paddy seedlings at 25 per cent of leaf extract. The germination percentage was reduced to 33 per cent and at 5 per cent, percentage leaf extract reduced 10 per cent of germination. The leaf extract of *A. viridis* also reduce the germination in 25 per cent concentration inhibit the 29 per cent of germination (Table 17.1).

Table 17.1: Allelopathic Effect of *A. viridis* and *A. caudatus* Plants on Seedlings Germination of Paddy

Sl.No.	*Concentration*	*Germination (%)*	
		A. caudates	*A. viridis*
1.	0 (Control)	100	100
2.	5 per cent	90	95
3.	10 per cent	78	82
4.	25 per cent	67	71

The seedling growth was measured with the parameters like shoot length, root length and number of rootlets in the paddy seedlings. The shoot length was measured in different days (10, 20 and 30 days) old paddy treated with 5, 10 and 25 per cent leaf extract of selected plants; inhibition of seedlings growth depends on the concentration and natural chemicals (Table 17.2).

The findings of Pandy (1997) supports the finding that the allelochemicals leached out of the parthenium residue appear to have inhibited the treated plants by affecting cellular membrane integrity, similar results are seen by Padhy *et al.* (2000) from Eucalyptes leaf litter leachates in germination and seedling growth of finger millet. The root length studies also reveals that there is allelopathic effect in selected plant in studied (Table 17.2). On the 3rd day in all the three concentration (5, 10 and 25 per cent) the root length is lesser than the control. The fresh weight and dry weight of the treated plants were found to be decreased them in plants under control. Chellamuthu *et al.* (1997) and Indira *et al.* (2001) have observed similar results. Allelopathic effects of *A. viridis* and *A. caudatus* significantly act on the growth parameters of *Oryza sativa.* The growth parameters were observed that there is a gradual decrease in their 25 per cent concentration and increase in 5 per cent concentration. As the concentration of two species of leaf extracts increases there was a significant reduction in the germination and seedling growth parameters of *Oryza sativa.* It is observed that the inhibition rate is increased with increase in concentration of extract *Amaranthus caudatus* and *Amaranthus viridis* for the study.

Table 17.2: Effect of *A. viridis* and *A. caudatus* Extract on Seedling Growth Parameter of Paddy

Sl.No.	*Seedlings*	*Days*	*Control*	*A. caudatus*			*A. viridis*		
				5%	*10%*	*25%*	*5%*	*10%*	*25%*
1.	Shoot length (cm)	10	9	7	7	7	8	8	8
		20	11	9	9	8	11	11	9
		30	16	9	9	9	12	11	11
2.	Root length (cm)	10	5.8	4.4	4.2	4.2	4.8	4.3	4.4
		20	6.5	5.9	4.5	4.2	5.7	4.9	4.4
		30	6.7	6.0	5.2	5.0	5.9	5.1	5.1
3.	Root lets	10	12.2	8.5	8.5	8.2	9.00	8.6	8.4
		20	13.5	10.5	10.1	8.2	11.8	10.2	8.7
		30	14.5	12.0	10.5	8.3	14.0	11.0	8.5
4.	Dry weight (mg)	10	12.2	9.0	8.5	8.5	10.0	8.5	8.5
		20	13.8	10.5	10.5	8.2	11.5	10.5	9.0
		30	15.0	12.0	11.0	8.2	14.0	12.0	9.5

References

Chellamuthu, V., Balasubramanian, T.N., Rajan, A.R. and S.P. Palaniappan, 1997. Allelopathic influence of *Prosopis julifloraon* field crops. *Allelopathy*, 4(2): 291–302.

Ghosh, B., Sengupta, T. and S.M. Sircar, 1978. Allelopathic effects of Plants. *Ind. J. of Exp. Biol.*, 16: 411.

Indira, E.P., Basha, S.C., Chacko, K.C and C.N. Krishnankutty, 2001. Effect of different sowing methods and seed germination and growth of Teak seedlings. *Indian Journal of Forestry*, 24(1): 93–96.

Lowry, O.H., Rosebrough, N.J., Farr, A.L and R.J. Randall, 1951. Estimation methods of Biochemicals. *J. Biol. Chem.*, 193: 265.

Padhy, B., Patnaik, P.K. and K. Tripathy, 2000. Allelopathic potential of *Eucalyptus* leaf litter leachates an germination and seedling growth of finger millet. *Allelopathy*, 7(1): 69–78.

Panday, D.K., 1997. Inhibition of Najas (*Najas hysterophorus*). *Allelopathy*, 4(1): 121–126.

Rice, E.L., 1984. *Allelopathy*. Academic press, London, pp. 65–125.

Chapter 18

Screening of Chickpea Genotypes Against Fusarium Wilt

V.K. Mandhare, G.P. Deshmukh and A.V. Suryawanshi

Pulses Improvement Project, Mahatma Phule Krishi Vidyapeeth, Rahuri – 413 722 (M.S), India

ABSTRACT

Wilt of chickpea caused by *Fusarium oxysparum* f.sp. *ciceri* is an important and commonly occurring disease in Maharashtra. Studies were conducted on chickpea at Pulses Improvement Project, MPKV, Rahuri during the year 2003–2004 to 2004–2005 in artificially wilt sick soils for identification of sources of resistance to *Fusarium* wilt. Of the 52 genotypes screened, forty were exhibited resistance. There was 100 per cent mortality in susceptible check (JG-62).

Keywords: *Chickpea genotypes, Screening, Fusarium wilt.*

Introduction

Chickpea (*Cicer arietinum* L.) is one of the most important pulse crop grown in India. In India area under chickpea was 6.50 million ha. with the production of 5.77 million tones with an average yield of 888 kg/ha. (Anonymous 2004). In Maharashtra, the area under the cultivation of chickpea was 7.95 lakh ha. with 4.21 lakh tonnes production and average yield 529 kg/ha (Anonymous 2004). Chickpea suffers from many diseases like root rot, wilt, stunt. But the wilt caused by *F. oxysporum* sp. *ciceri* is an important and commonly occurring disease in Maharashtra. The fungus is soil and seed borne and survives in soil in the absence of host for at least 6 years (Haware *et al.*, 1986 b) causing losses upto 100 per cent. Thus, a resistant variety is a measure to overcome these problems.

Materials and Methods

Fifty two promising chickpea genotypes were screened in artificially wilt sick plot having plot size 5 × 3 m with two replication at the spacing 30 × 10 cm during *rabi* season 2003–2004 to 2004–2005 at Mahatma Phule Krishi Vidyapeeth, Rahuri. All the cultural practices except plant protection measures were carried out to raise the crop. The wilt susceptible check JG-62 was sown intermittantly after every two test entries. The observation on number of plants wilted in each line were recorded at 30, 45, 60, 75 and 90 days after sowing and mean per cent wilt incidence was calculated (Mayee and Datar, 1985). The genotypes were graded into different disease reaction categories. The categories were:

Resistance (R): 0.0 to 10.0 per cent

Moderately resistance (MR): 10.1 to 20.0 per cent

Susceptible (S): 20.1 to 50.0 per cent

Highly susceptible (HS): 50.1 and above

Results and Discussion

The results presented in Table 18.1 revealed that JG-62 was completed wilted within 30 days after sowing thereby indicating high and uniform level of sickness in the field. All the 53 genotypes tested showed varying degree of mortality.

Among the various genotype tested PG–9708–4–2–4, PG–9708–4–2–5, PG–9708–4–2–6, PG–9708–4–2–8, PG–96025–15–5, PG–960251–15–12, PG–960251–15–24, PG–960255–10–2, PG–960256–1–5, Vishal, Virat (K), Vihar (K), PG–9425–5, PG–9425–9, PG–96006, JAKI–9218, PG–960258–4–15, PG–960258–4–19, PG–960258–12–1, PG–960258–10–12, PG–960258–1–11, PG–95333(K), PG–97313 (K), PG–9621–8, PG–9621–14, PG–9759–2–10, PG–00108, PG–00109, PG–97030, PG–97117, PG–9409–1, PG–9426–2, PG–9421–1, PG–00110, PG–01103, PG–95138, PG–9409–1–1, PG–9758–6–2, PG–97033, PG–9621–12 were found to be resistant.

The genotypes *viz.*, SAKI–9516 and PG–97301 were moderately resistant whereas, PG–00402 (K), PKV–2 (K), PG–01310, PG–97308 exhibited susceptible reaction. Pawar *et al.* (1993), Gupta (1995) and Suryawanshi *et al.* (2003) reported similar type of reaction with respect to *Fusarium* wilt when they screened various genotypes.

Table 18.1: Field Screening of Chickpea Genotypes Against *F. oxysporum* f. sp. *ciceri*

Sl.No.	Disease Reaction Categories	Genotypes
1.	Resistant (R) (0–10.0% wilt)	PG–9708–4–2–4, PG–9708–4–2–5, PG–9708–4–2–6, PG–9708–4–2–8, PG–960251–15–5, PG–960251–15–12, PG–960251–15–24, PG–960255–10–2, PG–960256–1–5, Vishal, Virat (K), Vihar (K), PG–9425–5, PG–9425–9, PG–96006, PG–960258–4–15, PG–960258–4–19, PG–960258–12–1, PG–960258–10–12, PG–960256–1–11, PG–96333 (K), PG–97313 (K), PG–9621–8, PG--9621–14, PG–9759–2–1Q, PG–00108, PG–00109, PG–97030, PG–97117, PG–9409–1, PG–9421–1, PG–00110, PG–01103, PG–95138, PG–9409–1–1, PG–9758–6–2, PG–97033 and PG–9621–12
2.	Moderately resistant (MR) (10.1–20.0% wilt)	SAKI–9516 and PG–97301
3.	Susceptible (S) (20.1–50.0% wilt)	PG–00402 (K), Vijay, PKV–2(K), PG–01308, PG–01310, PG–97308 and JG–62
4.	Highly Susceptible (HS) (50.1% and above wilt)	AKG–46–1,Vikas, PG–5, PG–12 and JG–62

References

Anonymous, 2004. Area and production under chickpea in India. *Agricultural Situation in India.*

Anonymous, 2004. *District-wise General Statistical Information of Agricultural Department* (MS) Part-II.

Gupta, O., 1995. Identification of chickpea genotypes with dual resistance against wilt and root rots. *ICPN*, 2: 27–28.

Haware, M.P., Nene, Y.L. and Natrajan, M., 1986b. Survival of *Fusarium oxysporum* f.sp. *ciceri* in soil in the absence of Chickpea. In: *Abst. of Papers Presented at Seminar on Management of Soil Borne Diseases of Crop Plant*, held at Tamil Nadu Agricultural University, Coimbatore (India), on October 5–9.

Mayee, C.D. and Datar, V.V., 1985. *Phytopathometry*. Tech. Bull. 1, Marathwada Agricultural University, Parbhani, pp. 144.

Pawar, K.B., N.J. Bendre, R.B. Deshmukh and S.B. Bhor, 1993. Field reactions of some chickpea lines to *Fusarium* wilt. *J. Maharashtra Agric. Univ.*, 18(2): 327–328.

Suryawanshi, A.V., V.K. Mandhare, M.M. Sanap and B.M. Jamadagni, 2003. Reaction of chickpea entries to *Fusarium* wilt and gram pod borer. *J. Maharashtra Agric. Univ.*, 28(2): 213–214.

Chapter 19

Screening of Pigeonpea Genotypes Against Wilt and Sterility Mosaic Disease in Maharashtra

G.P. Deshmukh, V.K. Mandhare and A.V. Suryawanshi

Pulses Improvement Project, Mahatma Phule Krishi Vidyapeeth, Rahuri – 413 722 (M.S.), India

ABSTRACT

Pigeonpea [*Cajanus cajan* (L.)] is one of the important grain legume. Wilt caused by *Fusarium oxysproum* f.sp. *udum* and sterility mosaic virus are the important diseases responsible for great yield losses. *Fusarium* wilt is soil and seed borne nature Sterility mosaic virus is transmitted by an eriophyid mite. In present investigation, thirty one genotypes were screened against *Fusarium* wilt in artificially wilt sick soil and screened for sterility mosaic disease using infector row technique at Pulses Improvement Project, MPKV, Rahuri. Of the 31 genotypes screened, the genotypes *viz.*, PT–9230, PT–02–134, PT–1037, PT–8208–1, BSMR–736 and BSMR–853 exhibited combined resistance against both the diseases. The genotype *viz.*, PT–02–5, PT–02–140 and MAL-3 were found to be resistance only for *Fusarium* wilt. MAL–6, Bahar, BDN–2010 showed resistant against sterility mosaic disease.

Keywords: Pigeonpea genotypes, Screening, Wilt and Sterility mosaic.

Introduction

Pigeonpea [*Cajanus cajan* (L)] is most important *Kharif* pulse crop of Maharashtra covering an area of 10.46 lakh ha with the production of 6.93 lakh tones (Anonymous, 2004). *Fusarium* wilt and sterility mosaic are the two major diseases affecting pigeonpea production in India. The fungus is soil and seed borne and survives in soil in the absence of host for at least six years. In Maharashtra, *Fusarium udum* causes 15 to 20 per cent grain yield losses (Nene *et al.*, 1989). In nature, the sterility mosaic disease is transmitted by an eriophyid mite (*Aceria cajani*) which is host specific on Pigeonpea and few of its wild relatives causing yield losses upto 95 per cent in south India (Reddy and Nene, 1981). It's severity has been recorded more in Tamil Nadu, Karnataka and Maharashtra (Kannaiyan *et al.*, 1984 and Mali *et al.*, 1971). Therefore, the present studies were undertaken.

Materials and Methods

Thirty one pigeonpea genotypes were screened for *Fusarium* wilt in artificially wilt sick plot and for sterility mosaic disease using infector row technique having plot size 5 × 3 m with two replication at the spacing 45 × 15 cm during *Kharif* season 2003–2004 to 2004–2005 at Pulses Improvement Project, Mahatma Phule Krishi Vidyapeeth, Rahuri. All the cultural practice except plant protection major carried out to raise the crop. The wilt susceptible check ICP–2376 was sown intermittently after every two test entries. Similarly, the sterility mosaic susceptible check ICP–8863 was sown as an infector, alternatively after every two test entries to ensure high disease pressure and monitor the disease on various test entries. The observation on number of plants wilted in each line was recorded from 45 days after sowing up to harvesting. The incidence of sterility mosaic disease was recorded before flowering, at flowering and at physiological maturity of the crop (Nene and Reddy, 1976). The mean percent wilt incidence was calculated (Mayee and Datar, 1985).

Results and Discussion

The data presented in Table 19.1 indicated that there was cent percent wilting was noticed on ICP–2376 a susceptible check while incidence of sterility mosaic on ICP–8863 (SM check) in screening nursery was 100 per cent. Out of 31 genotypes screened for two years, six genotypes *viz;* PT–9230, PT–02–134, PT–1037, PT–8208–1, BSMR–736 and BSMR–853 exhibited combined resistance against *Fusarium* wilt and sterility mosaic. Zote and Dutraj (2002) reported similar results showing BSMR-853 cultivar of Pigeonpea as multiple disease resistant. The entries *viz.*, PT–02–5, PT–02–140, MAL–3 were found to be resistant only to *Fusarium* wilt. MAL–6, Bahar and BDN–2010 genotypes were resistant against sterility mosaic disease. But, ICPL–87119 and JG–65 were moderately resistant for the sterility mosaic and remaining entries were highly susceptible to sterility mosaic. Singh *et al.* (1995) screened 450 genotypes of which 36 were resistant, 22 were moderately resistant, 16 were tolerant and remaining were highly susceptible to sterility mosaic disease.

Table 19.1: Field Screening of Pigeonpea Genotypes Against Wilt and Sterility Mosaic Disease

Sl.No.	*Disease Reaction Categories*	*Genotypes*
1.	Resistant (R) (0–10.0% wilt and SM)	Wilt: PT–02–5, PT–9230, BSMR–736, BSMR–853, MAL–3 PT–02–134 PT–02–140, PT–1037 and PT–8208–1
		SM: Bahar, PT–9230, BDN–2010, BSMR–736, BSMR–853, MAL–6, PT–02–134, PT–1037 and PT–8208–1
2.	Moderately resistant (MR) (10.1–20.0% wilt and SM)	Wilt: BDN–2010 and BDN–2004 SM: JG–65, MAL–3, NDA–99–6 and ICPL–87119
3.	Susceptible (S) (20.1–50.0% wilt and SM)	Wilt: PT–02–9, PT–02–25, BDN–2009, AKT–9929, C–11, NDA–99–6, PT–11–39, PT–01–77, BDN–2003–1 and ICP–237–6
		SM: PT–02–9, PT–02–25, BDN–708, AKT–929, C–11, BDN–702, PT–02–140, PT–11–39, PT–01–77, BDN–2003–1 and BDN–2004
4.	Highly Susceptible (HS) (50.1% and above wilt and SM)	Wilt: Bahar, JG–65, BDN–708, BDN–702, AKT–8811 BDN–2 BDN–2001–6 and ICP–2376
		SM: PT–02–5, BDN–2009, AKT–8811, BDN–2, BDN–2001–6 and ICP–8863

References

Anonymous, 2004. *District-wise General Statistical Information of Agricultural Department* (MS) Part-II.

Kannaiyan, J., Nene, Y.L., Reddy, M.V., Ryan, J.G. and Rahu, T.N., 1984. Prevalence of Pigeonpea diseases and associated crop losses in Asia, Africa and the Americas. *Tropical Pest Management*, 30(1): 62-71.

Mali, V.R., Shirsat, A.M. and Godbole, G.M., 1971. Occurrence of Pigeonpea sterility mosaic in Marathwada. *Research Bulletin*, M.A.U., Parbhani, 1(10): 148–149.

Mayee, C.D. and Datar, V.V., 1985. Phytopathometry. *Tech. Bull. J.*, Marathwada Agric. University, Parbhani,

Nene, Y.L. and Reddy, M.V., 1976. A new technique to screen pigeonpea for resistant to sterility mosaic. *Trop. Gram Legumes Bulletin*, 5: 23.

Nene, Y.L., Kannaiyan, J., Reddy, M.V., Zote, K.K., Mathmood, M. Hiremath, R., Shukla, V.P., Kotasthane, S.R. Sen Gupta, K., Jahagirdar, D.K., Jaque, M.F., Grewal, J.S. and Pal, M., 1989. Multi-locational testing of pigeonpea from broad-based resistance to *Fusarium* wilt in India. *Indian Phytopath.*, 42(3): 449–453.

Reddy, M.V. and Nene, Y.L., 1981. Estimation of yield loss in pigeonpea due to sterility mosaic. In: *Proceeding of the International Workshop on Pigeonpeas*, 15–19, December, 1980, ICRISAT Center, India, Patncheru, India. 2: 305–312.

Singh, B.B., D.P. Singh, N.P. Singh and R. Kumar, 1995. Evaluation of pigeonpea germplasm for resistance to sterility mosaic. *ICPN*, 2: 69–70.

Zote, K.K. and Dhutraj, D.N., 2002. A white seeded multiple disease resistant pigeonpea BSMR-853. *J. Maharashtra Agric. Univ.*, 27(3): 252–254.

Chapter 20

Assessment of the Quality of Drinking Water in Outer Rural Delhi: Physico-chemical Characteristics

Vijender Singh*

Indian Institute of Ecology and Environment, New Delhi

ABSTRACT

A study was conducted in the outer rural Delhi to examine the physico-chemical properties of the drinking water being used by the residents there. The main objective was to assess the fitness of the potable water in view of the increased incidences of gastro-intestinal disorders and other water-borne disease prevalent there. The water samples were collected from three zones where the source of water was different, *i.e.*, only by DJB or DJB plus groundwater or only groundwater. The sampled water was analyzed for pH, electrical conductivity, turbidity, residual chlorides, nitrates, phosphates and total organic carbon. It was observed that water supplied by Delhi Jal Board (DJB) was safer to use and did not contain much amount of inorganic elements. In contrast the groundwater was found to be rich in mineral elements (nitrates and phosphates) that promotes the growth of pathogenic microorganisms and thus poses health threats to residents using it. At some sampling areas using only groundwater; the water supply contained significantly higher amounts of nitrates, phosphates and TOC that were above the prescribes limits.

Keywords: *Physico-chemical properties, Potable water, Groundwater.*

Introduction

India has a rich water resource including a wide network of rivers and vast alluvial basins to hold groundwater. In spite of the availability of approximately 1100 km^3 water for meeting all kind of

* *Present Address: Vice-President of India Secretariat, New Delhi – 110 011.

needs, there are severe problems of water scarcity, primarily due to non-uniformity in availability. The problem has been further aggravated by the rapid increase in population thereby increasing the demand of water supply for irrigation, human and industrial consumption. It has resulted in severe depletion of available water resources in many parts of the country. Surface water and groundwater are the major sources of drinking water in urban and rural India. In urban areas, water for drinking and various domestic purposes is supplied by municipal authorities who supply after a thorough treatment of river water. However, despite the treatment the water gets contaminated during the distribution process either due to leaking from water pipes or escape of protozoa and enteric virus through filters or formation of bio-film in storage and distribution systems or even seepage of soil nutrients and soil particles through breakage in supply system. As per a report of Central Pollution Control Board, most of the Indian rivers are contaminated due to discharge of untreated sewage, domestic and industrial wastes and pesticide runoff from the fields. In fact, industrial waste and the municipal solid waste have emerged as one of the leading cause of pollution of surface and groundwater. In many parts of the country, available water is rendered non-potable because of presence of iron, nitrate, arsenic or heavy metals in excess. The situation gets worsened during the summer season due to water scarcity and rain water discharge. Contamination of the water resources available for household and drinking purposes with heavy elements, metal ions and harmful microorganisms is one of the serious and major health problems. As a result huge amount of money is spent for chemical treatment of contaminated water so as to make it potable. Thus there is a need to look for some useful indicators, both microbiological and physical, which can be used to monitor both drinking water system operation and performance. The problem of water contamination and scarcity is very severe in Delhi where there has been a rapid increase in population due to increased urbanization and industrialization. The majority of potable water is supplied by Delhi Jal Board (DJB) in the inner regions of Delhi whereas in the outer regions water is supplied through two different types of resources. These are water supplied by Delhi Jal Board (*i.e.* DJB supply) and Groundwater. People residing in outer Delhi experience very frequent episodes of gastro-intestinal disorders and outrage of diseases like cholera and jaundice. Presumably most of the disorders are caused by the contamination of the drinking potable water. A study was planned to assess the physical and chemical properties of potable water being used in different parts of Outer Delhi where the supply is through either DJB or from the groundwater.

Methods

Sampling Sites

For the present study area of outer rural Delhi was selected and categorized into three zones. These were: Zone I (that received water supply from DJB); Zone II (that received mixed supply of DJB and Groundwater 1–tubewell); and Zone III (drinking water supply from groundwater 1–tubewell and groundwater 2–hand pump). Both the groundwater sources are treated with chlorine before distribution. Overall, 51 sampling sites including 13 in Zone I, 30 in Zone II and 8 in Zone III were selected for monitoring the physico-chemical status of water supply.

Sample Collection

Water samples from the selected sites were collected and taken in 500 ml phosphate free washed glass bottles. The sample after collection were immediately placed in dark cooling boxes and processed within 6 h of collection.

Physico-chemical Analysis

In order to determine the compounds which might support bacterial growth, a variety of physico-chemical parameters were monitored. These include electrical conductivity (EC), pH, temperature, turbidity, organic carbon, residual chlorine, phosphate, and nitrate. Electrical conductivity (EC), pH and temperature measurements were performed according to standard procedures (APHA, 1992). Turbidity was measured by Nephelometric method (Hart *et al.*, 1992) with Hach Model 2100A Turbidimeter. Organic carbon was analyzed by a high temperature combustion technique using a Shimadzu 5000 TOC analyzer as per the method of Van Hall *et al.* (1963). Residual chlorine was estimated by Iodometric method as per Marks and Chamberlin (1953). Phosphates were analyzed by ascorbic acid method (Murphy and Riley, 1962). Nitrates were determined by following the method of Nivone (1964).

Results and Discussion

It is clear from the Table 20.1 that pH of water in collected from various sites in the zone I was lower than that from source of DJB except at the sites 7, 8, 9, 10. Values of pH in the sites 2 to 6 ranged from 7.29 to 7.33. The lowest values of pH were, however, recorded in sites 12 and 13 (Table 20.1). In zone II, however, where the supply of water was from DJB as well as from groundwater the value of pH was higher (7.5). At consumer's end also the pH values exhibited a great variation (Table 20.2). It ranged from 7.08 to 7.9 (site 26). In zone III, the pH of water collected at source was found to be 4.43 and 7.45, respectively for groundwater–2 and groundwater–3, respectively. At consumers end however, the pH values ranged from 7.08 to 7.91 (Table 20.3).

Table 20.1: Physico-chemical Characteristics of Drinking Water in Zone I (DJB supply) of Outer Rural Delhi

Sampling Site	*pH*	*EC (mho/cm)*	*NO_3N (ppm)*	*Residual Cl (ppm)*	*PO_4P (ppm)*	*Turbidity (NTU)*	*TOC (mg/l)*
			Source of DJB Water				
Site-1	7.50	0.486	1.61	0.18	0.01	0.23	1.3
			Consumer End Point				
Site-2	7.29	0.410	1.63	0.16	0.01	0.12	1.6
Site-3	7.30	0.416	1.65	0.15	0.02	0.11	1.4
Site-4	7.30	0.403	1.60	0.18	0.01	0.16	1.9
Site-5	7.29	0.401	1.60	0.16	0.01	0.19	1.5
Site-6	7.33	0.419	1.56	0.14	0.01	0.20	1.1
Site-7	7.60	0.26	1.56	0.16	0.01	0.16	1.2
Site-8	7.60	0.441	1.53	0.14	0.01	0.16	1.5
Site-9	7.64	0.418	1.59	0.15	0.01	0.18	1.6
Site-10	7.51	0.420	1.56	0.14	0.01	0.13	1.9
Site-11	7.11	0.341	2.06	0.11	0.68	0.18	2.1
Site-12	7.04	0.386	3.13	0.11	0.02	0.14	2.0
Site-13	7.06	0.389	2.72	0.17	0.00	0.15	2.0
LSD (0.05)	NS	NS	NS	NS	NS	NS	NS

The conductivity of water in the zone I at source and sampling sites did not show much variation. It was around 0.486 at the source point whereas at the consumer's end these values differed from 0.386 to 0.441 mmhos/cm (Table 20.1). The conductivity of water at the source in zone II was measured to be 1.283 mmhos/cm. It was almost three times higher than the DJB water supply. At the distribution ends the conductivity ranged from 0.249 to 1.386 mmhos/cm (Table 20.2). In zone III the conductivity of water was higher at source as well as at consumers end point (Table 20.3).

Table 20.2: Physico-chemical Characteristics of Drinking Water in Zone II (DJB + Groundwater-1 supply)

Sampling Site	*pH*	*EC (mho/cm)*	*NO_3N (ppm)*	*Residual Cl (ppm)*	*PO_4P (ppm)*	*Turbidity (NTU)*	*TOC (mg/l)*
Source of DJB + Groundwater							
Site-14	7.68	1.283	2.5	0.18	0.03	0.42	2.1
End point DJB and Groundwater							
Site-15	7.29	0.321	3.2	1.18	0.20	0.68	2.9
Site-16	7.70	0.516	11.0	0.14	0.32	0.56	3.1
Site-17	7.7	0.701	10.8	0.12	0.34	0.49	2.6
Site-18	7.35	0.343	2.90	0.10	0.56	0.34	2.9
Site-19	7.35	0.413	3.30	0.17	0.39	0.82	3.1
Site-20	7.36	0.244	3.30	0.15	0.90	0.99	3.4
Site-21	7.61	0.430	6.40	0.17	0.53	0.82	3.9
Site-22	7.25	1.183	33.70	0.25	0.44	0.54	3.1
Site-23	7.36	0.421	3.70	0.19	0.36	0.55	3.6
Site-24	7.2	0.308	3.3	0.17	0.29	0.46	3.9
Site-25	7.68	0. 985	15.7	0.12	0.58	0.87	3.7
Site-26	7.90	0.399	2.6	0.16	0.08	0.89	3.1
Site-27	7.33	0.389	1.6	0.16	0.41	0.56	2.7
Site-28	7.80	1.566	18.3	0.29	0.01	0.46	3.1
Site-29	7.22	0.411	30.7	0.28	0.01	0.82	2.9
Site-30	7.72	0.419	10.9	0.15	0.13	0.65	3.7
Site-31	7.57	0.402	.12.1	0.16	0.06	0.84	2.9
Site-32	7.51	1.341	24.8	0.31	0.01	0.82	4.1
Site-33	7.70	0.815	15.5	0.17	0.44	0.85	4.7
Site-34	7.56	1.123	26.0	0.29	0.95	0.79	4.2
Site-35	7.51	0.846	16.4	0.18	0.45	0.79	5.1
Site-36	7.08	0.363	2.6	0.12	0.00	0.48	3.8
Site-37	7.60	1.323	17.3	0.17	0.01	0.54	3.5
Site-38	7.64	1.361	19.0	0.29	0.03	0.48	4.1
Site-39	7.66	1.368	17.8	0.32	0.05	0.52	4.2
Site-40	7.37	0.382	2.3	0.18	0.00	0.49	3.6
Site-41	7.70	1.241	2.7	0.12	0.02	1.01	5.8
Site-42	7.31	0.404	1.7	0.13	0.00	0.56	3.9
Site-43	7.81	1.216	2.7	0.11	0.01	0.99	6.1
LSD (P = 0.05)	NA	NA	NA	NA	NA	NA	NA

Table 20.3: Physico-chemical Characteristics of Drinking Water in Zone III (Groundwater only)

Sampling Site	*pH*	*EC (mho/cm)*	*NO_3N (ppm)*	*Residual Cl (ppm)*	*PO_4P (ppm)*	*Turbidity (NTU)*	*TOC (mg/l)*
			Source of Groundwater-1				
Site-44	7.43	1.396	1.73	0.267	0.02	0.22	3.6
			Consumer End Point				
Site-45	7.45	1.384	17.73	0.268	0.04	0.54	4.2
Site-46	7.10	1.389	36.80	0.312	0.05	0.39	5.8
Site-47	7.08	1.426	35.47	0.312	0.30	0.42	6.1
Site-48	7.63	1.339	18.47	0.284	0.03	0.60	3.1
Site-49	7.75	1.466	1.73	0.312	0.02	0.63	6.1
Site-50	7.91	1.509	2.65	0.124	0.01	0.82	5.6
			Consumers Source and End Point (Hand pump–groundwater 3)				
Site-51	7.75	1.218	2.78	0.027	0.02	1.22	6.2
LSD (0.05)	NA	NA	NA	NA	NA	NA	NA

In zone I, the values of turbidity also did not show much difference. It varies from 0.11 to 0.20 NTU in contrast to 0.23 NTU at the source point. The turbidity of zone II varied from 0.42 to 1 NTU in contrast to 0.42 measured as source point. At many sites it was appreciably more and even double than source values thereby indicating the suspended materials. In zone III, however, turbidity at source point was quite less (0.22 NTU) and it increased drastically at all the consumer end points. It was maximum at site 50 and 51.

There was not much change in the residual chlorine in the zones I and II between consumer use point and the source point (Tables 20.1 and 20.2). However, in the zone III the residual chlorine was higher at many sampling sites, *e.g.* sites 46, 47 and 49. There was very little of chlorine in the hand pump groundwater (Table 20.3).

The content of nitrate did not show much variation at many of the consumer sites from that of source in the zone I except at sites 11, 12 and 13 where it was significantly higher (Table 20.1). In the zone II there was a large variation and at many sites (22, 29, 34) it was increased by over 10–15 times of that at source point (Table 20.2). However, in the zone 3 where groundwater was from hand pump, there was a significant increase in the nitrate content of the water supply except at site 49, 50 and 51 (Table 20.3).

In zone I and II not much difference in total organic carbon was observed between source and user end point. The only exception was site 43 in zone II, where it differed from source value (Tables 20.1 and 20.2). However, in the zone III an appreciable and significant increase was observed in the TOC value (Table 20.3).

Discussion

From the results it is clear that there was not much variation in the physico-chemical characteristics of water supply to the outer rural Delhi by the DJB except site 11, 12, 13. Even in zone I there was variation in EC but these were well within the admissible levels as per the Indian standards for safe water supply.

In the zone II there was an appreciable variation in the physico-chemical characteristic of water supply. In this zone where DJB water was mixed with groundwater-1 and then supplied through pipes to consumers. Phosphate content was significantly more in this zone at a number of sites. In the zone 3 there was a significantly higher range of phosphate and it was even more than the permissible limits of WHO and APHA. Increased phosphate content generally results in higher growth of microbial growth (Mettinen *et al.*, 1997) and bacteria such as *Vibrio* and *Salmonella* spp. which make the water unfit and health hazardous.

Though the nitrate content was significantly higher at a number of sampling sites in zone 2 and 3, yet it was less than safer limit of 45 ppm as per WHO. Probably, the increase in nitrate was due to presence of more microbial growth as promoted by increased phosphates. Nitrates and phosphates are the main mineral elements that are required for the growth of pathogenic microorganism and bacteria in the storage water (Botzenhart and Kufferath, 1976). So the water containing higher amounts of nitrates and phosphates are not safer for human consumption and generally lead to water-borne diseases such as gastro-intestinal disorders and jaundice etc. The increased amount of TOC in the drinking water promotes growth of many coliform bacteria (Van der Kooij, 1990) and thus make it unfit for human consumption. The water supply in the zone 3 has higher content of TOC since it comes only from the groundwater that indicates high contamination with pathogenic microorganisms. Earlier CPCB (2000) has reported that at many places the groundwater in Delhi has been contaminated with nitrates, sulphates and contains coliform bacteria.

Based on the study it is concluded that those areas in the outer Delhi which are using only groundwater are at potential risk of getting water-borne diseases as this water is rich in minerals and inorganic chemicals that make it an ambient place for the growth of pathogenic microorganisms.

References

APHA, 1992. *Standard Methods for Analysis of Water and Wastewater*, 18th ed. American Public Health Association, Inc., Washington DC.

Botzenhart, K. and Kufferath, R., 1976. Uberdie Verhrung versciendener Enterobacteriaceae sowie *Pseudomonas aeruginosa* and *Alcaligenes* sp. In: *Destiliertem Wasser, entionisiertem Wasser, leitungswasser and Mineralsalzlosung*. Zentralbl. Bacteriol. Hyg. I. abt. Orign. B. 163: 470–485.

CPCB, 2000. *Water Quality Status of Yamuna River: Assessment and Development of River Basin*. Series ADSO RBS/32/1999–2000.

Hart, V.S., Johnson C.E. and Letterrnan, R.D., 1992. An analysis of the low-level turbidity measurement. *J. Amer. Water Works Assoc.*, 84: 40.

Marks, H.C. and Chamberlain, N.S., 1953. Determination of residual chlorine in metal finishing wastes. *Anal. Chem.*, 24: 1985.

Mettinen, I.T., Vartiainen, T. and Mrtikainen, P.J., 1997. Changes in water microbial quality during bank filtration of lake water. *Can. J. Microbiology*, 43: 1126–1132.

Murphy, J. and Riby, J., 1962. A modified single solution method for the determination of phosphates in nature waters. *Anal. Chem. Acta*, 27: 31.

Navone, R., 1964. Proposed methods for nitrate in potable water. *J. Am. Water Works Assoc.*, 58: 781.

Van der Kooij, D., 1990. Assimilable organic carbon (AOC) in drinking water. In: *Drinking Water Microbiology*, (Ed.) G.A. McFeters, Springer-Verlag, New York, pp. 57–87.

Van Hall, C.E., Safranko, J. and Stenger V.A., 1963. Rapid combustion method for the determination of organic substances in aqueous solution. *Anal. Chem.*, 35: 315.

Chapter 21

Toxic Effect of Malathion on Quantitative Alteration of Protein in Muscular Tissues of *Glossogobius giuris*

V. Sreenivasa, V. Aravindan, M.B. Nadoni and P.S. Murthy

Physiology of Reproduction Unit, Department of Zoology, Bangalore University, Bangalore – 560 056

ABSTRACT

The investigations were made on protein assay of cardiac, red and white muscle tissues of freshwater gobiid fish, *Glossogobius giuris* exposed to sub-lethal concentrations of malathion for 24 to 96 hours. The protein content increased in short duration and declined in prolonged period of exposure to the pesticide toxicity. The significant variations were recorded in the protein content with increased concentration of malathion as well as the period of exposure. The possible role of malathion toxicity towards the alteration in protein content of the fish is discussed in the present communication.

Keywords: *Malathion, Protein, Glossogobius giuris, Muscular tissue.*

Introduction

The biodegradable organophosphorous pesticide (malathion) is widely used to control the pests in agricultural field. This enters the aquatic media by rain wash leading to an adverse effect on the metabolites of fish (Holden, 1972 and Malla Reddy and Philip, 1991). The use of pesticide has direct impact on physiological and biochemical changes in fish (Bhaskaran, 1988, Dravyam Selvarani and Rajamanickam, 2003).

Protein is significant component of the body and play an important role in the body construction and energy metabolism (Shah and Dubale, 1983). Organophosphorous pesticides are more toxic for fishes and likely to alter the tissue protein content (Jha and Jha, 1995). Saravana Bhavan and Geraldine (1997) have also reported an alteration in the fish tissue protein under stressed conditions. The toxic effect of pesticides on various biochemical constituents in the fish tissues have been documented (Dixshit, *et al.*, 1978; Malla Reddy and Bashamohideen, 1988 and Vasanthi, *et al.*, 1990). But the information is meager about the effect of malathion on the protein level of muscle tissues in *Glossogobius giuris*. It is therefore the present investigations were conducted to assess the toxic effect of malathion on quantitative alterations of protein in the cardiac, red and white muscles of *Glossogobius giuris*.

Materials and Methods

The *Glossogobius giuris* (Ham) were brought alive to the laboratory from Kalavarapalli Dam near Bangalore. The fishes were acclimatized for 15 days in 50 liters glass aquaria CA containing aerated tap water and fed daily with earthworms prior to their use in the experiments. The malathion was dissolved in Acetone and added to the test water to obtain the desired concentrations by adopting the dilution technique as outlined in APHA, AWWA WPCF (1985).

The acclimatized fishes were divided into four experimental groups of eight each. The first three groups were placed in different sub-lethal concentrations (0.05, 0.25 and 0.5 ppm) of malathion and the fourth group in freshwater served as control.

The fishes treated with sub-lethal concentrations were vivisected at the intervals of 24, 48, 72 and 96 hours. The cardiac, red and white muscles are removed for the estimation of protein content by following the method of Lowry *et al.* (1951).

Results

Significant variations were recorded in the protein content in the cardiac, red and white muscles of experimental fish *Glossogobius giuris* exposed to different concentrations of malathion for 24 to 96 hrs of exposure (Table 21.1).

Table 21.1: Protein Content (mg/g Wet Weight) in the Muscles of *Glossogobius giuris*

Tissue	*Control*	*Conc. of Malathion*	*Exposure Period (hours)*				*CD at 5%*
			24 hours	*48 hours*	*72 hours*	*96 hours*	
Cardiac muscle	21.33±0.08	0.05 ppm	$20.26^{d,1}$±0.02	$16.53^{c,1}$±0.01	$14.13^{b,1}$±0.06	$13.86^{a,2}$±0.07	0.15
		0.25 ppm	$20.80^{d,1}$±0.05	$18.40^{c,2}$±0.02	$15.46^{b,2}$±0.20	$12.80^{a,1}$±0.05	0.42
		0.50 ppm	$25.86^{c,1}$±0.28	$21.06^{b,3}$±0.02	$13.86^{a,1}$±0.19	$13.60^{a,2}$±0.11	0.79
	CD at 5%		0.69	0.10	0.39	0.46	
Red muscle	14.66±0.09	0.05 ppm	$15.2^{b,1}$±0.11	$16.80^{c,1}$±0.17	$15.20^{b,1}$±0.12	$14.13^{a,2}$±0.21	0.74
		0.25 ppm	$16.0^{c,2}$±0.05	$17.73^{d,2}$±0.16	$15.40^{b,1}$±0.16	$13.00^{a,1}$±0.11	0.36
		0.50 ppm	$18.6^{c,3}$±0.10	$20.20^{d,3}$±0.02	$18.00^{b,2}$±0.19	$14.93^{a,2}$±0.23	0.42
	CD at 5%		0.20	0.47	0.41	1.11	
White muscle	18.40±0.14	0.05 ppm	$21.00^{d,3}$±0.19	$17.40^{c,3}$±0.07	$10.00^{b,1}$±0.32	$8.20^{a,1}$±0.33	0.85
		0.25 ppm	$16.40^{b,2}$±0.23	$16.20^{b,2}$±0.17	$15.00^{a,2}$±0.23	$14.60^{a,2}$±0.04	0.63
		0.50 ppm	$13.60^{c,1}$±0.09	$8.60^{a,1}$±0.16	$10.60^{b,1}$±0.02	$8.26^{a,1}$±0.13	0.55
	CD at 5%		0.45	0.86	0.90	0.89	

Values given in the table are means±standard error.
Alphabetical superscripts in the rows and numeral superscripts in the columns are significantly different.

Cardiac Muscle

The protein content was found to be 21.33±0.08 mg/g in the cardiac muscles of control fish. After 24 hours of exposure, the fish treated with 0.05 ppm and 0.25 ppm of malathion revealed a decline in the protein content when compared to that of control. The fishes treated with 0.5 ppm of pesticide recorded increased protein level (25.86±0.28 mg/g wet weight) than control. The protein content did not reveal any significant variation with increased concentration of malathion in 24 hours. Whereas the significant increase was noticed with increasing concentrations of malathion after 48 hours of exposure. The gradual decrease in the level of protein was recorded in all the sub-lethal concentrations of malathion after 72 to 96 hrs of exposure. However, the fishes with 0.25 ppm shows 12.80±0.05 mg/gm of protein after 96 hour. When the concentrations of malathion were compared for protein changes under different exposure period, an overall decline was recorded in the protein content with increase in the exposure period.

Red Muscle

An increase in the protein content was recorded in the red muscles after treatment with sub-lethal concentrations of malathion during 48 hours of exposure as compared to the control fish. Fishes treated with 0.05 ppm and 0.25 ppm concentration shows 15.20±0.12 and 15.40±0.16 mg/g wet weight of protein after 72 hours, which were significantly lower than the fishes treated with 0.5 ppm (18.00±0.19 mg/g wet weight). After 96 hours of exposure the fishes with 0.05 ppm and 0.5 ppm malathion, the protein levels were higher (14.13±021 and 14.93±0.23 mg/g wet weight) than the 0.25 ppm treated fish (13.0±0.11 mg/g wet weight). For each treatment a general decline was observed in protein content with increasing exposure period from 24 to 96 hours ($p \leq 0.05$).

White Muscle

The fishes exposed to 0.05 ppm of malathion for 24 hrs shows significant increase in the protein content and in 0.25 and 0.5 ppm the level of protein content was declined when compared to control. Further, the protein content in the white muscle tissue was significantly declined with increase in the malathion concentrations from 48 to per cent hours of exposure. Thereafter the fishes treated with 0.05 ppm and 0.5 ppm for 72 and 96 hrs shows similar level of protein content and were significantly lower than those of treated with 0.25 ppm malathion.

Discussion

Protein being involved in the architecture and physiology of the cell seems to occupy a key role in building blocks of muscles and energy metabolism. It is clearly indicated by the existence of high protein synthesis activity leading to leading to an increase in the protein content in the cardiac, red and white muscles of fishes exposed to short duration malathion and the results are harmony with the findings of Vijaya Mohan Nair (2000) and Vasanthi *et al.* (1990).

The decrement in the protein content with progress in the exposure period in the fishes is suggested that the high protein hydrolysis that could be due to high pesticide interfering and impairment as well as lowering of protein synthesis in fish muscle (Katti and Sathyanesan, 1983 and Ghosh and Chatterjee, 1985). Malla Reddy and Bashamohideen (1988) reported that the increase in the protease activity might increase protein breakdown in the tissue of the fish exposed to pesticide. The treated fishes draw their extra energy requirement from body proteins by the process of gluconeogenesis (Aruna Khare and Sudha Singh, 2002) and inhibition of glycogenesis through stress induction in catecholamine as also suggested by Kasthuri and Chandran (1997).

The decline in the protein level during toxic exposure may be due to increased catabolism and decreased anabolism, the results are on the line with Qayyum and Shaffi (1977), Jha and Pandey (1977).

In view of the above facts it is evident that malathion alters protein metabolism by decreasing the protein content due to proteolysis. Hence a decrease in protein content in the cardiac, red and white muscles of *G. giuris* was observed after exposure to malathion.

Acknowledgement

The authors are thankful to the Chairman, Department of Zoology, Bangalore University for providing the necessary facilities.

References

APHA, AWWA, WPCF, 1985. *Standard Methods for the Examination of Water and Wastewater*, Washington DC.

Aruna Khare and Sudha Singh, 2002. Impact of malathion on protein content in the freshwater fish, *Clarias batracus. J. Ecotoxicol. Environ. Monit.*, 12: 129–132.

Bhaskaran, R., 1988. Effect of DOT and methyl parathion on the mitochondrial respiration SDH and ATPase activity of an air-breathing fish, *Channa striatus. Environ. and Ecol.*, 6: 198.

Davng, P.M., Khillare, Y.K and J.P. Sarawade, 2005. Chronic effect of pesticides on biochemical constituents in *Thynichthys sandkhol. Him. J. Environ. Zool.*, 19(2): 127–130.

Dixshit, T.S.S., Tondon, S.K., Datta, S.K, Gupta, P.K. and Jayraj, B., 1978. Comparative response of male rats to parathion and Lindane: Histopathological and biochemical studies. *Environ. Res.*, 17: 1–19.

Dravyam Selvarani and Rajamanickam, C., 2003. Toxicity of PCB 1232 on mitochondria of fish *Arius caelatus* (Valenciennes), *I. J. Expt. Biol.*, 41: 336–340.

Ghosh, T.K. and Chatterjee, S.K., 1985. Effect of chromium on tissue energy reserve in a freshwater fish, *Sarotherodon mossambicus. Environ. Ecol.*, 3: 178–179.

Holden, A.V., 1972. The effect of pesticides on life in freshwater. *Proc. Roy. Soc.*, London, 180: 383–394.

Jha, B.S. and Panday, S., 1989. Alternation in the total carbohydrate level of intestine, liver and gonads indicated by lead nitrate in the fish, *Channa punctatus*. In: *Environmental Risk Assessment*, (Eds.) Y.N. Sahai, P.B. Deshmuck, T.A. Mathai and K.S. Pillai. The Academy of Environmental Biology, India, pp. 207–211.

Jha, B.S and Jha, A.K., 1995. Biochemical profiles of Liver, Muscle and Gonads of the freshwater Fish, *Heteropneustes fossilis*. In: *Chromium Stress, Toxicity and Monitoring of Xenobiotics*, pp. 127–137.

Kasthuri, J. and Chandran, M.R., 1997. Sub-lethal effect of lead on feeding energetic, growth performance, biochemical composition and accumulation of the estuarine fish, *Mystue gulio* (ham). *J. Environ. Biol.*, 18: 95–101.

Katti, S.R. and Sathyanesan, A.G., 1983. Lead nitrate induced changes in lipid and cholesterol levels in the freshwater fish, *Clarias batrachus. Toxicol. Letter*, 9: 93–96.

Lowry, D.H., Rosebrough, N.J., Fair, A.L. and Randall, R.L., 1951. Protein measurement with the folin-phenol reagent. *J. Biol. Chem.*, 193: 265–275.

Malla Reddy, P. and Bashamohideen, Md., 1988. Toxic impact of fenvalerate on the protein metabolism in the bronchial tissue of a fish, *Cyprinus carpio. Curr. Sci.*, 57: 211–212.

Malla Reddy, P. and Philip, G.H., 1991, Hepato toxicity of malathion on the protein metabolism in *Cyprinus carpio. Acta. Hydrochem. Hydrobiol.*, 19(1): 127–130.

Qayyum, M.A. and Shaffi, S.A., 1977. Changes in tissue glycogen of a freshwater cat fish, *H. fossils* (block) mercury intoxication. *Curr. Sci.*, 46(18): 652–658.

Saravana Bhavan and P. Geraldine, 1997. Alternation in concentration of protein, carbohydrates, glycogen, free sugar and lipids in the prawn *Macrobrachium malcolmsoni* on exposed to sub-lethal concentration of endosulfan. *Pesticide Biochem. and Physiol.*, 58: 89–101.

Shah, P.H. and Dubale, M.S., 1983. Biochemical changes induced by malathion in the body organs of *Channa punctatus. J. Anat. Morphol. Physiol.*, 30(1&2): 107–118.

Vasanthi, R., Baskaran, P. and Palanichamy, S., 1990. Influence of carbofuran on growth and protein conversion efficiency in some freshwater fishes. *J. Ecobiol.*, 2: 85–88.

Vijaya Mohan Nair, G.A., 2000. Impact of titanium dioxide factory effluents on the biochemical composition of fresh water fishes *O. mossambicus* and *Etroplus maculates. Poll. Res.*, 19(1): 67–71.

Chapter 22

Morphological, Cultural, Physiological and Nutritional Studies of *Fusarium* Wilt Pathogen of Chickpea

V.S. Shinde, V.K. Mandhare and A.V. Suryawanshi

Pulses Improvement Project, Mahatma Phule Krishi Vidyapeeth, Rahuri – 413 722 (MS)

ABSTRACT

The fungus *Fusarium oxysporum* f.sp. *ciceri* produces hyaline, septate, profusely branched and cottony white to pink mycelium. Macro conidia were fusiform, curved or falcate shaped with three to four septa while micro conidia were oval, mostly non septate and rarely with single septa. The pathogens produce abundant pale brown chlamydospores either singly or in chains. Host extract agar, Richard's agar, Coon's agar as well as Potato dextrose agar medium was basal medium for growth and sporulation of fungus. It utilize mannitol, glucose, sucrose and dextrose as a carbon sources while Ammonium tartarate and potassium nitrate as a nitrogen sources. The optimum temperature and pH for the growth of fungus was 30°C and 6.7 respectively.

Keywords: Fusarium, Wilt pathogen, Chickpea.

Introduction

Chickpea (*Cicer arietinum* L.) is one of the most important pulse crop grown in India. In India area under chickpea was 6.50 million ha with the production of 5.77 million tonnes with an average yield of 888 kg/ha. In Maharashtra, the area under the cultivation of chickpea was 7.95 lakh ha with 4.21 lakh tonnes production and average yield 529 kg/ha (Anonymous, 2004). Wilt of chickpea caused by *F. oxysporum* Schl. Snyd. and Hans. f.sp. *ciceri* (Padwick) is most destructive seed and soil borne disease (Haware *et al.*, 1986b) and is widely distributed in 32 countries (Haware *et al.*, 1986a and Nene *et al.*, 1996). The present investigation were therefore, undertaken to study the morphological, cultural, physiological and nutritional aspects of *F. oxysporum* f. sp. *ciceri*.

Materials and Methods

The laboratory experiment of morphological, cultural and physiological studies on chickpea against *Fusarium* wilt was conducted in the laboratory of Pulses Improvement Project, Mahatma Phule Krishi Vidyapeeth, Rahuri during 2003–04.

Morphological Studies

The *Fusarium* culture used under study were obtained from single spore isolation. To determine the size of conidia and chlamydospores, one hundred spores were measured under high power (45 x) using micrometer.

Cultural Studies

Culture media *viz.*, Potato dextrose agar, Richard's agar, Host extract agar, Sobouraud's agar, Ashby's agar, Sacch's agar, Kirchoff's agar, Coon's agar, Czapeck's agar and Leonian agar were tested for the growth and sporulation of the fungus.

Physiological Studies

Temperature requirement and Hydrogen ion concentration (pH) required for the fungus were studied by routine methods.

Nutritional Studies

The utilization of carbon *viz.*, sucrose, dextrose, glucose, lactose, mannitol, dextrin, sorbitol, fructose, galactose, xylose and nitrogen *viz.*, ammonium nitrate, ammonium tartarate, ammonium sulphate, potassium nitrate, urea, calcium nitrate, tyrosine, peptone and ammonium acetate by the fungus was studies by adopting routine procedures.

Results and Discussion

Morphological Studies

The growth of the fungus was noticed to be cottony white to pinkish in colour with hyaline, septate and profusely branched mycelium and measured 0.9–1.35 µm (average 1.12 µm) in diameter. The conidia were variable in size. Macroconidia were fusiform, fulcate/curved shaped, mostly 3–4 septate and measured 45.0–22.5 × 5.4–4.0 µm. Micro-conidia were oval, hyaline, mostly non-septate or rarely with single septa and measured 13.5–5.4 × 2.70–1.80 µm. Abundant chlamydospores (9 µm in dia.) were produced which were hyaline to pale brown either single or in chains.

Cultural Studies

Profuse growth, sporulation and dry mycelial weight of *F. oxysporum* f. sp. *ciceri* were obtained on Host extract agar, Richard's agar, Coons agar and Potato dextrose agar (Table 22.1). Kewate (1986) reported Richard's and Potato dextrose agar as best medium.

Physiological Studies

The fungus *F. oxysporum* f. sp. *ciceri* preferred to grow in the temperature range of 10–40° C (Table 22.2). Below and above this range, there was no growth. Maximum growth (on the basis of dry mycellial weight) and sporulation was observed at 30°C. The fungus could grow in the presence of wide range of pH *i.e.* 4.4 to 9.0. However, the maximum mycelial growth was harvested at pH 6.7.

Table 22.1: Cultural Characters of *F. oxysporum* f. sp. *ciceri* on Different Media

Sl.No.	*Cultural Media*	*Colony Dia (mm)**	*Dry Mycelial Wt.* (mg) in broth*	*Sporulation*	*Growth Characters*
1.	Host extract agar	90	98	++++	Circular colony, mycelium loose, submerged and dirty white
2.	Potato dextrose agar	83	90	++++	Circular colony, dirty white mycelium, compact with concentric rings
3.	Richard's agar	86	92	++++	Circular colony, mycelium compact, cottony white
4.	Sobouraud's agar	75	82	++++	Circular colony, mycelium compact, submerged, pinkish
5.	Ashby's agar	73	80	+++	Circular colony, mycelium submerged, loose, dirty white
6.	Sacch's agar	54	69	++	Circular colony, mycelium submerged, raised only at center, poor growth, cottony white
7.	Kirchoff's agar	76	82	++	Mycelium spreaded, showed concentric rings, whitish
8.	Coon's agar	84	90	++++	Circular colony, mycelium compact and raised at center
9.	Czapeck's agar	78	85	++	Flat growth, mycelium compact and raised at center
10.	Leonian agar	72	78	++	Circular colony, mycelium smooth

*: Average of three replication.

+: Poor/Scanty sporulation; ++: Moderate sporulation; +++: Good sporulation; ++++: Profuse sporulation.

Table 22.2: Effect of Temperature and pH on Growth and Sporulation of *F. oxysporum* f. sp. *ciceri*

Sl.No.	*Temperature*				*pH*			
	Incubation Temperature (°C)	*Colony Dia. (mm)**	*Dry Mycelium Wt. (mg)**	*Sporulation*	*Before Sterilization*	*After Sterilization*	*Dry Mycelium Wt. (mg)**	*Sporulation*
1.	0	0	0	–	4.0	4.4	98	+
2.	5	0	0	–	5.2	4.8	138	+
3.	10	14	66	+	5.0	5.5	220	++
4.	15	28	116	+	6.8	6.2	245	+++
5.	20	48	152	++	7.2	6.7	274	++++
6.	25	70	248	+++	7.6	7.0	205	++
7.	28	80	262	++++	8.2	7.7	180	+
8.	30	90	278	++++	8.6	8.3	156	+
9.	35	60	200	++	9.3	9.0	60	+
10.	40	20	82	+				
11.	45	0	0	–				
	SE±	2.14	2.36				2.98	
	CD at 5%	6.33	6.96				9.04	

*: Average of three replications.

+: Poor/Scanty sporulation; ++: Moderate sporulation; +++: Good sporulation; ++++: Profuse sporulation; –: No sporulation.

Nutritional Studies

The fungus utilized all the carbon sources (Table 22.3). However, good growth, maximum dry mycelial weight and sporulation was recorded on mannitol and sucrose.

Table 22.3: Utilization of Carbon Sources by *F. oxysporum* f.sp. *ciceri* on Solid and Liquid Media

Sl.No.	Source of Carbon	Colony Diameter (mm)*	Dry Mycelial Wt. (mg)*	Sporulation
1.	Sucrose	88	244	++++
2.	Dextrose	86	200	+++
3.	Glucose	88	212	+++
4.	Lactose	82	192	++
5.	Mannitol	90	288	++++
6.	Dextrin	82	190	+++
7.	Sorbitol	80	180	++
8.	Fructose	74	176	++
9.	Galactose	56	98	+
10.	Xylose	80	190	+++
11.	Control (without sugar)	0	0	–
	SE±	2.14	6.17	
	CD at 5%	6.31	18.33	

*: Average of three replications.

+: Poor/Scanty sporulation; ++: Moderate sporulation; +++: Good sporulation; ++++: Profuse sporulation; –: No sporulation.

Table 22.4: Utilization of Nitrogen Sources by *F. oxysporum* f. sp. *ciceri* on Solid and Liquid Media

Sl.No.	Nitrogen Sources	Colony Diameter (mm)*	Dry Mycelial Wt. (mg)*	Sporulation
1.	Ammonium nitrate	70	212	+++
2.	Urea	72	259	+++
3.	Ammonium sulphate	32	124	+
4.	Ammonium tartarate	86	279	++++
5	Potassium nitrate	82	264	++++
6.	Calcium nitrate	78	252	+++
7.	Ammonium acetate	48	166	+
8.	Tyrosine	66	178	++
9.	Peptone	70	200	++
10.	Control (without N)	0	0	–
	SE±	2.24	3.09	
	CD at 5%	6.67	9.10	

*: Average of three replications.

+: Poor/Scanty sporulation; ++: Moderate sporulation; +++: Good sporulation; ++++: Profuse sporulation; –: No sporulation.

Among the nitrogen sources, good growth, maximum dry mycelial weight and sporulation were recorded on ammonium tartarate and potassium nitrate. Calcium nitrate, urea and ammonium nitrate found to be good and tyrosine and peptone was found to be moderately effective source of nitrogen (Table 22.4). Kewate (1986) reported above nitrogen sources to be the best for *F. oxysporum*.

References

Anonymous, 2004. Area and production under chickpea in India. *Agricultural Situation in India.*

Anonymous, 2004. *District-wise General Statistical Information of Agricultural Department* (MS) Part-II.

Haware, M.P., Nene, Y.L. and Mathur, S.B. 1986a. Seed-borne diseases of chickpea. *Technical Bulletin* I, Danish Government, Institute of Seed Pathology, Copenhagen, Denmark, pp. 32.

Haware, M.P., Nene, Y.L. and Natrajan, M., 1986b. Survival of *Fusarium oxysporum* f.sp. *ciceri* in soil in absence of chickpea. In: *Management of Soil-borne Diseases of Crop Plants*. Proc. Natl. Sem., 8–10 January, 1986, Tamil Nadu Agric. University, Coimbatore, India.

Kewate, S.B. 1986. Studies on wilt of Gram (*Cicer arietinum* L) caused by *Fusarium oxysporum* f.sp. *ciceri*. *M.Sc. (Agri.) Thesis* (Unpublished) Submitted to MPKV, Rahuri.

Nene, Y.L., Shelia, V.K. and Sharma, S.B., 1996. *A World List of Chickpea and Pigeonpea Pathogens*, 5th edn. Patancheru, A.P. India, ICRISAT, pp. 27.

Chapter 23

Ecological Study of Soil Microarthropods in Banana (*Musa sp.*) Plantation of Cachar District, Assam

Ranabijoy Gope and D.C. Ray
Department of Ecology and Environmental Science, Assam University, Silchar – 788 011

ABSTRACT

Ecology of microarthropods under banana (*Musa* sp.) plantation was investigated from March 2004 to October 2004 at Dorgakona Village, Silchar, Assam. The diversity of microarthropod was found to be high during the period of investigation whereas evenness showed stability of community. Moisture content and organic carbon showed positive correlationship ($p < 0.001$) with the microarthropod population. However, environmental factors (aerial and soil temperature), pH did not show any influence on the microarthropod population.

Keywords: *Musa sp., Microarthropods, Edaphic factor, Diversity, Correlation co-efficient.*

Introduction

Soil inhabiting microarthropods are one of the major components of total soil dwelling faunas. These arthropods belongs to different taxonomic groups and every taxonomic groups are having their own niche. These arthropods are responsible for the litter breakdown and decomposition (Abbot and Crossley, 1982; Seastedt 1984; Hagvar, 1988). They also help in the maintenance of physical and chemical properties of soil. Reddy, 1984 reported that the soil inhabiting arthropods have crucial role in soil aeration and turnover. Some progressive attempt has been made to understand the effect of land use pattern on the ecology of soil microarthropods world wide but in India the scenario is reasonable though some extensive works have been made by Hazra and Choudhuri (1983), Reddy (1984b), Sengupta and Sanyal (1991), Mitra *et al.* (1983), Hatter and Alfred (1986), Singh and Yadava (1998),

Bhattacharya *et al.*, 1981. These works were mainly based on agriculture, forest, jhum land and plantation. *Musa* sp. (banana) is extensively cultivated in this district by farmers in the field as well as in the home gardens. Soil fauna plays an important role in decomposition process and nutrient release in the banana field. No work has been carried by any workers in this district so far. Considering the importance of micro and mesofauna of the soil the present investigation was undertaken to know the diversity and edaphic characteristics of *Musa* sp. field. Cachar, Assam. The selected study site having a distance of 22 km from Silchar town. This area situated at latitude of 92°45′25.9″E and longitude of 24°41′29.9″N. The climatic condition of this area was warm humid and monsoon dominating here from March to May and annual rainfall was received over 2500 mm. The selected banana field was 5-years old.

Experimental Design

For the study of soil inhabiting microarthropods soil was sampled monthly basis from March 2004 to October 2004 by using a iron made rectangular sampler (5 cm × 5 cm) at a depth of 10 cm. After sampling soil sample were brought to the laboratory. Microarthropods were extracted by modified Tullgren funnel extractor with 40 watt electric bulb (as a heat producer) for 72 hours. (The extracted arthropods were preserved in 80 per cent ethanol. Microarthropods were sorted upto group level by using stereoscopic binocular microscope (10x × 6x). Apart from sampling one composite sampling occasion for the study of physico-chemical parameters of soil. Among physico-chemical parameters moisture content, pH, water holding capacity, bulk density, porosity, organic carbon content and soil colour were analysed by using standard methods. The obtained data from the experiment were further interpretated by using statistical software (Biodiversity Pro).

Results and Discussion

Monthly variation in diversity and evenness of soil microarthropod population were incorporated in Table 23.1. From this table it was found that Shannon diversity index (0.903) and evenness (0.903) were maximum in April and minimum diversity found in July (0.479). On the other hand evenness found low in October (0.667).

Table 23.1: Shannon Diversity Index and Evenness Index (Pielou) of Microarthropod Population During March to October, 2004

Month	*Shannon Diversity Index*	*Evenness*
March	0.794	0.828
April	0.903	0.903
May	0.708	0.838
June	0.587	0.754
July	0.479	0.685
August	0.758	0.795
September	0.773	0.742
October	0.720	0.667

Similarity matrix of microarthropod population was studied with their monthly variation. Dendogram indicated that there were high similarity of microarthropod abundance with March–April, May–August, June–July, and September–October as a single cluster analysis (Figure 23.1).

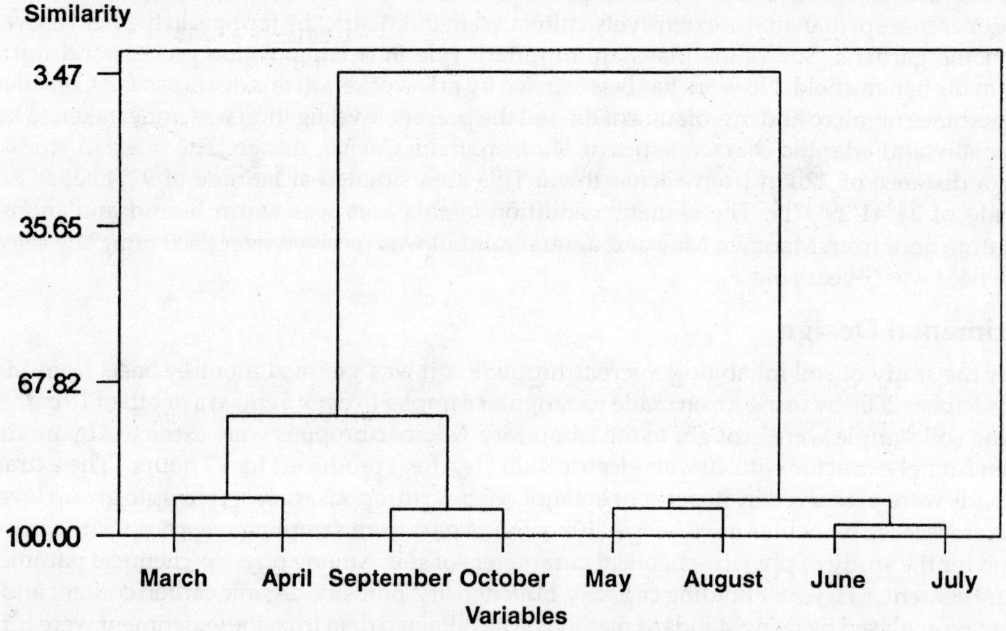

Figure 23.1: Showing the Dendrogram

Among physico-chemical properties it was found that the soil was very pale brown in colour (10YR 7/4). Bulk density showed 1.27gm cm^{-3} whereas porosity recorded 52.07 per cent. The water holding capacity was found to be 45.11 per cent. The soil was clay-loam in character having 55.50 per cent sand, 26.00 per cent silt and clay was recorded 18.50 per cent (Table 23.2).

Table 23.2: Edaphic Characteristics of *Musa* sp. Field During March to October 2004

Parameters	*Values*
Soil colour	Very pale brown (10YR 7/4)
Bulk density (gm cm^{-3})	1.27
Porosity (per cent)	52.07
Water holding capacity (per cent)	45.11
Texture Sand (per cent)	55.50
Silt (per cent)	26.00
Clay (per cent)	18.50

Correlations co-efficient of environmental and edaphic factors were employed with microarthropod population (Figure 23.2). Microarthropod population showed positive correlationship ($r = 0.725$ $p < 0.001$) with moisture content and organic carbon content ($r = 0.768$ $p < 0.001$) only whereas pH and environmental factors did not show any correlationship. Diversity of microarthropod population in this banana field is very high (0.903). Maximum diversity is noticed during April whereas in our finding the faunal diversity is recorded less during the same period (Wallwork, 1967) which is not in agreement with the present investigation. Evenness is more, which corroborates the findings of Hattar *et al.*, 1986. Cluster analysis shows single linkage clustering. Soil moisture shows positive correlationship with microarthropod population which is in agreement with the findings of Christiansen, 1964; Choudhuri and Roy, 1972; Hazra and Choudhuri, 1983; Singh and Yadava, 1998. Soil organic carbon content of banana fields also shows positively significant correlationship with microarthropod population which corroborates the findings

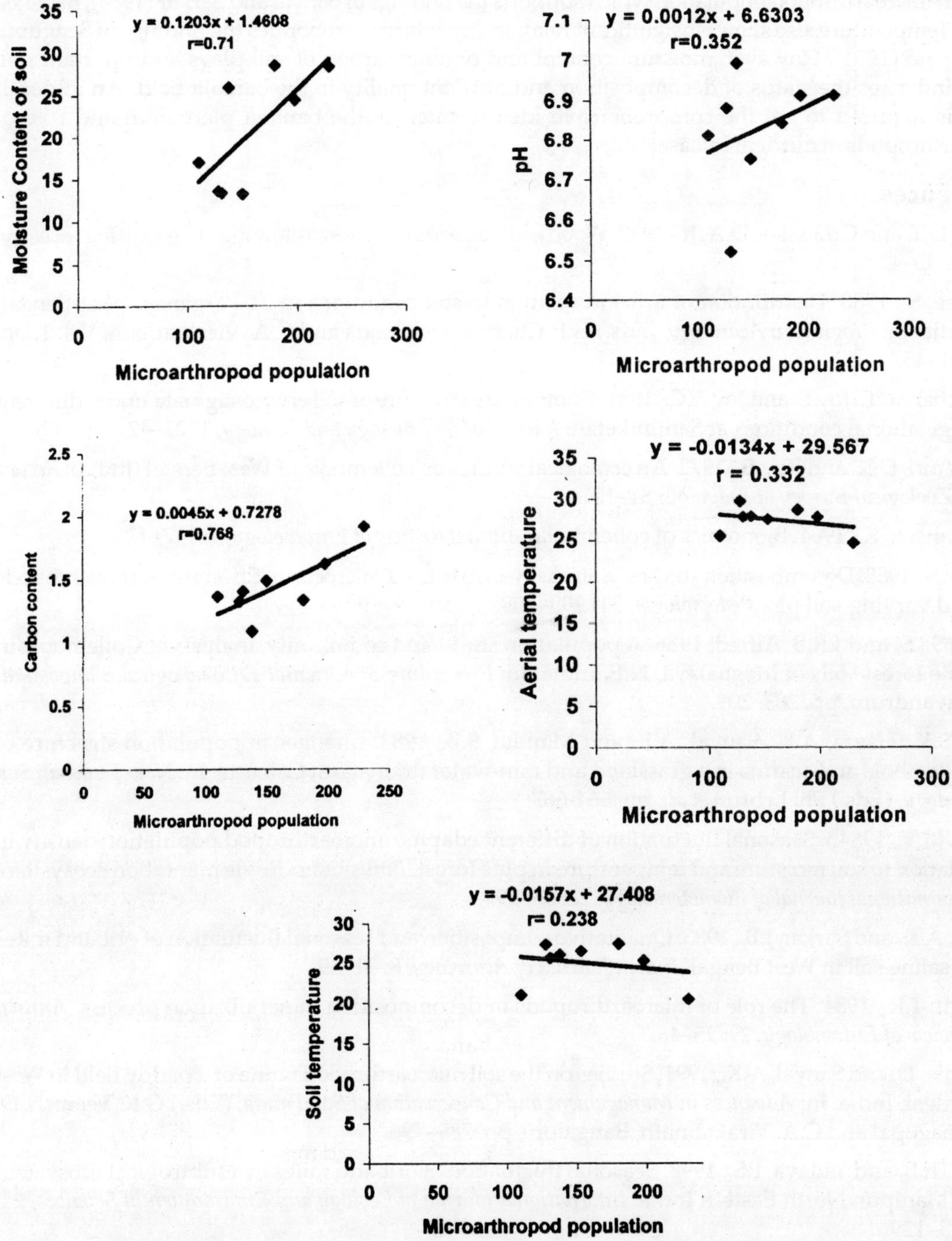

Figure 23.2: Correlation Co-efficient Between Microarthropod Population and Various Edaphic and Environmental Factors

of Joy and Bhattacharya (1981), Banerjee (1988). The insignificant correlationship shown by pH of the soil with microarthropod population which supports the findings of Sanyal and Sarkar (1983), Banerjee (1988). Temperature also shows insignificant relationship which corroborates the findings of Sengupta and Sanyal (1991). However, moisture content and organic carbon of soil plays an important role which indicates the status of decomposition and nutrient quality in the banana field. An in-depth study is required to get the comprehensive idea to manage the banana plantation and role of microarthropods in nutrient release.

References

Abbot, D.T. and Croossley, D.A. Jr., 1982. Woody litter decomposition following clear-cutting. *Ecology,* 63: 35–42.

Banerjee, S., 1988. Distribution of acari in relation to soil conditions in 24-Parganas, West Bengal, India. In: *Progress in Acarology,* (Eds.) G.P. Channa Basavanna and C.A. Viraktamath, Vol. 1, pp. 451–457.

Bhattacharya, T., Joy, S. and Joy, V.C., 1981. Community structure of soil cryptostigmata under different vegetational conditions at Santiniketan. *Journal of Soil Biology and Ecology*, 1: 27–42.

Choudhuri, D.K. and Roy, S., 1972. An ecological studies on collembola of West Bengal (India). *Record of Zoological Survey of India*, 66: 81–101.

Christiansen, K., 1964. Bionomics of collembola. *Annual Review of Entomology*, 9: 147–178.

Hagvar, S., 1988. Decomposition studies in an easily-constructed microcosm: Effects of microarthropods and varying soil pH. *Pedobiologia,* 31: 293–303.

Hattar, S.J.S. and J.R.B. Alfred, 1986. A population study and community analysis of Collembola in pine forest soils of Meghalaya, N.E. India. In: *Proceeding 3rd Oriental Entomological Symposium,* Trivandrum, pp. 203–209.

Mitra, S.K., Hazra, A.K., Sanyal, A.K. and Mandal, S.B., 1983. Changes in population structure of collembola and acarina in a grassland and rain-water drainage at Calcutta. In: *New Trends in Soil Biology,* (Eds.) Ph. Lebrun *et al.*, pp. 664-667.

Reddy, M.V., 1984b. Seasonal fluctuation of different edaphic microarthropod population density in relation to soil moisture and temperature in pine forest, *Pinus kesiya* Royle plantation ecosystem. *International Journal of Biometereology*, 28: 55–59.

Sanyal, A.K. and Sarkar, J.B., 1983. Qualitative composition and seasonal fluctuation of oribatid mites in saline soil in West Bengal. *Indian Journal of Acarology*, 8: 31–39.

Seastedt, T.R., 1984. The role of microarthropods in decomposition mineralization process. *Annual Review of Entomology*, 29: 25–46.

Sengupta, D. and Sanyal, A.K., 1991. Studies on the soil microarthropod fauna of a paddy field in West Bengal, India. In: *Advances in Management and Conservation of Soil Fauna,* (Eds.) G.K. Veeresh, D. Rajagopal and C.A. Viraktamath, Bangalore, pp. 789–796.

Singh, Th.B. and Yadava, P.S., 1998. Seasonal fluctuation of oribatid mites in a subtropical ecosystem of Manipur, North Eastern India. *International Journal of Ecology and Environmental Sciences*, 24: 123–129.

Wallwork, J.A., 1967. Some oribatei (Acari : Cryptostigmata) from Techa (3rd series). *Review of Zoology and Botany*, Africa, 75: 35–45.

Chapter 24

Food Preferences of the Brown Trout (*Salmo trutta* L.) in Relation to the Benthic Macroinvertebrates of River Sindh, Kashmir Valley

Haroon Ul Rashid and Ashok K. Pandit

Aquatic Ecology Laboratory, P.G. Department of Environmental Science, The University of Kashmir, Srinagar –190 006, J&K, India

ABSTRACT

The food preferences of the brown trout in relation to the density of benthic macroinvertebrates have been studied in River Sindh during August, 2005. The fish was found to be a typical benthophagic carnivore, eating on almost every type of small animal present in its habitat. A correlation between the organisms found in the gut contents and the river macroinvertebrates showed that there was not any specificity for a particular food item as the average number of macroinvertebrate individuals collected per m^2 of the river and the average number of macroinvertebrate individuals found in the gut of fish specimens showed a significant positive correlation (r = 0.905).

Keywords: Brown trout, Benthic macroinvertebrates, Gut contents, Food preferences, River Sindh.

Introduction

Brown trout, an exotic fish, has been introduced into many streams of Kashmir including River Sindh which runs in the north of the Srinagar city from Baltal to Shadipora (Narayan Bagh). The site selected (Mamar) for the collection purposes is 50 km away to the north-east of the Srinagar city and lies at an altitude of about 1,900 m (a.s.l). The bottom substrate of the river is primarily made of boulders and gravel. The trout, being a cold water fish, has adapted well to the conditions of Kashmir streams and some high altitude lakes and has become a very successful and dominant fish of River Sindh. Though a lot of work has been done worldwide on the food and feeding of trout (Hynes, 1970; Sagar and Eldon, 1983; Dedual and Collier, 1995; Merz, 2002; Kara and Alp, 2005), yet very little information is available on the food preferences of brown trout in Kashmir streams inspite of the fact that other fishes including *Schizothorax* spp. have been studied thoroughly as regards their food and feeding (Das and Subla, 1963; Jan and Das, 1970; 1971; Subla and Das, 1970; Yousuf and Firdous, 1997; 2001; Pandit, 1999). In order to investigate the food preferences of brown trout vis-à-vis the macroinvertebrate density in the River Sindh, the present study was undertaken.

Materials and Methods

35 trout specimens were captured from 7th–14th of August, 2005 with the help of French angling spoon (No. 02) over a river stretch of about 3 km between 4:00 pm and 7:00 pm. The fish were dissected at the collection site and the gut contents were preserved in 5 per cent formalin in 250 ml plastic containers (APHA, 1998). The gut contents were weighed on a micro-analytical balance (Dhona 200D). The length of the fish specimens was also measured at the site. The angling spoon was used in order to catch only the carnivorous brown trout and hence the unnecessary damage, by electrofishing, to other herbivore fishes was thus avoided.

The benthic macroinvertebrates were collected with the help of Surber type sampler (30 cm × 30 cm square frame) by lifting the river bed substrate (stones, boulders etc.) after placing the sampler at right positions (Surber, 1936). The substrates so disturbed were carried to the shore and manual picking of the macroinvertebrates both from within the sampler and from the substrate surface (undetached macroinvertebrates) was done with the help of forceps and preserved in 5 per cent formalin (APHA, 1998). The macroinvertebrates were collected after a stretch of every kilometer and the sample was taken in triplicate at all the points of collection.

For the identification of both the gut contents and the benthic macroinvertebrates, stereoscopic microscopes were used and the identification was facilitated by the standard works of Edmondson (1959), Pennak (1978) and Engblom and Lingdell (1999). The percentage of the gut contents was done as per the number of organisms found in the gut.

Statistical Analysis

To assess the relative importance of food items, an index of relative importance (IRI) (Hyslop, 1980) was calculated for each food category.

$$IRI = (\%N + \%W) \times \%O$$

where,

%N: A food item's percentage of the total number of organisms ingested

%W: A food item's percentage of the total weight of the food ingested, and

%O: A food item's percentage frequency occurrence in all stomachs that contained food.

In order to find out a correlation between the per cent density of benthic macroinvertebrates in the river, per cent gut contents of the fish and %IRI of the gut contents the Radar diagrams and Bar diagrams were used.

Pearson's coefficient of correlation 'r' was used to correlate the average number of macroinvertebrates found in the river per m^2 and the average number of macroinvertebrates found in the gut contents of fish. The value of 'r' was calculated at *d.f* = 6 (*d.f* = Degree of freedom) at 0.01 level. 'r' was calculated by the formula:

$$r = \frac{\sum dx \cdot dy}{\sqrt{\left(\sum x^2 \cdot \sum y^2\right)}}$$

where,

r: Correlation coefficient

dx: Deviations of x variable (average number of macroinvertebrates per m^2 of the river)

dy: Deviations of y variable (average number of macroinvertebrates found in the gut of each fish)

Σ dx . dy: Sum of multiplications of deviations x and y

$\Sigma dx^2 . \Sigma dy^2$: sum of the multiplications of squares of the deviations x and y.

Results and Discussion

Only 35 fish specimens were collected as the fish stock is dwindling in River Sindh because of the uneven discharges (courtesy of Upper Sindh Hydel Power Projects, stages I and II), and also the over fishing by the local inhabitants.

A careful examination of the macroinvertebrate samples revealed that the River Sindh was dominated by the insects in the macroinvertebrate fauna at the site of collection (Mamar). Macroinvertebrates play a very important role as consumers at the intermediate levels of lotic food webs and are influenced by both bottom-up and top-down forces in streams (Wallace and Webster, 1996). On the basis of the dominance pattern the most abundant order of macroinvertebrates was found to be the Trichoptera contributing 38.36 per cent of the total organisms collected, followed by Diptera (28.77 per cent) and Ephemeroptera (24.70 per cent). During sampling, snails and *Gammarus* sp. were also collected and on the basis of their sporadic occurrence they were placed in "others" category (Table 24.1). Amongst the macroinvertebrates, the abundant genus were found to be *Triaenodes* sp., followed by Diamesinae larvae, *Hydropsyche* sp., *Mediopsis* sp. (water mite), *Nigrobaetis* sp., *Baetis* sp., *Epeorus* sp., *Ecdyonurus* sp. etc. in a decreasing order.

Table 24.1: Density (m^2) of Macroinvertebrates in River Sindh at Mamar

Order	*Average Number or Individuals Collected (Density)/m^2*	*% Density*
Ephemeroptera	103	24.70
Trichoptera	160	38.36
Diptera	120	28.77
Coleoptera	000	00.00
Plecoptera	005	01.19
Hydracarina (Acarina)	025	05.99
Odonata	000	00.00
Others	004	00.99
Total	417	100.00

Note: Hydracarina is only a term of convenience and not a taxonomic term (Pennak, 1978).

The gut contents of the fish (Table 24.2) showed that the food composition had the major contribution of trichopteran larvae (27.00 per cent), followed by ephemeropteran larvae (23.95 per cent) and dipteran larvae (22.75 per cent), which is in consonance with the findings of Merz (2002). Dedual and Collier (1995) found the proportion of Diptera in the diet of small trout specimens to be higher than in large specimens and the reverse was observed for the proportion of Trichoptera. Further, Allan (1981) examined the composition (by numbers) of brook trout diet at different times of the year and predicted that relative abundance and then body size seemed to explain the food choice of the fish. Since the fish specimens collected during the present study were of intermediate size (average length = 165 mm), no huge difference was seen in the trichopteran and dipteran food quantity utilized.

Table 24.2: Macroinvertebrate Food Items Found in the Gut Contents

Order	*Average Number of Individuals Found in the Gut of Each Fish*	*% Contribution of Food Composition*
Ephemeroptera	5.33	23.95
Trichoptera	6.00	27.00
Diptera	5.07	22.75
Coleoptera	1.86	08.38
Plecoptera	0.53	02.38
Hydracarina (Acarina)	0.27	01.19
Odonata	0.27	01.19
Others	2.93	13.16
Total	**22.26**	**100.00**

The per cent index of relative importance (%IRI) depicted that the most preferred and important food item for the brown trout was the order Trichoptera (%IRI = 37.416 per cent), followed by Ephemeroptera and Diptera with the values of 22.591 per cent and 21.459 per cent respectively (Table 24.3). The fourth item as per %IRI was the "others" category including terrestrial ants, *Gammarus* sp., snails, sand etc., contributing 15.685 per cent to the %IRI.

Table 24.3: The Index of Relative Importance of the Food Items Found in the Gut of the Fish

Order	*%N*	*%W*	*%O*	*%IRI*
Ephemeroptera	23.95	20.40	70	22.591
Trichoptera	27.00	33.49	85	37.416
Diptera	22.75	14.11	80	21.459
Coleoptera	08.38	09.36	20	02.581
Plecoptera	02.38	01.00	05	00.122
Hydracarina (Acarina)	01.19	00.22	05	00.051
Odonata	01.19	01.42	05	00.095
Others	13.16	20.00	65	15.685
Total	**100.00**	**100.00**		**100.00**

Note: The IRI values of each food item were converted to percentage based on total IRIs to make the comparisons.

Table 24.4: Comparing the %IRI of Food Items with the Per cent Density of the Macroinvertebrates in the Habitat

Order	*% IRI*	*% Density*
Ephemeroptera	22.591	24.70
Trichoptera	37.416	38.36
Diptera	21.459	28.77
Coleopteran	02.581	00.00
Plecoptera	00.122	01.19
Hydracarina (Acarina)	00.051	05.99
Odonata	00.095	00.00
Others	15.685	00.99
Total	**100.000**	**100.00**

Although no coleopteran species was found in the macroinvertebrate sample of the habitat, yet the group contributed 8.38 per cent to the gut contents. Similarly no Odonata were found in the macroinvertebrate sample in the habitat, but its contribution in the gut contents was found to be 1.19 per cent (Table 24.5) thereby indicating that the source of the food item must be a place other than the sampled river stretch. The distribution of fish and macroinvertebrates was reported by Straskraba (1965) showing the abundance of a common amphipod to decrease in a 100m stretch, coincidental with a high weir and the occurrence of large numbers of brown trout (*c.f.* Allan, 1983). Williams *et al.* (2003) also believe that the utilization of a particular food item directly affects the assemblage structure of macroinvertebrates. Further, the presence of the terrestrial ants in the gut contents revealed that the brown trout had a greater tendency to eat larger prey from wherever available. The above assumption gains further foothold from the fact that even though the small sized water mites (Hydracarina) were present in comparatively greater numbers (5.99 per cent) in the niche, yet their overall contribution to the gut contents was only 1.19 per cent.

Table 24.5: Comparing the Per cent Density of the Macroinvertebrates and their Percentage in the Trout Gut Contents

Order	*% Density*	*% Food Composition (Gut Contents)*
Ephemeroptera	24.70	23.95
Trichoptera	38.36	27.00
Diptera	28.77	22.75
Coleopteran	00.00	08.38
Plecoptera	01.19	02.38
Hydracarina (Acarina)	05.99	01.19
Odonata	00.00	01.19
Others	00.99	13.16
Total	**100.00**	**100.00**

Sagar and Eldon (1983) reported that the major part of the brown trout food were chironomid larvae, trichopteran larvae and adult dipterans with almost no clear seasonal differences in the food utilized. It was also reported that the gut food items of trout and the composition of benthos was similar, a fact also corroborated by the present study. However, according to Sagar and Eldon (1983) the gut food items of torrent fish, common bully and quinnat salmon, differed considerably from the composition of benthos in the habitat.

In the present study flying terrestrial ants were also found in the gut contents in addition to some snails and *Gammarus* sp. Merz (2002) in another study found bird feathers, mammalian hair and terrestrial ants in the stomachs of steelhead trout. These findings indicate that the brown trout like steelhead trout eats everything that is smaller than the trout itself and looks alive. In the present study a good quantity of sand was also found in the gut of the fishes thereby indicating the benthophagic feeding habits of the fish.

The correlations between the per cent density of rnacroinvertebrates in the river stretch, per cent gut contents and the per cent IRI of the gut contents showed that there was no specific selectivity in the food items (Table 24.4). It was seen that the prey items were consumed in almost similar proportions to their occurrence in the river (Figures 24.1–24.4). Laggarigue *et al.* (2002) further opined that the brown trout chiefly feeds on the prey that is the most available in the habitat as is true for the present investigation.

The calculated value of the Pearson's coefficient of correlation between the average number of macroinvertebrates collected per m^2 of the river and the average number of macroinvertebrates found in the gut contents of each fish was r = 0.905. On observing the value of 'r' at *d.f* 6, the value came to be 0.834 at 0.01 level. Therefore, it is clear that 'r' is very significant at 0.01 level and hence there is a significant positive correlation between the two variables.

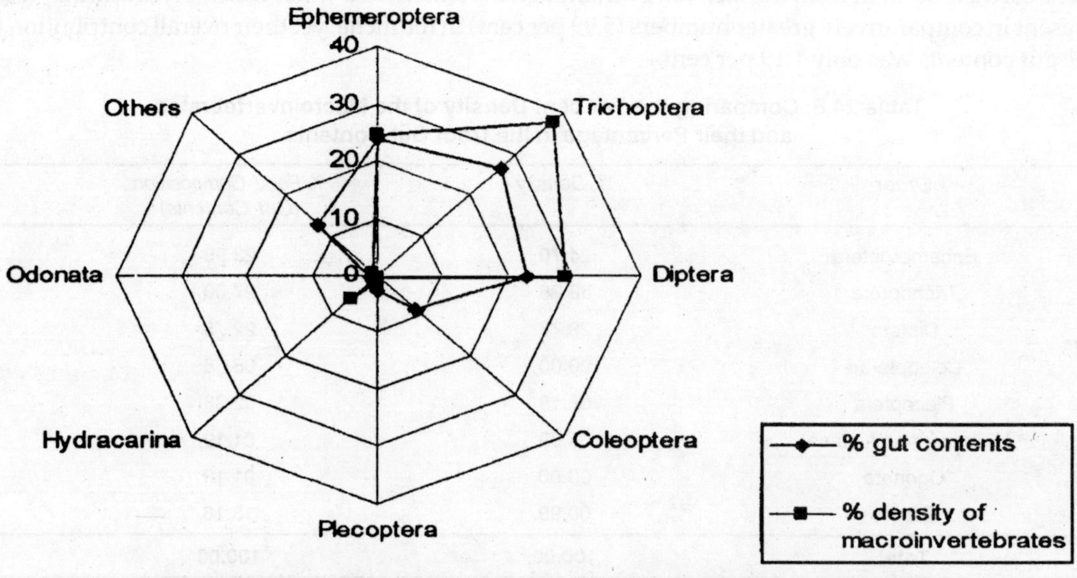

Figure 24.1: Radar Diagram Showing the Overlap of Macroinvertebrates as Per cent Gut Contents and their Per cent Density in the River

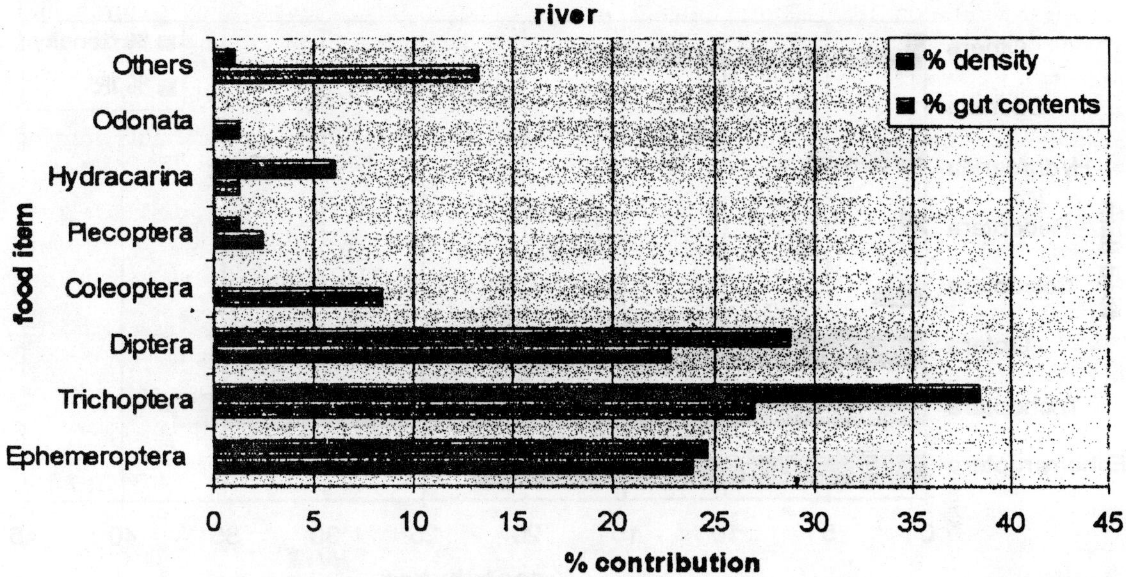

Figure 24.2: Bar Diagram Correlating the Presence of Macroinvertebrates as Per cent Gut Contents and their Per cent Density in the River

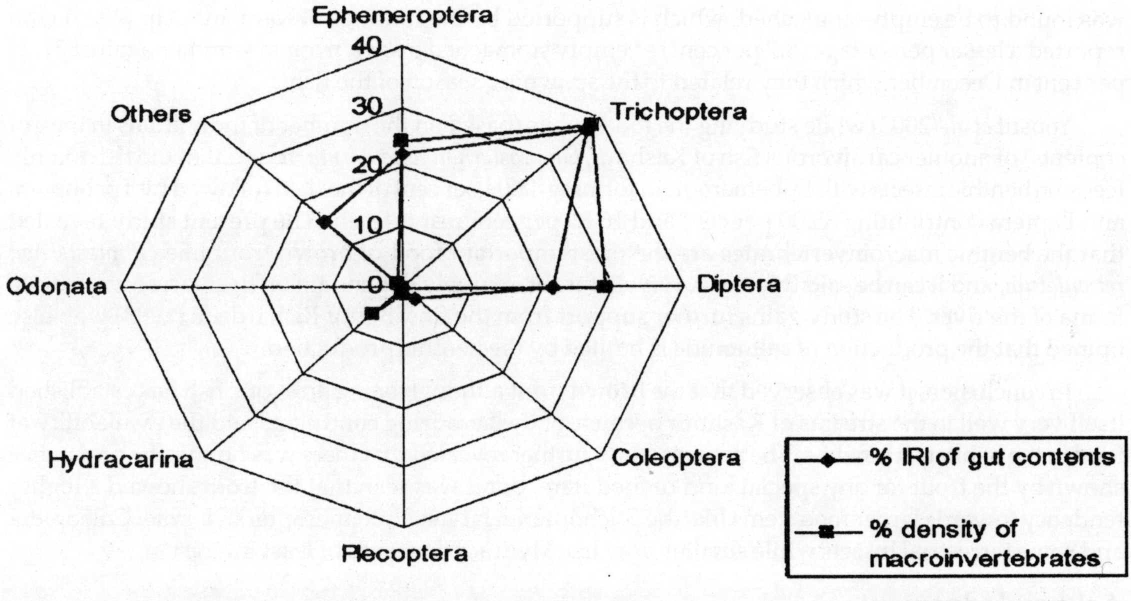

Figure 24.3: Radar Diagram Showing the Overlap of Per cent Density of Macroinvertebrates in the River and their %IRI as Trout Food

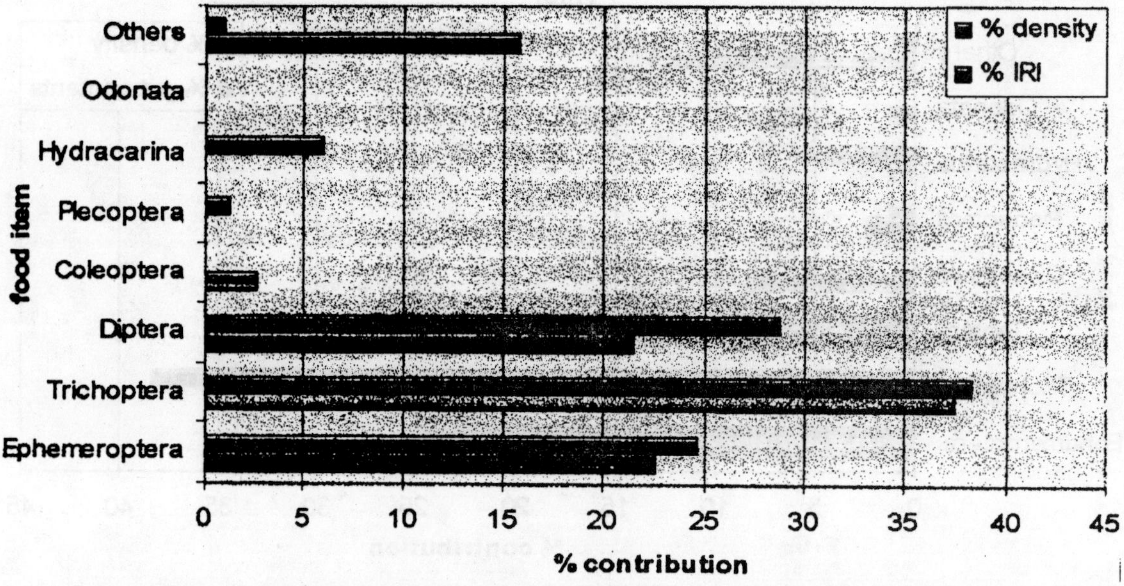

Figure 24.4: Bar Diagram Correlating the Per cent Density of the Macroinvertebrates in the River and their %IRI as the Trout Food

Since the collection of fish and macroinvertebrates was done in the first half of August, no fish was found to be empty stomached, which is supported by the findings of Kara and Alp (2005) who reported a lesser percentage (8.7 per cent) of empty stomached brown trout in summer against 24.24 per cent in December which they related to the spawning season of the fish.

Yousuf *et al.* (2003) while studying the food habits (based on the number of individuals in the gut contents) of another carnivorous fish of Kashmir, *Glyptosternon reticulatum,* found that the fish mainly feeds on benthic insects with Ephemeroptera forming 48.09 per cent of the diet, followed by Trichoptera and Diptera contributing 35.00 per cent and 16.86 per cent respectively. The present study revealed that the benthic macroinvertebrates are the most important food of brown trout like *Glyptosternon reticulatum,* and it can be said that the survival of trout is largely dependant on the macroinvertebrate fauna of the river. The study gains further support from the findings of Richardson (1993) who also opined that the production of salmonids is limited by the benthic production.

In conclusion, it was observed that the brown trout although being an exotic fish has established itself very well in the streams of Kashmir because of the favouring conditions and the availability of food in the habitats it inhabits. The present study further revealed that there was no specific preference shown by the trout for any special kind of food item, but it was seen that the trout showed a higher tendency towards larger food items like the Trichopteran larvae, Ephemeropteran larvae, Coleoptera and some terrestrial insects while smaller prey like Hydracarina were of least interest.

Acknowledgement

The authors highly acknowledge Mr. Shakeel Ahmad and Mr. Mukhtar Ahmad, the local anglers, for their help in collection of the trout specimens.

References

Allan, J.D., 1981. Determinants of diet of brook trout (*Salvelinus fontanalis*) in a mountain stream. *Can. J. Fish. Aquat. Sci.*, 38: 184–192.

Allan, J.D., 1983. Predator-prey relationships in streams. In: *Stream Ecology: Application and Testing of General Ecological Theory*, (Eds.) J.R. Barnes and G.W. Minshall. Plenum Press, New York and London, pp. 191–229.

A.P.H.A., 1998. *Standard Methods for Examination of Water and Wastewater*, 20th Ed. American Public Health Association, Washington, D.C.

Das, S.M. and Subla, B.A., 1963. The Ichthyofauna of Kashmir, Part I. History, topography, origin, ecology and general distribution. *Ichthyologica*, 2: 87–106.

Dedual, M. and Collier, K.J., 1995. Aspects of Juvenile rainbow trout (*Oncorhynchus mykiss*) diet in relation to food supply during summer in the lower Tongariro River, New Zealand. *New Zealand Journal of Marine and Freshwater Research*, 29: 381–391.

Edmondson, W.T., 1959. *Freshwater Biology*. John Wiley, N.Y.

Engblom, E. and Lingdell, P.E., 1999. Analysis of benthic invertebrates. In: *River Jhelum, Kashmir Valley: Impacts on the Aquatic Environment*, (Ed.) Lennart Nyman. Swedmar, Sweden, pp. 39–77.

Hynes, H.B.N., 1970. *The Ecology of Running Waters*. Liverpool University Press, UK

Hyslop, J.E., 1980. Stomach contents analysis: A review of methods and their application. *Journal of Fish Biology*, 17: 429–441.

Jan, N.A. and Das, S.M., 1970. Qualitative and quantitative studies on the food of eight fish species of Kashmir. *Ichthyologica*, 10: 20–26.

Jan, N.A. and Das, S.M., 1971. Studies on the food and seasonal variation in four fishes of Kashmir valley. *Kashmir Sci.*, 8: 102–107.

Kara, C. and Alp, A., 2005. Feeding habits and diet composition of brown trout (*Salmo trutta*) in the upper streams of River Ceyhan and River Euphrates in Turkey. *Turk. J. Vet. Anim. Sci.*, 29: 417–428.

Laggarigue, T., Cereghino, R., Lim, P., Reyes-Merchant, P., Chappaz, R., Lavandier, P., Belaud, A., 2002. Diel and seasonal variations in brown trout (*Salmo trutta*) feeding patterns and relationship with invertebrate drift under natural and hydropeaking conditions in a mountain stream. *Aqua. Living Res.*, 15: 129–137.

Merz, J.E., 2002. Seasonal feeding habits, growth and movement of steelhead trout in the lower Mokelumne River, California. *California Fish and Game*, 88(3): 95–111.

Pandit, A.K., 1999. Ecology of fishes in wetland ecosystems of Kashmir with emphasis on their management. In: *Inland Water Resources: India*, (Eds.) M.K. Durga Prasad and P.S. Pitchaiah. Discovery Publications, Delhi, India, pp. 188–201.

Pennak, R.W., 1978. *Freshwater Invertebrates of United States*. John Wiley and Sons, London.

Richardson, J.S., 1993. Limits to productivity in streams: Evidence from studies of macroinvertebrates. *Can. Spec. Publ. Fish. Aquat. Sci.*, 118: 9–15.

Sagar, P.M. and Eldon, G.A., 1983. Food and feeding of small fish in the Rakaia River, New Zealand. *New Zealand Journal of Marine and Freshwater Research*, 17: 213–226.

Straskraba, M., 1965. The effect of fish on the number of invertebrates in ponds and streams. *Mitt. Internat. Verein. Limnol.*, 13: 106–127.

Subla, B.A. and Das, S.M., 1970. Studies on the feeding habits, the food and the seasonal fluctuations in feeding in nine Kashmir fishes. *Kashmir Sci.*, 7: 25–44.

Surber, E.W., 1936. Rainbow trout and bottom fauna production in one mile of stream. *Trans. Am. Fish. Soc.*, 66: 193–202.

Wallace, J.B. and Webster, J.R., 1996. The role of macroinvertebrates in stream ecosystem function. *Ann. Rev. Entomol.*, 41: 115–139.

Williams, L.R., Taylor, C.M. and Warren Jr., M.L., 2003. Influence of fish predation on assemblage structure of macroinvertebrates in an intermittent stream. *Trans. Am. Fish. Soc.*, 132: 120–130.

Yousuf, A.R. and Firdous, G., 1997. Food spectrum of crucian carp, *Carassius carassius* (Linnaeus) in Anchar lake, Kashmir. *Oriental Sci.*, 2: 35–40.

Yousuf, A.R. and Firdous, G., 2001. Food spectrum of mirror carp in a deep mesotrophic Himalayan lake. *J. Res. Dev.*, 1: 60–67.

Yousuf, A.R., Bhat, F.A., Mehdi, D. Ali, S. and Ahangar, M.A., 2003. Food and feeding habits of *Glyptosternon reticulatum* Mc Clelland and Griffith in torrential streams of Kashmir Himalayas. *J. Res. Dev.*, 3: 123–133.

Chapter 25

Aquatic Insects as Biological Indicators of Water Pollution

S. Paul Sebastian, R. Kavitha and A. Christopher Lourduraj

Department of Environmental Sciences, Tamil Nadu Agricultural University, Coimbatore – 641 003

There has been serious concern about declining water quality in streams and rivers since the 1960s. Initially, concerns were centered on releases of point source pollutants such as heavy metals, sewage, and other chemical wastes from industrial and municipal origins. These were harmful both to human and stream ecosystem health. High non point source runoff may change stream water colour and turbidity (clarity), increase the amount of organic matter and nutrients (usually nitrogen and phosphorus), and increase sediment suspended in the water. Once in streams, these materials separately or in combination can seriously degrade water quality for humans and aquatic life. Unlike point source pollution, non point source pollution is much more difficult to detect and control because runoff does not come from a few easily identifiable sources, but instead stems from a number of locations scattered across a watershed. Non point source pollution is among the most important sources of water quality impairment.

The community of benthic macroinvertebrates in a water body is affected by the habitat structure, water quality, and other environmental factors. Factors that negatively impact the benthic community population and diversity are called stressors. For water quality management purposes, it is important to distinguish between natural and anthropogenic stressors. Natural stressors include but are not limited to high winds, low and high rainfall, frost action, snowfall and intense sunshine. Anthropogenic stressors include hydraulic alterations; the impact of point and non-point sources of pollution such as sediment, organic, and chemical (*e.g.*, heavy metals and pesticides) loads; changes in pH and water temperature; and predation or competition by introduced species.

The traditional water quality monitoring approach has been to collect stream water samples and analyze them in a laboratory for suspected physical and chemical pollutants. Unfortunately, because sampling and analysis are expensive and because concentrations of pollutants vary greatly with time and location, physical and chemical monitoring alone often cannot detect non point source pollution problems. In the mid-1960s, it was recognized that physical and chemical monitoring of water do not adequately describe the adverse effects of water contamination on aquatic life (Jackson and Brungs, 1966). John Cairns, a pioneer in aquatic biology, suggested that biological monitoring could be a useful supplement to (but not a substitute for) physical and chemical monitoring of water (Cairns *et al.*, 1973). Benthic macroinvertebrate bio assessments use the number, type, and sensitivity to certain pollutants (or tolerance) of benthic macroinvertebrates to calculate a series of metrics and provide an overall water quality rating. The presence, absence, abundance, and diversity of these organisms in a water body, when compared to a regional reference condition are used to classify the status of the water body as "impaired" or "non impaired" owing to the benthic condition.

A biological approach to water quality monitoring–biomonitoring–incorporates the use of stream organisms themselves as a basis for pollution detection. Europeans first adopted this strategy in the early 1900s to identify organic pollution in large rivers. In the United States, the use of stream organisms as biological indicators or "sentinels" has become widespread only over the last two decades. Several agencies including the Environmental Protection Agency (EPA), the Natural Resources Conservation Service (NRCS), and the U.S. Geological Survey (USGS) now employ biologists whose main task is to implement biomonitoring in streams and rivers across the country. The underlying concept of biomonitoring is simple certain types of stream animals occur or thrive only under certain water quality conditions. When conditions change, such as when a stream receives significant nonpoint source runoff, the abundance and distribution of animals in the affected site change as well.

Although fish and algae have been used in stream biomonitoring programs, benthic invertebrates are the most commonly used organisms. Benthic invertebrates are widely used as bioindicators because

1. They constitute the majority of species present in streams.
2. The numerous species present often show a wide range of sensitivity to pollution.
3. They are relatively easy to sample and identify.
4. The short life cycles and high movement of many species may provide reliable and rapid evidence of the return of favourable water quality after a pollution event.

The combined use of benthic invertebrate bioindicators and traditional stream water quality monitoring allows a comprehensive means of assessing water quality pollution from nonpoint sources within forested watersheds.

What Are Benthic Invertebrates?

Benthic invertebrates are small animals that live on the bottom of a pond, lake, stream, or river for at least part of their lives. They inhabit tiny spaces between submerged stones, within organic debris, on logs and aquatic plants, or within fine sediments (silt, clay). Technically, invertebrates are animals that do not have backbones like the larger animals (vertebrates) such as fishes, amphibians, reptiles, birds, and mammals.

Benthic "macroinvertebrates" are bottom-dwelling invertebrates large enough to be seen with the naked eye. They are usually greater than 1 mm or 1/32 inch long. Most species of stream macroinvertebrates are aquatic insects, although crustaceans (crayfish, side swimmers, aquatic pillbugs), molluscs (snails, mussels, clams), oligochaetes (earthworms, leeches), and arachnids (aquatic mites) also occur commonly (http://www.aces.edu/pubs/docs/A/ANR-1167/).

Major Groups of Aquatic Insects

There are so many different kinds of Aquatic Insects, it is difficult to appreciate their biological diversity without considering some of the individual kinds. The following section provides a brief summary of the eight major groups (Reese Voshell, 2003).

Mayflies (Ephemeroptera)

Larvae of mayflies live in a wide variety of flowing and standing waters. Most of them eat plant material, either by scraping algae or collecting small pieces of detritus from the bottom. Larvae breathe dissolved oxygen by means of gills on the abdomen. They have incomplete metamorphosis. Most mayflies are sensitive to pollution, although there are a couple of exceptions. The most unusual feature of mayflies is that the adults only live a few hours and never eat.

Dragonflies and Damselflies (Odonata)

Larvae of dragonflies and damselflies are most common in standing or slow-moving waters. All of them are predators. Larvae breathe dissolved oxygen with gills, which are located either inside the rear portion of the abdomen (dragonflies) or on the end of the abdomen (damselflies). They have incomplete metamorphosis. Many kinds are fairly tolerant of pollution, but some kinds only live in unique habitats, such as bogs high in the mountains. The most unusual feature of this group is the way the larvae catch their food with an elbowed lower lip, which they can shoot out in front of the head.

Stoneflies (Plecoptera)

Larvae of stoneflies live only in flowing waters, often cool, swift streams with high dissolved oxygen. Some feed on plant material, either by shredding dead leaves and other large pieces of detritus, while others are predators. Larvae breathe dissolved oxygen. Some have gills on their thorax, but others just obtain dissolved oxygen all over their body. They have incomplete metamorphosis. Almost all of the stoneflies are sensitive to pollution. The most 'Unusual feature of this group is that some kinds are programmed to emerge only during the coldest months; hence, they are called the winter stoneflies.

True Bugs (Hemiptera)

Most of the true bugs live on land, but the aquatic kinds are most common in the shallow areas around the edge of standing waters. Both the adults and the larvae of the aquatic kinds live in the water. Both stages are usually found on submerged aquatic plants. Almost all of them are predators. They breathe oxygen from the air, either by taking a bubble underwater or by sticking a breathing tube up into the air. They have incomplete metamorphosis. Most kinds are tolerant of pollution. The most unusual feature of this group is the way they kill and eat their prey. True bugs have a sharp beak that they stick into the body of their prey, and then they pump in poison to kill their prey, after which they suck out the body fluids. Some of the larger kinds feed on small fish and tadpoles.

Dobsonflies and Alderflies (Megaloptera)

Larvae of different kinds live in flowing or standing waters. They are all predators. They breathe dissolved oxygen by means of gills and their overall body surface. They have complete metamorphosis. Mature larvae the water and dig out a protected space under a rock or log for the pupa stage. Different kinds are either sensitive or tolerant to pollution. Larvae of some of the larger kinds are called hellgrammites, which are popular as live bait for small mouth bass and other warm-water fish species.

Water Beetles (Coleoptera)

There are more species of beetles than any other insects, but most of them live on land. Most of the Water beetles are more common in standing or slow-moving waters, but a few kinds are only found in swiftly flowing waters. Both the adults and the larvae of the aquatic kinds live in the water. Water beetles feed in different ways, primarily by preying on other animals, scraping algae, or collecting small particles of detritus from the bottom. All of the adults breathe air by taking a bubble underwater, while most of the larvae breathe dissolved oxygen by a combination of gills and their overall body surface. They have complete metamorphosis and leave the water for the pupa stage. Water beetles range from sensitive to somewhat tolerant of pollution. The most unusual feature of water beetles is that some of the adults live for several years.

Caddis Flies (Trichoptera)

Larvae of different caddis flies live in a wide variety of flowing and standing waters. They also have a wide range of feeding habits, including scraping algae, collecting fine particles of detritus from the bottom or from the water, shredding dead leaves, and preying on other invertebrates. They breathe dissolved oxygen by means of gills and their overall body surface. Caddis flies have complete metamorphosis and remain in the Water for the pupa stage. Most kinds are sensitive to pollution, but a few kinds are somewhat tolerant of moderate levels of pollution. The most distinctive feature of caddis flies is their ability to spin silk out of their lower lip. They use this material to glue together stones or pieces of vegetation into a small house for their protection during the larva and pupa stages. Some also use strands of silk to make a net for filtering particles of food from the water.

True Flies (Diptera)

This group has more kinds on land, but there are also many aquatic kinds. They have a wide range of feeding habits, including scraping algae, collecting fine particles of detritus from the bottom or from the water, shredding dead leaves, and preying on other invertebrates. They breathe dissolved oxygen by means of gills and their overall body surface. True flies have complete metamorphosis and remain in the water for the pupa stage. The most distinctive feature of this group is their ecological diversity. Some kinds live in the cleanest habitats (*e.g.*, swift, cool, mountain streams), while others live in some of the harshest natural habitats on the earth (*e.g.*, arctic tundra ponds, geothermal springs, alkaline lakes, mucky swamps). They have equally diverse responses to pollution, with some kinds being exceptionally sensitive, while other kinds endure the worst imaginable water quality (*e.g.*, raw sewage or acid mine drainage).

Factors Affecting Aquatic Insect Diversity

Hydraulic Alteration

Changes in stream water depth and velocity (flow volume and rate) can negatively impact the benthic community. These changes can be caused by increased surface-runoff from adjacent lands (including flooding), sustained drought, discharges from municipal and industrial outfalls, and pooling behind man-made dams. Low flows reduce the available habitat capacity for aquatic organisms and have higher water temperatures. High flows can scour the substrate, move rocks and other valuable habitat areas downstream, and often carry higher loads of sediment and other pollutants.

Sediment Load

Sediment from non-point sources in surface-runoff and solids in municipal and industrial discharges can have multiple effects on the stream water environment. These effects include increased turbidity that reduces light penetration, increased water temperature, and reduced dissolved oxygen levels. The solids can interfere with the respiration of benthic macroinvertebrates by injuring the gills. Deposited sediment that fills interstitial spaces of the substrata reduces the available habitat for some macroinvertebrate species. Solids in water reduce the photosynthesis of aquatic plants, which are the food source for some benthic macroinvertebrates, and may clog the feeding nets of other benthic macroinvertebrates. Solids reduce the visibility in the water and can thus lower the success rate of predatory macroinvertebrates in capturing prey.

Organic Load

Organic enrichment, which leads to high dissolved organic carbon levels and associated high biochemical oxygen demand, results in low dissolved oxygen concentrations in water. The consequences of low dissolved oxygen (DO) levels include: a decrease in the number of oxygen-sensitive organisms, an increase in low-DO tolerant organisms, and therefore changes in the macroinvertebrate community composition. Deposited organic sludge forms a blanket over the benthic macroinvertebrates resulting in the loss of interstitial' organisms. Sludge deposition in low velocity waters can release methane and hydrogen sulfide into the water and may result in the elimination of an entire community of benthic macroinvertebrates.

Chemical Load

Chemical inputs into water bodies can originate from industrial, agricultural, and urban sources. The effects of chemical inputs on benthic organisms vary and can be detrimental to some species. For example, high ammonia concentrations in water can be toxic to certain benthic organisms and may cause a total elimination of or a decrease in the population of these species. Other toxins may have lethal effects on different species and cause changes in the benthic community structure and diversity.

Changes in pH

Stream water pH can be affected by long-term acid precipitation, acid mine drainage, and acids in effluent from industrial plants. Under acidic conditions, benthic macroinvertebrates that employ carbonate shell structures (*e.g.*, crayfish, snails, clams, and mussels) are unable to properly build shell material. Acid conditions also indirectly affect benthic macroinvertebrates because the toxicity of some pollutants (*e.g.*, ammonia) is more intense under acidic conditions. Toxic metals such as aluminum, manganese, and mercury become more mobile under acidic conditions and therefore become more likely to impact the benthic community.

Water Temperature

The riparian canopy and stream bank vegetation influence stream water temperatures. For instance, the cutting of trees and tall vegetation along the stream bank allows in more direct sunlight, which heats the water. The discharge of high temperature waters from industrial sources also influences the stream water temperatures. Changes in the benthic macroinvertebrate community can be expected under conditions of higher water temperatures. One of the most obvious effects of water temperature change concerns its effect on the amount of available dissolved oxygen, with warmer waters containing less dissolved oxygen. Water temperature regime also effects developmental and growth characteristics of benthic macroinvertebrates.

How are Benthic Invertebrates Used in Biomonitoring?

Aquatic Insects are excellent overall indicators of both recent and long-term environmental conditions (Patrick and Palavage, 1994). The immature stages of Aquatic Insects have short life cycles, often several generations a year, and remain in the general area of propagation. Thus, when environmental changes occur, the species must endure the disturbance, adapt quickly, or die and be replaced by more tolerant species. These changes often result in an overabundance of a few tolerant species, and the communities become destabilized or "unbalanced." Members of the order Diptera, or true flies, are especially good "bioindicators" of Aquatic environmental conditions because, in addition to the attributes of other Aquatic Insects, they occupy the full spectrum of habitats and conditions (Paine and Gaufin, 1956; Roback, 1957; Mason, 1975; Hudson *et al.*, 1990).

Typically, benthic invertebrates are sampled from streams using dip nets or kick screens for qualitative collections or quadrat samplers such as a Surber or Hess sampler for more precise, quantitative collections. After collection, the samples are examined either in the field or in the laboratory. The invertebrates are removed from stones and organic debris, identified as belonging to a particular taxonomic group, and counted. Once counted, invertebrates can be compared to samples taken in the same stream but at different times, such as before and after a suspected pollutant has entered a stream. They also can be compared to samples taken from two or more streams at approximately the same time, such as from a stream suspected of receiving a pollutant and a nearby undisturbed reference stream.

Invertebrates may be quantified by species richness (number of unique types of invertebrates present in a sample), abundance (total number of invertebrates in a sample), relative abundance (number of invertebrates in the sample from one species relative to another), and species diversity (distribution of total individuals across species in the sample).

Three main parameters (MBI, KBI, and %EPT) are analyzed to address the biology impairment. The macroinvertebrate Biotic Index (MBI) rates the nutrient and oxygen demanding pollution tolerance of large taxonomic groups (order and family). Higher values indicate greater pollution tolerances. Along with the number of individuals within a rated group, a single index value is computed which characterizes the overall tolerance of the community. The higher the index values, the more tolerant the community is of organic pollution exerting oxygen demands in the stream setting. Index values greater than 5.4 are indicative of non-support of the aquatic life use; values between 4.51 and 5.39 are indicative of partial support and values at or below 4.5 indicate full support of the aquatic life use. The Kansas Biotic Index (KBI) is similar to the MBI in that it indicates the impact of nutrient and oxygen demanding pollutants. The EPT index is the proportion of aquatic taxa present within a stream belonging to pollution intolerant orders: Ephemeroptera, Plecoptera and Trichoptera (mayflies, stoneflies and caddis flies). Higher percentages of total taxa comprising these three groups indicate less pollutant stress and better water quality.

Bioassessment Protocols

Rapid Bioassessment Protocols (RBPs) for using benthic macroinvertebrates as indicators of biological integrity of waters were initially developed in the 1980s as a cost-effective screening tool. Since then RBPs have evolved into three protocols (RBPI, RBPII, RBPIII) that incorporate different levels of rigorousness. The benthic macroinvertebrate protocols (RBPI, RBPII, RBPIII) differ in the level of effort, taxonomic identification level, expertise required to perform them, and in the usefulness of obtained data. RBPI Sampling procedures are not standardized; family-level taxonomic identification is carried out in the field; assessment decision is based on "best professional judgment." RBPII Sampling procedures are standardized; family-level taxonomic identification is carried out in the field or in the laboratory; assessment decision is based on numerical data. RBPIII Sampling procedures are standardized; genus/species level taxonomic identification is carried out in the laboratory; assessment decision based on numerical data. RBPI and RBPII are useful approaches for setting priorities but are less rigorous than RBPIII. Because RBPIII involves organism identification to the lowest practical level (genus or species), it is the most labour intensive approach but gives relatively detailed information for trend analysis. The RBPII and RBPIII data are used to calculate a variety of values of metrics. A metric is a calculated term or enumeration that represents some aspect of the biological assemblage structure, function, or other measurable characteristics that changes in some predictable way in response to environmental influences (including human influences). A multimetric approach aggregates metrics into an overall assessment of the biological condition. Each calculated metric is assigned a score based on a comparison to the reference (undisturbed) condition. Scores for all metrics are then summed and compared to the total metric score for the reference condition. The per cent comparison between the total scores provides a final evaluation of the biological condition. The U.S. EPA has suggested minimum requirements for state biological assessment programs as part of the Section 305(b) reporting requirements. These requirements are based upon existing state programs and when followed ensure greater accuracy and consistency in state biological assessment and criteria development efforts. Suggested requirements include the use of multiple assemblages multiple metric indices, habitat structure assessment; regional reference conditions, index periods, standard operating procedures, and a quality assurance program.

Advantages of Using Benthic Bioassessments

Advantages of using benthic macroinvertebrates as a bioassessment tool are listed as follows (Barbour *et al.*, 1999)

1. Macroinvertebrates assemblages are good indicators of localized environmental conditions. Because many benthic macroinvertebrates have limited migration patterns or a sessile mode of life, they are particularly well suited for assessing site specific impacts.
2. Macroinvertebrates integrate the effects of short term environmental variations. Most species have a complex life cycle of approximately one year or more. Sensitive species respond quickly to environmental stress while the overall community responds more slowly.
3. Macroinvertebrates are relatively easy to identify to family; many taxa can be identified to lower taxonomic levels with ease. An experienced biologist can easily detect a degraded condition with an examination of the benthic macroinvertebrates assemblage.
4. Benthic macroinvertebrates assemblages are made of species that constitute a broad range of trophic levels and pollution tolerances, thus providing strong and graded information for interpreting cumulative effects.
5. Sampling benthic macroinvertebrates is relatively easy, requires few people and inexpensive equipment, and has minimal detrimental effect on the resident biota.
6. Benthic macroinvertebrates serve as a primary food source for fish, including many recreationally and commercially important species.
7. Benthic macroinvertebrates are abundant in most streams. Many small streams (1st and 2nd order), which normally support a diverse macroinvertebrate fauna, only support a limited fish fauna.
8. Many state water quality agencies have more expertise with invertebrates than fish. Therefore, most state water quality agencies that routinely perform biosurveys focus on macroinvertebrates.

The aquatic insects are like the canaries in the coal mine for water quality. Many roles performed by macroinvertebrates in aquatic ecosystem under sores the importance of their conservation. Macroinvertebrates have served as valuable indicators of degradation of aquatic ecosystem, and as increasing demands are placed on our water resources, their value in assessments of there impacts is necessary.

References

Barbour, M.T., J. Gerritsen, B.D. Snyder and J.B. Stribling, 1999. *Rapid Bioassessment Protocols for Use in Streams and Wadeable Rivers: Periphyton, Benthic Macroinvertebrates, and Fish*, 2nd Edition. EPA 841-B-99-002. U.S. EPA.

Cairns, Jr. J., J.W. Hall, E.L. Morgan, R.E. Sparks, W.T. Waller, and G.F. Westlake, 1973. The development of an automated biological monitoring system for water quality. *VWRRC Bull.*, Virginia Water Resources Research Center, Virginia Tech, Blacksburg, Virginia, 59: 50.

http://www.aces.edu/pubs/docs/A/ANR-1167/.

Hudson, P.L., D.R. Lenat, B.A. Caldwell, and D. Smith, 1990. Chironomidae of the southeastern United States: a checklist of species and notes on biology, distribution, and habitat. *U.S. Fish and Wildlife Service Fish and Wildlife Res.*, 7: 46.

Jackson, H.W. and W.A. Brungs, Jr., 1966. Biomonitoring of industrial effluents. In: *Proceedings, Industrial Waste Conference*, Purdue University, L. (1): 117–124.

Mason, W.T., Jr., 1975. Chironomidae (Diptera) as biological indicators of water quality. In: *Organisms and Biological Communities as Indicators of Environmental Quality*, (Eds.) C.C. King and L.E. Elfner. Circular 8. Ohio Biological Survey, Columbus, pp. 40–51.

Paine, G.W. and A.R. Gaufin, 1956. Aquatic diptera as indicators of pollution in a midwestern stream. *The Ohio Journal of Science*, 56: 291–304.

Patrick, R. and D.M. Palavage, 1994. The value of species as indicators of water quality. In: *Proceedings of the Academy of Natural Sciences*, Philadelphia, 145: 55–59.

Reese Voshell, 2003. http://www.ext.vt.edu/pubs/fisheries/420-531/420-531.

Roback, S.S., 1957. The immature tendipedids of the Philadelphia area. *Academy of Sciences of Philadelphia Monograph*, Washington, D.C., 9: 152.

Chapter 26

Diversity and Composition of Insecta in Rice Agroecosystem in Barak Valley of Assam (N.E. India)

D.C. Ray and Partha P. Bhattacharjee

Department of Ecology and Environmental Science, Assam University, Silchar – 788 011

ABSTRACT

Diversity and density of insects were studied in 5 (five) sites of Cachar district of Assam during 2003–04. Order-wise study of insect showed maximum abundance by Coleoptera (35–53 per cent) followed by Hemiptera (9–37 per cent). Site-wise and season-wise study of Shannon-diversity index (H) proved moderate to high diversity. Sorensen Co-efficient of similarity (Cs) showed more similarity of insect population distribution. Diurnal activity of insect revealed maximum (159.0) activity at 17–18 hrs during Boro season followed by Amon season (115.0) at 15–16 hrs 't'–test proved insignificant regarding the activities of insects at peak hrs in all the three seasons. Density was found to be more by *Coccinelid sp.* (659/m^2) in Boro season followed by *Oxya hyla hyla* (136/m^2) in Amon season.

Keywords: Diversity index, Sorensen Co-efficient, Insects, Agroecosystem, Diurnal activity, Density.

Introduction

Insects are the dominant components of most ecosystems and agroecosystems are no exception. They play a key role in pollination and natural pest control, but it is a fact that most of the agricultural pests are insects (LaSalle, 1999). Thus, insect biodiversity form a vital component of a variety of ecosystem functioning.

According to the status report of India' s faunal wealth, large proportion of Indian fauna constitute Insecta (more than 40 per cent of the total fauna) (Anonymous, 1991). But this faunal diversity is under estimated for improper surveys and lack of systematic identifications, particularly insects whose only larval and pupal stages occur in aquatic habitats (Brij Gopal, 1997). A number of workers worked on insect diversity on various ecosystem levels (Engels and Wood, 1999; Wood and Lenne, 1999; Chesalmah *et al.*, 2000; Ogbeibu, 2001).

Barak valley-southern part of Assam belongs to North-east hotspot area and Himalayan foot hills where the study is being carried out. A great insect diversity was noticed from rice agroecosystem. Considering the importance of insects in agroecosystem an attempt has been made to study the diversity and composition of insects during different seasons.

Description of the Study Sites

The present investigation was undertaken in Cachar district southern part of Assam. The geographical location of study sites are situated at longitudes of 24°41′29.9″ N and latitude of 92°45′25.9″ E. Altitude of the study areas are about 36 meter above msl. The entire study areas are low plain land and wetland with small hillocks. Rainfall is moderate to high. Paddy is cultivated in medium extent and mostly depends upon climatic conditions. The region has a plenty of faunal biodiversity. Vegetation is evergreen semi-evergreen and subtropical. The overlying soil cover is mostly sandy laterite. In most of the areas land has poor nutrient status except the riverine plains of Barak River. Most of the hillocks are covered by tea plantation. Paddy is cultivated in the low plains between the hillocks.

Materials and Methods

Five sites *viz.*, Irongmara, Machghat, Bariknagar, Dudpatil and Binnakandi were selected for the study purpose during 2003–04. A quadrat. (1m × 1m) was used to collect the insect sample as a standard device (Ghosh, 1985). The diversity of insects were determined by counting the numbers and kept in 70 per cent alcohol for preservation. The known insects were released to the field in living conditions and unknown specimens were kept in preservative for identification and further study. The sample was drawn at an interval of 10 (ten) days from all the sites. From each site 5 (five) quadrats were drawn randomly from 0.5-hectare are cropland during the Boro season.

Diurnal studies were conducted in Irongmara paddy field during Amon, Aus and Boro seasons on Sataki, Mayamoti and Basful varieties, respectively. The period of study were confined to 7–17 hrs at Amon season, from 6 to 18 hrs at Aus and Boro seasons at an interval of 1 (one) hour. Here also 5 (five) quadrats of same size were randomly drawn from 0.5 hectare of cropland.

Shannon-diversity index (H) was employed with the following formula (Shannon and Weaver, 1949)

$$H = -\Sigma \, Pi \, . \, \text{Log } Pi$$

Sorensen Co-efficient (Similarity) (Cs) was employed for the determination of similarity and dissimilarity of insects between communities of different seasons by the following formula (Sorensen, 1948)

$$Cs = \frac{2a}{(2a + b + c)}$$

't'–test was employed to assess the significance of diurnal activities of insects at various peak hours.

Results and Discussion

Group-wise abundance of insects was incorporated in Table 26.1. As regards the abundance it was found that more dominant group was Coleoptera (35–53 per cent) followed by Hemiptera (9–37 per cent) in all the sites out of seven available orders. Lowest group was considered as Odonata (4–14 per cent) (Table 26.1). Shannon diversity index indicated moderate diversity of insects in all the study sites during Boro season only (Table 26.2). Whereas season-wise diversity index (H) indicated also moderate diversity of insects between the community (Table 26.3).

Table 26.1: Group-wise Diversity of Insects on Rice Ecosystem During Boro Season (2003)

Orders	*Sites*				
	Irongmara	*Machghat*	*Bariknagar*	*Dudpatil*	*Binnakandi*
Coleoptera	35.55%	52.14%	47.37%	53.40%	36.49%
Lepidoptera	19.05%	3.11%	1.75%	4.85%	0
Hymenoptera	3.81%	3.11%	3.51%	0	5.40%
Diptera	3.49%	19.07%	0	0	2.70%
Hemiptera	12.38%	9.34%	0	13.01%	37.84%
Orthoptera	21.59%	5.06%	33.33%	3.88%	12.16%
Odonata	4.13%	8.17%	14.03%	4.85%	5.40%

Note: '0' means nil.

Table 26.2: Showing the Shannon-diversity Index (H) in Five Sites of Barak Valley (2003)

Study Sites	*Boro Season*	
	Diversity Index	*Paddy Varieties Studied*
Irongmara	H = 1.72547	Basful
Machghat	H = 1.9050	Basful
Barik nagar	H = 1.5441	Tupa Sali
Dudpatil	H = 1.6339	Asatti
Binnakandi	H = 1.6757	Jaya

Table 26.3: Shannon-diversity Index (H) Showing Insect Communities During Three Seasons (2003–04)

Diversity Index	*Seasons*		
	Aus	*Amon*	*Boro*
Shannon diversity index (H)	1.8694	1.8765	1.8581

As regards the Sorensen Co-efficient (Cs) of similarity between the two seasons it was found more similarity (0.75-0.92) of insect population distribution (Table 26.4). Diurnal activities of insect population was studied during three season (Tables 26.5–26.7 and Figure 26.1). During Aus season study was undertaken at hourly interval from 6–18 hrs. Maximum activities were recorded at 6–7 hrs

(67.0) and 17–18 hrs (66.0). Most part of the day the activities of insects were moderate (Table 26.5). Diurnal activities were studied during Amon season, which indicated maximum activity (115.0) at 15–16 hrs followed by 14–15 hrs (100.0). Least activity 57.0 was noticed at 7–8 hrs (Table 26.6). In Boro season the peak activity was recorded 159.0 at 17–18 hrs followed by 16–17 hrs where the insect population was recorded 132.0. The lowest activity (78.0) was recorded at 7–8 hrs (Table 26.7). However, a 't'–test was employed between the peak hour activity of insects between two seasons. Although the peak hour activities varied in different seasons it did not significantly different from Aus vs. Amon, Aus vs. Boro and Amon vs. Boro (P < 0.05) (Tables 26.5–26.7).

Table 26.4: Sorensen Co-efficient of Insect Communities During Three Seasons (2003–04)

	Aus/Amon	*Amon/Boro*	*Boro/Aus*
Sorensen Co-efficient (Cs)			
Similarity	0.758	0.846	0.928
Dissimilarity	0.242	0.154	0.072

Table 26.5: Diurnal Variation of Insects from Paddy Field of Irongmara During Aus Season (2003)

No. of Obs.	*Time of Obs.*	*A*	*B*	*C*	*D*	*E*	*F*	*G*	*H*	*I*	*J*	*K*	*L*	*M*	*N*	*O*	*Total*
1.	6–7	18	5	2	2	5	3	7	6	1	1	0	0	0	0	0	67
2.	7–8	13	2	3	2	5	0	3	4	5	2	2	0	0	0	0	54
3.	8–9	17	1	2	2	8	0	2	7	3	0	0	1	1	1	0	57
4.	9–10	17	1	4	2	4	0	4	4	3	0	0	2	1	0	0	47
5.	10–11	10	1	3	7	11	1	7	6	3	0	0	1	1	1	2	64
6.	11–12	13	1	2	4	10	0	6	5	3	0	0	0	0	1	2	54
7.	12–13	12	0	1	1	8	1	4	7	0	1	0	1	0	2	0	43
8.	13–14	8	0	2	0	10	0	5	5	2	0	0	0	0	1	1	38
9.	14–15	13	0	2	0	7	0	3	3	2	0	0	0	0	1	1	37
10.	15–16	15	2	1	2	6	0	3	6	0	1	0	0	0	0	0	45
11.	16–17	15	0	2	2	10	0	3	6	2	0	0	0	0	1	2	52
12.	17–18	16	1	4	0	7	0	7	7	2	3	0	0	0	1	2	66
Total		**167**	**14**	**28**	**24**	**91**	**5**	**54**	**66**	**26**	**8**	**2**	**5**	**3**	**9**	**10**	**624**

Note: '0' means nil.

A: *Coccinelid* sp., B: *Bagrada crusifararum;* C: Demsel fly; D: *Neurothemis* sp.; E: *Oxya hyla hyla;* F: *Tryporyza incertula;* G: *Nephotettix virescens* (Dist); H: *Musca domestica;* I: *Alstonae* sp.; J: *Attractomorpha crenulata;* K: *Dicladispa armigera;* L: *Apis* sp.; M: *Vespa orientalis;* N: *Demoleus* sp.; O: *Leptocoryza acuta.*

Aus vs. Boro at 15–16 hrs: t = 1.24 (insignificant: P < 0.05)

16 17 hrs: t = 1.72 (insignificant: P < 0.05)

17 18 hrs: t = 0.91 (insignificant: P < 0.05)

Table 26.6: Diurnal Variation of Insects from Paddy Field of Irongmara During Amon Season (2003)

No. of Obs.	*Time of Obs.*	*A*	*B*	*C*	*D*	*E*	*F*	*G*	*H*	*I*	*J*	*K*	*L*	*M*	*N*	*Total*
1.	6–7	12	22	6	0	2	4	0	8	2	10	2	0	0	0	68
2.	7–8	5	21	7	0	0	0	4	2	0	13	5	0	0	0	57
3.	8–9	16	39	0	0	0	4	2	0	5	13	2	0	0	0	81
4.	9–10	20	44	2	1	0	3	4	0	0	10	3	0	0	1	88
5.	10–11	13	51	2	1	0	0	0	7	3	12	4	0	0	0	93
6.	11–12	12	31	1	1	0	0	5	2	5	10	3	0	1	0	71
7.	12–13	10	23	3	2	0	3	2	8	0	4	3	1	0	0	59
8.	13–14	14	39	3	4	0	0	3	1	8	22	5	0	0	0	99
9.	14–15	20	39	1	14	1	3	0	3	3	16	0	0	0	0	100
10.	15–16	14	51	8	19	1	6	1	5	3	7	0	0	0	0	115
Total		**136**	**360**	**33**	**42**	**4**	**23**	**21**	**36**	**29**	**117**	**27**	**1**	**1**	**1**	**831**

Note: '0' means nil.

A: *Oxya hyla hyla*; B: *Coccinelid* sp.; C: *Neurothemis* sp; D: *Attractomorpha* sp.; E: Demsel fly; F: *Leptocoryza acuta*; G: *Sitophilus oryzae*; H: *Musca domestica*; I: *Bagrada; crusiferarum*; J: *Tryporyza incertula*; K: *Galerucella* sp.; L: *Demoleus* sp; M: *Dicladispa armigera*; N: *Apis* sp.

Aus vs. Amon at 15–16 hrs: t = 1.19 (insignificant: $P < 0.05$).

Table 26.7: Diurnal Variation of Insects from Paddy Field of Irongmara During Boro Season (2003–04)

No. of Obs.	*Time of Obs.*	*A*	*B*	*C*	*D*	*E*	*F*	*G*	*H*	*I*	*J*	*K*	*L*	*M*	*Total*
1.	6–7	49	11	4	1	3	9	12	2	2	0	0	3	0	96
2.	7–8	43	9	3	2	5	4	7	3	0	1	0	1	0	78
3.	8–9	60	8	6	4	5	11	14	2	3	0	0	3	1	117
4.	9–10	55	9	3	5	12	6	9	4	3	1	2	2	0	111
5.	10–11	63	5	2	3	12	4	17	1	1	1	2	4	0	115
6.	11–12	47	5	2	9	6	4	8	1	2	1	3	2	0	90
7.	12–13	42	8	5	3	8	3	10	1	5	0	0	1	1	87
8.	13–14	41	8	6	3	7	3	6	2	2	1	2	1	0	82
9.	14–15	51	8	5	7	6	6	10	0	4	3	0	5	1	106
10.	15–16	59	8	5	10	13	9	9	3	5	1	2	6	0	130
11.	17–18	61	8	9	14	12	11	6	7	0	0	0	4	0	132
12.	18–19	88	7	11	7	11	4	16	2	2	1	4	6	0	159
Total		**659**	**94**	**61**	**68**	**100**	**74**	**124**	**28**	**29**	**10**	**15**	**38**	**3**	**1303**

Note: '0' means nil.

A: *Coccinelid* sp.; B: *Leptocoryza acuta*; C: *Neurothemis* sp; D: *Oxya hyla hyla*; E: *Tryporyza incertula*; F: *Musca domestica*; G: *Dicladispa armigera*; H: *Bagrada crusiferarum*; I: Demsel fly; J: *Attractomorpha crenulata*; K: *Nephotettix virescens* (Dist); L: *Alstonae* sp.; M: *Demoleus sp.*

Amon vs. Boro at 15–16 hrs: t = 0.0043 (insignificant: $P < 0.05$)

Table 26.8: List of Insects with Their Density (m^{-2}) in Paddy Field During Three Crop Seasons (2003–04)

Taxa Scientific Name	Common Name	Aus	Amon	Boro
Coccinelid sp.	Lady bird beetle	167	360	659
Alstonae sp.	Blue beetle	26	–	38
Dicladispa armigera Ol.	Rice hispa	2	1	124
Sitophilus oryzae Linn.	Rice weevil	–	21	–
Galerucella sp.	–	–	27	–
Tryporyza incertula	Yellow rice borer	5	117	100
Demoleus sp.	Butter fly	9	1	3
–	Demsel fly	28	4	29
Neurothemis sp.	Dragon fly	24	33	61
Oxya hyla hyla	Grasshopper	91	136	68
Attractomorpha crenulata	Grasshopper	8	42	10
Bagrada crusiferarum	–	14	29	28
Leptocoryza acuta	Mahua	10	23	94
Nephotettix virescens (Dist)	Green leafhopper	54	–	15
Musca domestica	Fly	66	36	74
Apis sp.	Bee	5	1	–
Vespa orientalis	Wasp	3	–	–

–: Means nil.

During three crop seasons the insect density was studied m^2 area basis (Table 26.8). Study indicated that maximum numbers were *Coccinelid* sp. in all the seasons. Peak population was recorded as 659/m^2 during Boro season followed by 360/m^2 during Amon season. Least numbers (167/m^2) were recorded in Aus season. The second dominant insect was *Tryporyza incertulus* was found in all the study seasons. Some insects *viz., Sitophilus oryzae, Gallerucella* sp. and *Vespa orientalis* were not found in all the seasons. Among the crops, rice cultivation is predominant in Barak Valley of Assam. Insect diversity and abundance are very rich in rice agroecosystem. It is perhaps tropical area, moderate to high humidity and rainfall prevails during the rainy season which helps in multiplication of insect population. Group-wise study reveals that out of seven orders Coleoptera occupies the dominant: group. Ogbeibu (2001) studied group-wise insect on aquatic ecosystem in four sites in Nigeria. Site-wise study of Shannon diversity index (H) reveals moderate to high diversity in all the sites but in season-wise it also proves moderate to high diversity. Diversity index also reported by Ogbeibu (2001) and Nautiyal *et al.* (2000) on aquatic system and found higher diversity index.

Sorensen Co-efficient of similarity reveals high similarity of insect population among the three seasons. Ogbeibu (2001) reported significant dissimilarity among the four study sites. This may be because of aquatic ecosystem where the work is performed. As regards the diurnal study it reveals out of three seasons maximum activities (159.0) reported during Boro season (2003–04) at 17–18 hrs followed by Amon season (115.0) at 15–16 hrs (Figure 26.1). This may be because of less temperature preferred by insects which prevails at that time. 't'–test was employed to assess the activities of insect

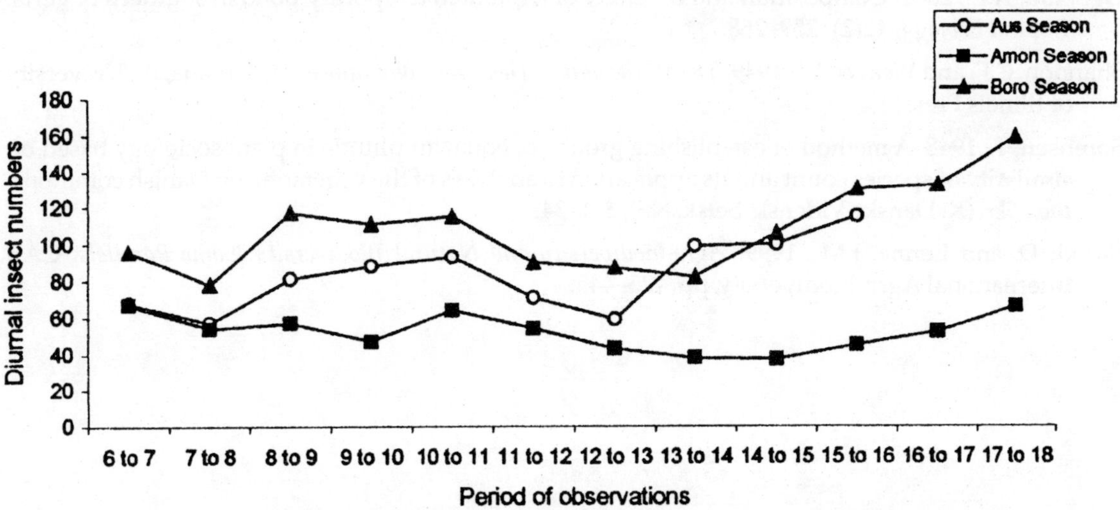

Figure 26.1: Showing Diurnal Insect Abundance During Three Seasons (2003–04)

among the peak hour period between Aus vs. Boro, Aus vs. Amon and Amon vs. Boro seasons. It proves the difference of activities insignificant. ($P < 0.05$) among all the seasons although the numbers of insect abundance is varying at different hour intervals. Density of insect taxa is found to be maximum in case of *Coccinelid* sp. (659/m^2) in Boro season followed by *Oxya hyla hyla* (136/m^2) in Amon season. It may be noted that in all the three seasons *Coccinelid* sp. is found to be dominating. Dominating numbers recorded may be because of species which can sustains in all types of environmental conditions. *Galerucella* sp., *Vespa orientalis* species may not available because of either seasonal insect or non-availability of food in off-season.

References

Anonymous, 1991. *Animal Resources of India: Protozoa to Mammalia*. State of the Art, Zoological Survey of India, Calcutta, 694 pp.

Chesalmah, M.R., Hassan, S.T.S. and Abu Hassan, A., 2000. Local movement and feeding pattern of adult *Neurothemis tullia* (Drury) (Odonata : Libellulidae) in a rainfed rice field. *Tropical Ecology*, 41(2): 233–241.

Engels, J.M.M. and Wood, D., 1999. *Conservation of Agro-biodiversity*. CAB International Agro-biodiversity, pp. 355–385.

Ghosh, A.K., 1985. *Sampling Techniques in Rice Research*. ICAR Publication, New Delhi, 623 pp.

Gopal, B., 1997. Biodiversity in Inland Aquatic Ecosystems in India: An overview. *International Journal of Ecology and Environmental Sciences*, 23(40): 305–313, International Scientific Publications, New Delhi.

LaSalle, J., 1999. *Insect Biodiversity in Agroecosystems: Function, Value and Optimization*. CAB International Agro-biodiversity, pp. 155–182.

Nautiyal, C., Nautiyal, P. and Singh, H.R., 2000. Species richness and diversity of epilithic diatom communities on different natural substrates in the cold water river Alaknanda. *Tropical Ecology*, 41(2): 255–258.

Ogbeibu, A.E., 2001. Composition and diversity of Diptera in temporary pond in southern Nigeria. *Tropical Ecology*, 42(2): 259–268.

Shannon, C.E. and Weaver, W., 1949. *The Mathematical Theory of Communication*. Urbana, IL, University of Illinois Press.

Sorensen, T., 1948. A method of establishing groups of equal amplitude in plant sociology based on similarity of species count and its application to analyses of the vegetation on Danish commons. *Biol. Skr.* (K. Danske Vidensk. Selsk. NS), 5: 1–34.

Wood, D. and Lenne, J.M., 1999. *Agro-biodiversity and Natural Biodiversity: Some Parallels*. CAB International Agro-biodiversity, pp. 425–445.

Chapter 27

Physico-chemical Analysis of the Soil Modified by *Coptotermes heimi* (Wasmann) (Rhinotermitidae : Isoptera : Insecta)

*C.B. Arora and H.R. Pajni**

Department of Biology, D.A.V. College, Hoshiarpur, Punjab

**Retired, Department of Zoology, Punjab University, Chandigarh.*

ABSTRACT

Coptotermes heimi (Wasmann) is a termite species which belongs to a primitive family Rhinotermitidae of order Isoptera of the class Insecta. This species constructs a type of semi-carton nest and is very peculiar in its feeding habits where the workers consume the core matter and leaving only the outer hard portion of the attacked cellulosic material untouched. The hollow central portion of it is filled with soil mixed with saliva, excreta and remains of the dead insects. It is further surrounded by semi-carton fillings. *C. heimi* also constructs external and internal gallery system made up of soil, excreta etc to locate the new feeding and nesting sites. For this, the soil which is transferred from deeper layers to the surface layer for the construction of the nest and galleries and for filling the empty space of the attacked material gets enriched with nutrients. The soil also gets altered in its various physical properties. The quality of the modified soil is improved in respect of water holding capacity, moisture contents and porosity. The particle size ratio of the soil is also increased towards fine sand and clay particles. Due to the presence of higher contents of organic matter, the soil also shows higher plastic limit than the liquid limit. The modified soil also becomes more alkaline and shows its total nitrogen, potassium and available nitrogen contents increased appreciably. However, there is no change in the sodium contents. In the present observations the extent, up to which the *C. heimi* alters the physico-chemical properties of the soil used, has been discussed.

Keywords: *Physical, Chemical, Modified soil and properties.*

Introduction

As a general rule, the termites transfer a lot of soil from varying depths and use it for the construction of their nests (Plate 27.3), mud mounds and the superficial mud galleries (Plate 27.2) through which they shift to new attacking sites (Pajni and Arora 1989; Pajni and Arora, 1994; Pajni *et al.*, 1996 and Arora, 2004). During this process the soil is not only treated with the saliva as a carrying material but it also undergoes other physical and chemical changes on account of the addition of faecal matter, discarded food particles and dead bodies of the young and adults of different castes. In this manner, the resulting soil is modified greatly and perhaps has an improved quality (Lee and Wood, 1971).

C. heimi is one of the most common and serious pest of trees and being a plant feeding species, confines its activities largely to the stems of the attacked trees. It transfers and changes the soil which is ultimately packed in the hollow spaces in the consumed stem (Plate 27.1) or is used in the preparation of semi-carton nest (Plates 27.3 and 27.4) (Beeson, 1941 and Sen-Sarma *et al.*, 1975). In the present study the role played by *C. heimi* in altering the physico-chemical properties of the soil has been discussed.

Plate 27.1: Vertical Section Through Different Sections of the Stem of a *Populus* sp. Tree Completely Damaged by *C. heimi*

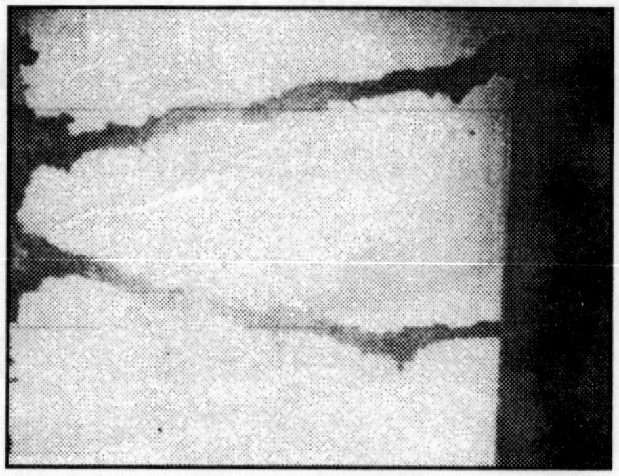

Plate 27.2: Two Exposed Galleries of *C. heimi* on the Wall of a Building

Plate 27.3: A Semi-carton Nest in its Original Position with Surrounding Layer of Soil Removed from the Basal Part

Plate 27.4: Semi-carton Nest after Removal from the Host Tree

Materials and Methods

In order to assess the degree of soil modification by *C. heimi*, the soil was collected from three different heights of a termite-damaged popular plant. The soil was subjected to physical and chemical analyses in the Regional Engineering College, Chandigarh and in Central Soil and Water Conservation Research Training Institute, Chandigarh respectively. As regards physical and chemical analysis of the modified soil, several techniques were employed. The particle size of the soil was determined by wet sieve analysis as well as sedimentation analysis methods. The liquid limit was estimated by Casagrande's (1932) cup and blow method. The plastic limit was observed by rolling and remoulding the soil into threads till the same started crumbling at a diameter of 3 mm. Void ratio was calculated from the difference between moisture contents and dry density. Water holding capacity and moisture were determined by following Chopra and Kanwar (1982) and Roonwal (1962) respectively. In the chemical estimations, pH was measured with an electric digital pH meter by using 1 gm of soil in 2.5 cc of distilled water. For organic matter or organic carbon contents of the soil, dry combustion and wet digestion methods were used. Total and available nitrogen were estimated by modified Kjeldahl's and alkaline permangnate methods respectively. Total sodium and potassium cations were determined with a flame photometer by burning their leachate prepared in ammonium acetate. Amount of phosphorous was estimated by following Chopra and Kanwar (1982). The attributes of the modified soil were compared with the control sample of soil taken from near the base of the experimental tree. The data are included in Table 27.1.

Table 27.1: Physico-chemical Analysis of the Unmodified Soil and Soil Modified by *C. heimi*

Property		*Unmodified Soil*	*Modified Soil*	*P <*
1.	**Physical**	$\overline{X}$±S.D.	$\overline{X}$±S.D.	
1.1	Mechanical Analysis/ Soil Grain Size Analysis			
1.1.1	% Coarse Sand	07.89±01.26	09.66±01.02	0.010
1.1.2	% Fine Sand	30.00±02.81	41.03±01.49	0.001
1.1.3	% Silt	32.93±02.29	25.56±00.41	0.001
1.1.4	% Clay	29.38±02.43	23.74±00.82	0.001
1.2	% Moisture Contents	10.12±03.79	54.02±11.29	0.001
1 3	% Liquid Limit	26.87±02.03	35.24±06.73	0.001
1.4	% Plastic Limit	20.50±01.50	38.47±19.73	0.025
1.5	% Void Ratio	40.12±07.47	82.75±19.27	0.001
1.6	% Water Holding Capacity	45.91±05.70	67.95±08.70	0.001
2.	**Chemical**			
2.1	pH	08.68±00.09	09.93±00.09	0.001
2.2	Total Nitrogen in (ppm)	531.67±39.86	1940.44±129.34	0.001
2.3	Available Nitrogen (ppm)	054.54±04.44	174.75±29.36	0.001
2.4	Total Phosphorus (ppm)	9.425±00.59	61.28±08.54	0.001
2.5	Total Potassium (ppm)	141.24±16.98	525.33±30.29	0.001
2.6	Total Sodium (ppm)	94.36±03.40	93.28±13.85	NS
2.7	% Organic Matter	01.80±00.66	29.67±04.25	0.001

NS: Non-significant; N: 8 in all the cases.

Observations and Discussion

It is evident from the table that the modified soil contains much more amount of fine sand and a little more amount of coarse sand. However, the silt and clay components undergo sufficient decrease in the modified soil. Apart from the changes in these standard components of the soil, the modified soil also contains a very large amount of organic matter which is very poorly represented in the control soil. All these variations in the components of the modified soil help in increasing the void ratio, the liquid limit and the plastic limit of the soil and consequently the fertility of the soil. It has, however, to be noted that the moisture contents value for the plastic limit unusually exceeds the moisture contents value of the liquid limit of the modified soil. This is rather an exceptional case because by its very definition, the moisture content value at the plastic limit of the soil should be less than at the liquid limit. It is possible that very high amount of organic matter drastically changes the plastic limit in such a manner that it defies the expected value of moisture content and soil no longer remains plastic.

The huge addition of organic matter in the modified soil is also expressed by the increased quantity of elements like total nitrogen, available nitrogen, phosphorus and potassium. There is however very little change in sodium contents in the two types of soil. The significant increase in the minerals of the modified soil supports the on going process of mineralization. In fact, both the physical and chemical qualities of the soil show useful changes in the modified soil which is greatly enriched and fertilized as a result of these changes.

There is no other information on the changes accompanying the modification of soil in a tree dwelling species. *C. acinaciformis* and *C. lacteus*, which have been investigated for this purpose (Lee and Wood, 1971), have also been taken from a habitat where they form termitaria. Other species about which data on the modified soil is available come from mound forming categories. In all these species, the changes in the percentage of different components occur in such a manner that they improve the porosity and water holding capacity of the soil. The porosity and water holding capacity of the soil are known to be dependent on the increase in organic matter, silt or clay components (Lee and Wood, 1971; Sharma and Misra, 1985). In all these cases also the improvement of the soil is due to the increase of one or more of these components. However, a minor decrease or no change in any of these components can be compensated by a correspondingly excessive increase in other components, as is the case in *C. heimi*. It can thus be concluded that the soil in the nests of both the tree dwelling and ground dwelling species of termites, is always modified to the beneficial side, largely due to the addition of the organic matter. This is amply supported by the noted increase in the total nitrogen, phosphorus and potassium in all types of modified soil (Lee and Wood, 1971; Mishra and Sen-Sarma, 1979; Golley, 1983; Salick *et al.*, 1983).

Acknowledgements

Authors are sincerely thankful to Shri M.S. Jangir, Superintendent Engineer, P.W.D. (B&R), Haryana and Shri L. Thakur, Scientist (B), Central Road Research Institute (CRRI), New Delhi for helping to analyse the physical properties of the soil and, Dr. S.S. Grewal, Scientist-2 (Soils) and Shri M.L. Juneja, Sr. Technical Assistant, Central Soil and Water Conservation Research Training Institute, Research Centre, Chandigarh for helping in Chemical analysis of the soil.

References

Arora, C.B., 2004. Termites and Human Welfare. In: *Biospectrum*, 7(1,2): 1–3.

Beeson, C.F.C., 1941. *The Ecology and Control of the Forest Insects of India and the Neighbouring Countries.* Controller of Publication, Govt. of India, Delhi, pp. 403–425.

Casagrande, A., 1932. Research on the Atterberg limits of soils. *Public Roads*, 13(8): 121–136.

Chopra, S.L. and J.S. Kanwar, 1982. *Analytical Agricultural Chemistry*. Kalyani Publishers, Ludhiana, pp. 518.

Golley, F.B., 1983. Decomposition. In: *Tropical Rainforest Ecosystem: A Structure and Function*, (Ed.) F.B. Golley. Elsevier Scientific Publishing Company, Amsterdam-Netherland, pp. 157–166.

Lee, K.E. and T.G. Wood, 1971. *Termites and Soils*. U.S. Edition, Published by Academic Press Inc., New York, pp. 251.

Mishra, S.C. and P.K. Sen-Sarma, 1979. Studies on deterioration of wood by insects. V. Influence of temperature and relative humidity on wood consumption and digestibility in *Neotermes bosei* Snyder (Isoptera : Kalotermitidae). *Material u. Organismen*, 14: 278–286.

Pajni, H.R. and C.B. Arora, 1989. A study of the termites of Chandigarh and their control. Final progress report of Science and Technology Council, U.T. Chandigarh Research Project, pp. 1–27.

Pajni, H.R. and C.B. Arora, 1994. Studies on the bionomics and management of termites of Haryana. Final progress report of Haryana Council for Science and Technology Research Project, pp. 1–82.

Pajni, H.R., C.B. Arora, A.K. Girdhar and I.S. Bhatti, 1996. Biology of termites. *Ann. Entomol.*, 14(1): 15–32.

Roonwal, M.L., 1962. Biology and ecology of Oriental termites (Isoptera). No. 5: Mound structure, nest and moisture content of fungus combs in *Odontotermes obeses* with a discussion on the association of fungi with termites. *Rec. Ind. Mus.*, 58: 131–150.

Salick, J., R. Herrera and C.F. Jordan, 1983. Termitaria: Nutrient patchiness in nutrient-deficient rain forests. *Biotropica*, 15(1): 1–7.

Sen-Sarma, P.K., M.L. Thakur, S.C. Misra and B.K. Gupta, 1975. Wood destroying termites of India. Final technical report under PL-480 project of U.S.A. (1968-1973), F.R.I. and Colleges, Dehradun, pp. 187.

Sharma, P.D. and R. Misra, 1985. *Elements of Ecology*. Rastogi Publ., Shivaji Road, Meerut, pp. 422.

Chapter 28

Treatment Studies on Pthalogen Blue Dye Waste from a Dye House in Tiruppur

K. Sadhana[1], K. Revathi[2], Suman Gulati[2], V. Rekha[2], N. Uma Chandra Meera Lakshmi[2] and R. Kungumapriya[2]

[1]*Department of Civil Engineering, Government College of Technology, Coimbatore – 641 013*

[2]*P.G. and Research Department of Zoology, Ethiraj College for Women, Chennai – 600 008*

ABSTRACT

Industrial waste is the most objectionable pollutant causing serious threat to the environment today. In the present study the raw waste from a Vat dye house in Chennai was collected and subjected to physico-chemical characterization, based on which suitable treatment strategies for the dye house waste have been suggested.

***Keywords**: Dye house waste, Pthalogen blue dye, Treatment plants.*

Introduction

Industrial waste of varying kinds and concentrations are contained in the spent process waters of manufacturing establishments. They are derived from washing, flushing, extracting and impregnations processes. They contain various kinds of suspended, colloidal and dissolved solids of mineral and organic origin. It is a dire need today to treat the effluent derived from the industrial sector.

The pollutants of the industrial wastes include raw materials, process chemicals, final products, process intermediates and process-by-products. Organic substances of the industrial wastes deplete the oxygen content of the streams, lakes and other water bodies. Acids and alkalis make the water bodies unfit for human use and survival of aquatic organisms. Inorganic salts present in the effluents make the water hard. Oil, grease and dyes are also amongst some of the pollutants from the industrial sector.

The industrial waste also causes severe damage to the sewage disposal system. It is only natural for the industries to pressurize that its wastes can best be disposed of in the domestic sewer system. However, before accepting any industrial waste, it is very important to understand the effects of the wastes upon all components of the city disposal system. The dissolved and colloidal organic matter imposes a load on the biological units of the treatment plants. All the organic matter present in the industrial waste may not oxidize at the same rate. The rate of decomposition of the industrial waste may be faster or slower than for the domestic sewage organic matter. The toxic metal ions interfere with biological oxidation by tying up enzymes required to oxidize organic matter. Acid and alkalis corrode the treatment units causing severe damage. Inflammables in the effluent may lead to explosions and pieces of fat may clog nozzles. Also the settling rate of the suspended solids present in large quantities in the industrial effluent varies considerably from the domestic sewage. These differences must be considered in the design and operation of the disposal systems.

In order to meet the existing terminal standards, the industries have to install capital-intensive and more sophisticated treatment systems, with heavy recurring costs. In addition to set up of waste water treatment plants another feasible alternative is to implement various pollution prevention measures using a 'methods approach', with the objective of not only optimizing chemical usage but also reducing the pollution load to the subsequent wastewater treatment system (UNEP Industry and Environment, 1995).

Dyes are obtained from animal and vegetable sources. Pthalogen blue dye belongs to the vat and naphtha dye. As far as the waste of pthalogen blue dye is concerned, the materials that can cause pollution include colour, acids, suspended solids, organic matter and temperature. In this study the feasibility of plain sedimentation and chemical treatment to treat pthalogen blue dye waste was examined.

Materials and Methods

The effluent sample used in this study was collected in polythene carbuoys from the dye house and was subjected to physico-chemical analysis in the laboratory. During the characterization the following analyses were performed, pH, Acidity, Solids (Total, Suspended and Dissolved Solids), Chemical Oxygen Demand (COD), Biological Oxygen Demand (BOD) and Colour. During the treatment study solids, COD and colour in terms of luminescence were the major parameters employed to determine the treatment efficiency. The analyses were performed according to procedures outlined in the standard methods and the results were tabulated.

Based on the effluent characteristics it was subjected to two different treatment methods namely: Plain sedimentation and Sedimentation aided with coagulation.

Plain Sedimentation

As soon as the turbulence was retarded by offering storage to water, the impurities tend to settle down at the bottom of the tank. This was the principle behind sedimentation. The purpose of sedimentation was to separate the settleable solids so that they do not form sludge banks when discharged into water courses. One litre measuring jars were used in order to evaluate the optimum detention period. Six one litre measuring jars were taken and filled with the sample. They were given a detention period of 30, 120, 180, 240, and 360 minutes. After each of the detention periods, 50 ml of the sample was collected from the jar and parameters such as suspended solids, volatile suspended solids and COD were determined. From the results the optimum detention period was determined.

Sedimentation Aided with Coagulation

The dye waste was also subjected to a chemical treatment method (*i.e.*) coagulation using aluminium sulphate as a coagulant. An optimum coagulant dosage was determined performing a jar test as follows to determine the optimum dosage of the coagulant required to remove the Total solids, Suspended solids and COD: 500 ml of waste sample already subjected to plain sedimentation treatment was taken in six one litre beakers. The beakers were positioned under the stirrers of a multiunit stirrer. The coagulant in solution dosage of 200, 300, 400, 500, 600, and 700 mg/l was added to each of the six beakers and flash mixed for 60 seconds at 90 rpm and then flocculated for 30 minutes at 25 rpm. The samples were then allowed to settle for 30 minutes. The clarified sample was then siphoned out into a clean flask and subjected to further analysis of Total solids, Suspended solids and COD.

With the results obtained from the laboratory tests, various treatment units such as sedimentation tanks and chemical coagulation units were designed.

Results and Discussion

In order to design the treatment units such as sedimentation tanks and chemical coagulation units for the dye house waste, experiments were conducted in the laboratory. The physico-chemical characters of the dye house waste are represented in Table 28.1. The percentage removal of suspended solids, volatile suspended solids and COD by the plain sedimentation procedure is represented in Tables 28.2–28.4. From these results the optimum detention period for removal of suspended solids, volatile suspended solids and COD was found to be 3 hours. After giving the detention period of 3 hours in the plain sedimentation, the percentage removal of total solids was 47.4 per cent, suspended solids was 87.4 per cent, COD was 19 per cent and the colour removal was 32 per cent. Using this detention period a sedimentation unit can be designed for the dye house waste. The plain sedimentation technique was found to be efficient in the removal of colloidal solids present in the dye house waste. But the color and COD were not satisfactorily removed by the plain sedimentation method. Hence, a coagulant was employed for subsequent treatment of the dye house waste. Storage reservoirs may also serve as sedimentation basins for treatment of effluent from the industrial sectors. However, they cannot affect proper sedimentation because of factors such as, density currents, turbulence caused by winds etc. Special basins therefore have to be constructed to purify the waste water by the process of sedimentation.

In the chemical treatment method, the optimum dosage of the coagulant was determined employing the standard jar tests. The results are represented in Tables 28.5–28.7. The optimum dosage of the coagulant was found to be 400 mg/l at which the percentage removal of total solids, suspended solids, and COD from the dye house waste was found to be 3.9 per cent, 29.7 per cent and 71 per cent respectively. This optimum dosage of the coagulant can be taken into consideration for the design of the chemical treatment plants.

Table 28.1: Physico-chemical Characteristics of the Dye House Waste

Sl.No.	*Parameters*	*Concentration*
1.	pH	5.13–6
2.	Acidity	150 mg/l
3.	Solids (mg/l)	
	(*a*) Total solids	3460
	(*b*) Suspended solids	1110
	(*c*) Dissolved solids	2350
4.	COD, (mg/l)	525
5.	BOD, (mg/l)	120
6.	Colour	
	(*a*) Luminance	94 per cent
	(*b*) Dominant wavelength	490–495
	(*c*) Hue	Blue

Table 28.2: Percentage Removal of Total Solids by Plain Sedimentation Treatment

Sl.No.	*Time in Min.*	*TS (mg/l)*	*Per cent Removal*
1.	Influent	3460	–
2.	30	2056	40.58
3.	120	1912	44.7
4.	180	1820	47.39
5.	240	1780	48.55
6.	360	1748	49.48

Table 28.3: Percentage Removal of Suspended Solids by Plain Sedimentation Treatment

Sl.No.	*Time in Min.*	*Suspended Solids (mg/l)*	*Per cent Removal*
1.	Influent	1110	–
2.	30	250	76.9
3.	120	168	84.86
4.	180	140	87.40
5.	240	132	88.10
6.	360	120	89.20

Table 28.4: Percentage Removal of COD by Plain Sedimentation Treatment

Sl.No.	*Time in Min.*	*COD (mg/l)*	*Per cent Removal*
1.	Influent	525	–
2.	30	460	12.38
3.	120	460	12.38
4.	180	428	19.04
5.	240	400	23.80
6.	360	400	23.80

Table 28.5: Optimum Dose of Alum for Removal of Total Solids from Dye House Waste by Chemical Treatment Method

Sl.No.	Alum Dose (mg/l)	Total Solids (mg/l)	Per cent Removal
1.	Influent	1888	–
2.	200	1840	2.5
3.	300	1838	2.6
4.	400	1814	3.9
5.	500	1812	4.0
6.	600	1803	4.5
7.	700	1798	4.8

Table 28.6: Optimum Dose of Alum for Removal of Suspended Solids from Dye House Waste by Chemical Treatment Method

Sl.No.	Alum Dose (mg/l)	Suspended Solids (mg/l)	Per cent Removal
1.	Influent	148	–
2.	200	120	18.91
3.	300	118	20.27
4.	400	104	29.73
5.	500	102	31.08
6.	600	98	33.78
7.	700	96	35.13

Table 28.7: Optimum Dose of Alum for Removal of COD from Dye House Waste by Chemical Treatment Method

Sl.No.	Alum Dose (mg/l)	COD (mg/l)	Per cent Removal
1.	Influent	450	–
2.	200	210	53.33
3.	300	190	57.77
4.	400	130	71.11
5.	500	120	73.33
6.	600	115	74.44
7.	700	110	75.55

This technique of treating waste water from the dye houses has been followed predominantly today. Due to the presence of several dyes, particularly reactive dyes, the biological treatment is often found less effective in the case of waste water from dye houses. Therefore, applicability of various physico-chemical treatment methods needs to be investigated in pursuit of an alternative to biological treatment of textile wastewater. A physico-chemical treatment scheme, involving chemical coagulation-sedimentation, dual media filtration, activated carbon adsorption followed by chemical oxidation serves to be effective (Pathe *et al.*, 2005). The results of this study have been employed for the set up of the treatment plants for the dye house based on the physico-chemical treatment scheme for the waste water.

References

_____1995. Pollution Prevention Strategy at an H-Acid Manufacturing Unit, *UNEP Industry and Environment*.

Pathe, P.P., Biswas, A.K., Rao, N.N. and Kaul, S.N., 2005. Physico-chemical treatment of wastewater from clusters of small scale cotton textile units. *Env. Tech.*, 26(3): 313–328(16).

Chapter 29

Preliminary Study on the Seasonal Distribution of Plankton in Irai River at Irai Dam Site, District Chandrapur, Maharashtra

A.P. Sawane[1], P.G. Puranik[2] and A.N. Lonkar[3]

[1]*Anand Niketan College, Warora*

[2]*Former Principal, Sevadal Mahila Mahavidyalaya, Nagpur*

[3]*Nutan Adarsh Mahavidyalaya, Umrer.*

ABSTRACT

The qualitative and quantitative estimate of Zooplanktons provides good indices of water quality and the capacity of water to sustain heterotrophic communities. Zooplanktons act as primary consumers and are important organisms in aquatic food chain. The study of seasonal distribution or fluctuations of zooplanktons have the greater importance. Ferneska and Lewkowein (1966) and Schindler and Noven (1971) have noted the enormous growth of rotifers in lake and reservoirs of Ontario. Important studies are also made by Prasadan (1971), Dalela *et al.* (1984), Biswas and Konar (2000). In the present study the water samples were collected from Irai dam site of Irai river. The zooplanktons noted are *Difflugia, Cyclops, Diaptomus, Chydorus, Moina, Brachionus calciflorus* and *Brachionus fulcatus* and *Cypris*. The quantitative zooplankton analysis shows that the total zooplankton density is more in the winter seasons and less at the end of summer. The rotifers are found more in number and Ostracodes are least. The results are discussed in relationship with the seasons and lentic condition of dam site of Irai river.

Keywords: *Zooplanktons, Seasons, Irai river, Irai dam site.*

Introduction

Zooplanktons are heterogeneous assemblage of minute floating animal forms found in water. Their qualitative and quantitative estimate provides good indices of water quality and the capacity of water to sustain heterotrophic communities. Zooplanktons usually act as primary consumers and constitute an important link between phytoplankton's (primary producers) and the consumers like fishes in aquatic food chain. The physico-chemical condition of water influences the distribution of Zooplanktons. The relevant studies on various aspects of zooplanktons are made by Ferneska and Lewkowiz (1966), Schindler and Noven (1971), Verma and Dalela (1975), Prasadam (1977), Nasar (1977), Penak (1978), Baker (1979), Jyoti and Sehagal (1979), Dalela *et al.* (1984), Balki *et al.* (1984), Laal and Karthikeyan (1993), Kumar (1994), Abdus and Altaff (1995), Edmondson (1996), Kaur *et al.* (1999), Biswas and Konar (2000), Pathak and Mudgal (2002), Shankar (2002) and Patra and Datta (2004). For fulfilling the water requirement of Chandrapur city and Thermal Power Station, the Irai dam have the greater importance. So considering the importance of water quality the study about the seasonal distribution of Zooplankton is attempted at Irai darn site of Irai river.

Materials and Methods

The water sample was collected from Irai river at Irai Dam site near the water supply pumping station (supplies water to Chandrapur thermal power station). The Irai dam having the water submerged area of about 5800 hectares and having the catchments area of about 500 km^2. The samples of zooplanktons were collected at an interval of 15 days in each month from September 2003 to August 2004. The water sample of maximum 40 liters was collected and was passed through the plankton collecting net made up of silk bolting cloth No. 25. The concentrated sample collected at the bottom tube of plankton net was preserved in 5 per cent formalin. The preserved sample was gently stirred to obtain the uniform suspension and with the help of wide mouth pipette the sample was quickly drawn and transported to the Sedgewick Rafter cell. The zooplanktons were counted in entire Sedgewick Rafter cell as per the methodology of Michael (1973) and Michael (1986).

Results and Discussion

The water sample for zooplankton analysis was collected from dam site of Irai river. The zooplankton noted are *Diffulgia* (protozoa), *Cyclops*, *Diaptomus* (copepoda), *Chydorus* and *Moiana* (cladocera), *Brachionus calciflorus* and *Brachionus fulcatus* (rotifera) and *Cypris* (ostracoda).

The quantitative zooplankton analysis shows that the total plankton density is more in winter season and less in rainy season. At this site the quantitative relationship amongst groups of zooplankton is rotifera > copepoda> protozoa > cladocera > ostracoda (Table 29.1).

Quantitative analysis shows the presence of larger numbers of rotifers at this station. Which may be due to lentic condition of water in Irai dam The zooplanktons noted at this station are rotifers, copepods, protozoans, cladocerans and ostracodes. Edmondson (1996) and Baker (1979) have observed that the high rotifer population in winter could be attributed with the favourable temperature and availability of abundance of food material in the form of bacteria, nanoplankton and suspended detritus. Hutchinson (196 7) observed that the Brachionus species are very common in temperate and tropical waters. In the present study maximum numbers of zooplanktons belongs to the group rotifera. The similar finding has been reported by Ferneska and Lewkowiez (1966) and Schindler and Noven (1971) they have noted the enormous growth of rotifers in lake and reservoirs at Ontario. Prasadam (1977) has reported the abundance of cladocerans during winter and summer months. Penak (1978) noted to the tolerance of wide range of ecological factors by ostracodes. Baker (1979) have studied the seasonal variation in zooplankton community and recorded the abundance of rotifers during winter due to availability of food and favourable temperature. Jyoti and Sehagal (1979) have recorded the most diversified species and rich rotifera group in reservoirs.

Table 29.1: Quantitative Analysis of Zooplanktons in Water Samples of Irai River at Dam Site of During the Year 2003–2004

Forms	*Winter (Oct, Nov, Dec, Jan)*	*Summer (Feb, Mar, Apr, May)*	*Rainy (Jun, Jul, Aug, Sep)*
Difflugia	36	30	26
Cyclop	22	24	19
Diaptomus	25	20	17
Moina	10	08	08
Brachionus calciflorus	36	31	28
Brachionus fulcatus	13	20	12
Cypris	6	14	29

Figures in table indicates total number of zooplanktons.

Datta *et al.* (1984) reported the abundance of zooplanktons during winter and summer months. Balki *et al.* (1984) studied the rotifera in reservoir and recorded the species diversity. Kumar (1994) has recorded the species diversity of rotifera in Pepara reservoir.

Laal and Karthikeyan (1993) have reported the maximum rotifers at polluted zone of different rivers. Abdus and Altaff (1995) have recorded the more quantity of zooplanktons during winter season. High population of rotifers in winter was attributed to favourable temperature and abundance of food (Edmondson 1996). Kaur *et al.* (1999) reported the diversity of zooplanktons in Kunj lake. Biswas and Konar (2000) have more rotifers at station of mixing zone of waters in Damodhar river water. Pathak and Mudgal (2002) have noted 19 species of zooplankton in Virla reservoir (M.P.). Shankar (2002) have studied the 300 planktons in freshwater bodies of Dharvar and reported the higher quantity of rotifers. In hydrological study of Monar river, the high dissolved oxygen content in water indicates the dominance of rotifera and copepoda, (Mone and Madlapure 2003). Patra and Datta (2004) have noted the seasonal fluctuation of zooplankton is governed by abiotic and biotic factors.

It is also noted that the protozoan species *Difflugia* is present at this site. Verma and Dalela (1975) have reported the presence of *difflugia* species and Arcella species in eutrophic water in Kalindi at Mansurpur.

In the present observation the rotifers are found in maximum numbers and ostracods are in less numbers. Similar observations are noted by Nasar (1977) in the freshwater pond at Bhagalpur the maximum number of rotifera and least number of ostracoda, the rotifera were the dominant group in the present study at this station. The quantity of zooplanktons was noted more in winter season may be due to static or lentic condition and favourable temperature of water in Irai Dam.

Acknowledgements

The authors are thankful to Dr. K.M. Kulkarni, Director of higher education, Pune for his help and for his precious guidance. The authors are also grateful to Dr. Raman, NEERI, Nagpur for his valuable suggestions. The authors express appreciation to Dr. Miss A.M. Bhate, Department of Zoology, Dharampeth Science College, Nagpur for her help during research work.

References

Abdus Saboor and Altaff, 1995. Qualitative and quantitative analysis of zooplankton population of tropical pond during summer and rainy season. *J. Ecobiol.*, 7(4): 269–275.

Baker, R.L., 1979. Specific status of *Keratella cochlearis* (Goose) and *K. Eurlinare* Ahlaofrom (Rotifera and Brachionidae) morphological and Ecological consideration. *Can J. Zoo.*, 57(9): 1719–1722.

Balki, M.H., Yosuf, A.R. and Quadri, M.Y., 1984. Rotifera of Anchor lake during summer and winter. *Geobios.* New Reports, 3: 163–165.

Biswas, B.K. and Konar, S.K., 2000. Influence of Nunia Nallah (canal) discharge on plankton abundance and diversity in the river Damodar at Narankuri (Rani Ganj) in West Bengal. *Indian J. Environ. and Ecoplan.*, 3: 209–217.

Edmondson, W.T. (Ed), 1996. *Freshwater Biology*. John Wiley and Sons Inc., New York.

Ferneska, M. and Lewkowiez, S., 1966. Zooplankton in pond in relation to certain chemical factory. *Acta. Hydrobiol.*, 8: 127–153.

Hutchinson, E.G., 1967. *A Treatise on Limnology,* Vol. II: Introduction to take biology and the limnoplankton. John Wiley and Sons, New York, p. 1115.

Jyoti, M.K. and Sehgal, M., 1979. Ecology of Rotifers of Surinsar-a-sub tropical freshwater lake in Jammu (Jammu and Kashmir) India. *Hydrobiology,* 27: 160–180.

Kaurh, K.S., Bath, G., Monder and Dhilon, S., 1999. Aquatic invertebrate diversity of Kunj lake, Punjab. *Indian J. Env. and Ecoplan.*, 2(1): 37–41.

Kumar, A., 1994. Periodicity and abundance of rotifers in relation to certain physico-chemical characteristics of two ecologically different wetland of Santal Pargana, Bihar. *Indian J. Ecol.*, 21: 54–59.

Laal, A.K. and Karthikeyan, M., 1993. Rotifer population or productivity indicators. *Current Science,* 5: 874–875.

Mone, A.M. and Madlapure, V.R., 2003. Hydrological studies on Manar River. Souvenir. Abstract, *National Conference on Management of Water Resources: Emerging Challenges Trends and Solutions,* Nanded, February, 21–22, pp. 71.

Michael, P., 1986. *Ecological Methods for Field and Laboratory Investigations.* Tata Mc-Hill Publishing Co. Ltd., pp. 131–145.

Michael, R.G., 1973. A guide to the study of freshwater organisms. *Journal Madurai Univ. Suppl.*, 1: 185.

Nasar, S.A.K., 1977. Investigation on the seasonal productivity of zooplankton in the freshwater pond in Bhagalpur, India. *Acta. Hydrochem. Hydrobiol.*, 5: 577–584.

Pathak, S.K. and Mudgal, L.K., 2002. A preliminary survey of zooplankton of Virla reservoir and Khargone (M.P.), India. *Indian J. Environment and Ecoplan.*, 6(2): 297–300.

Patra, S.B. and Datta, N.C., 2004. Seasonal fluctuations of different zooplanktonic group of a rainfed wetland in relation to some abiotic factors. *Indian J. Environ. and Ecoplan.*, 8(1): 7–12.

Pennak, R.W., 1978. *Freshwater Invertebrates of United States*, 2nd Edition. John Wiley and Sons, New York, pp. 803.

Prasadam, R.D., 1977. Observation of zooplankton population of some freshwater impoundments in Karnataka. In: Proc. Synop. Warm Wat. *Zooplanktons*, NIO 609, pp. 214–225.

Schindler, D.W. and Noven, B., 1971. Vertical distribution and seasonal abundance of zooplankton in two shallow lakes at Ontario. *J. Fish. Res. Bd., Can.*, 28: 245–256.

Shankar, P. Hosmani, 2002. Phytoplankton–zooplankton relationship in four fresh water bodies of Dharwar. *Indian J. Environ. Ecoplan.*, 6(1): 23–28.

Verma, S.R and Dalela, R.C., 1975. The pollution of Kalindi by industrial waste near Mansurpur, Part II: Biological index of population and biological characteristics of the river. *Acta. Hydrochem. Hydrobiol.*, 3: 256–214.

Welch, P.S., 1952. *Limnology*, 2nd Edition. Mc-Graw Hill Book Company Inc., New York, pp. 538.

Chapter 30

Studies on the Effect of Variation in Sweep Line Length of Bottom Trawls Over Fish Catch Along Mangalore Coast

Jaya Naik, B. Hanumantahppa, C.V. Raju and Shashidhar H. Badami

Department of Fishery Engineering, College of Fisheries,
Mangalore Karnataka Veterinary, Animal and Fisheries Sciences University, Bidar, India

ABSTRACT

Studies on comparative fishing was carried out to know the effect of sweep line, which is being used along Mangalore coast, and was divided into two categories *viz.*, 24–26 m and 26–28 m sweep line lengths, and were tested for a relative evaluation. Among these two sweep line 26–28 m was found to be more suitable, because of its wide horizontal mouth opening of net fishes herd into the direction of the net. Hence this category has resulted more catch, and it was the determental effect when compared to other category.

Keywords: *Bottom trawls, Trawl catchability, Trawl efficiency, Trawl performance, Selectivity.*

Introduction

In India a substantial portion of mechanized catch of fish and shellfish, which have very much demand. Hence, to make use of these resources we should chose a proper designed gear, selection of head rope, footrope, bridle and sweep line lengths. The status of marine capture Fisheries in India is in crucial phase now. During the earlier phase, the capture of Marine fisheries development and the fisheries resources remained rather underutilized, which is in the later phases, most of the resources have been either fully exploited or feared to be exploited. Consequently the present status of marine fisheries sector, which is thus for enjoyed free access to the resources is not prepared to face the stringent restrictive management.

The arrangement of sweep line lengths in trawl gear is an important factor, which influences the mouth opening of the trawl gear. As the mouth opening increased the amount of catch will also increases because of the swept area and herding effect. Crew (1964) reported that the importance of sweep line lengths to improve the performance of the trawl system by herding the fish into the direction of the trawl net. Mathai *et al.* (1984) and Rajan *et al.* (1990) undertook the studies on the optimization of sweep line lengths for the demersal trawls. Naik *et al.* (2004) conducted the survey on bridle lengths to know the effect of catch and found that a 19–21 m bridle length is more effective. Paine and Gruver (1996) have been found that the design of trawl bridle rigging, footrope attachment, alternative maintenance and their technique enables the net to fish slightly across the dwelling fishes. Sreekrishna (1995) reported that Mullets and to some extent prawns are known to have habits of jumping out of water, when encountered with obstacles and disturbed. Keeping all these views in picture attempts have been made to know the effect of HOBT with two different categories of sweep line lengths.

Materials and Methods

The experimental survey was carried out along Mangalore Coast wherein fishing operation were carried out by the private owners trawlers of Mangalore coast. Keeping the bridle lengths constant (19–21 m), sweep line lengths were divided in to two categories *i.e.* 24–26 m and 26–28 m, which is made up of 12 mm dia polythene ropes. The lengths of sweep line were slightly shorter than that of head rope and footrope. The rigging of the sweep line is in between the two otter boards. Each day these two categories of sweep line lengths with a regular rotation of the net, otter boards, floats, sinkers and sweep line lengths have been collected each day for these two-surveyed categories, in each month at least 10 boats and maximum of 50 boats were chosen for the study and based on information collected, which has been divided and catch has been averaged for their respective categories and found the more suitable category of sweep line lengths to catch the demersal fishes.

Results and Discussion

Results of investigation with respect to two categories 24–26 m and 26–28 m of sweep line lengths are given in Table 30.1 and 30.2 respectively. The total catch in case of 26–28 m sweep line length was 57,025 kg and it was 59,348 kgs for 26–28 m sweep line length during the fishing season September 2001 to May 2002. From the two Tables it can be observed that, the sweep line length having 26–28 m yielded higher catch when compared to 24–26 m sweep line length because of its vertical mouth opening and its swept area. Not much information is available on this aspect of getting more catch by increasing or decreasing sweep line lengths of the net. (Not much work has been done on this aspect to get a higher catch by increasing the Vertical mouth opening of the net.

A higher catch was observed in both the sweep line net during colder season, especially in the month of January and December. This may be due to the region that the fish would have moved to bottom region because of the higher temperature existing there. This is also proved from the Table that there was a lower catch during the month of April, it could be due to higher temperature existing on the surface.

Among all the species caught by both the categories of sweep line Pink perch contributed highest catch in 24–26 m sweep line length followed by 26–28 m sweep line lengths. And the region being Pink perch is a bottom dwelling fish. The prawn catch was higher in case of 24–26 m sweep line length followed by 26–28 m sweep line lengths of the net. As squilla enters the Net while hauling and it is a mid water dwelling by-catch.

Table 30.1: Monthly Average Catch (kg) per Boat Obtained with 24–26 m Sweep Line During September 2001–2002

Sl.No.	*Species*	*Months*									
		Sept.	*Oct.*	*Nov*	*Dec.*	*Jan.*	*Feb.*	*Mar.*	*Apr.*	*May*	*Total*
1.	Sharks	208	56	64	112	88	25	55	208	108	924
2.	Seer fish	–	–	695	670	560	–	450	–	508	2883
3.	Tuna	75	485	475	230	235	130	28	75	220	1953
4	Lactarius	–	75	25	220	70	75	206	–	105	776
5	Carangids	10	40	75	125	27	10	13	10	23	333
6.	Pomfrets	100	70	250	270	230	–	–	100	123	1143
7	Silver bellies	20	79	333	347	70	80	85	20	321	1355
8.	Sciaenids	35	620	–	50	48	–	22	35	480	1290
9.	Ribbonfish	110	910	650	330	220	1000	950	110	890	5170
10.	Flat fish	–	300	50	60	55	300	35	–	200	1000
11.	Soles	–	250	155	145	125	75	25	–	725	1500
12.	Pink perch	220	1050	2050	6400	6600	550	200	220	1200	18490
13.	Lizard fish	–	85	55	235	265	100	110	–	115	965
14.	Crabs	100	85	30	140	160	150	100	100	85	950
15.	Prawns	300	693	850	820	550	680	720	300	780	5693
16.	Squilla	1000	1600	2000	1500	1600	1600	800	1000	1500	12600
	Total	**2178**	**6398**	**7757**	**11654**	**10903**	**4775**	**3799**	**2178**	**7383**	**57025**

Table 30.2: Monthly Average Catch (kg) per Boat Obtained with 26–28 m Sweep Line During September 2001–2002

Sl.No.	*Species*	*Months*									
		Sept.	*Oct.*	*Nov*	*Dec.*	*Jan.*	*Feb.*	*Mar.*	*Apr.*	*May*	*Total*
1.	Sharks	104	156	34	100	55	25	155	212	218	1059
2.	Seer fish	200	400	395	695	650	305	455	308	285	3693
3.	Tuna	95	585	575	230	180	128	175	225	300	2493
4.	Lactarius	220	–	75	25	80	95	196	130	110	931
5.	Carangids	45	10	85	140	30	35	25	100	123	593
6	Pomfrets	80	90	220	230	220	30	40	75	85	1070
7.	Silver bellies	60	59	210	245	170	180	65	85	300	1374
8.	Sciaenids	55	420	200	150	32	–	42	55	165	1119
9.	Ribbonfish	120	1050	900	270	340	660	920	130	240	4630
10.	Flat fish	–	200	120	90	45	280	55	75	180	1045
11.	Soles	250	–	135	165	130	35	65	625	150	1555
12.	Pink perch	1000	1220	1100	6700	6300	320	450	325	185	17600
13.	Lizard fish	55	85	–	240	270	200	220	550	700	2320
14.	Crabs	140	35	85	100	280	70	100	86	120	1016
15.	Prawns	660	170	830	860	650	780	720	780	300	5750
16.	Squilla	1500	1750	1800	1750	1600	900	800	2000	1000	13100
	Total	**4584**	**6230**	**6764**	**11990**	**11032**	**4043**	**4483**	**5761**	**4461**	**59348**

Acknowledgement

I am grateful to the Director of Instruction (Fisheries) for their help and cooperation during the period of our study and also wish to thanks who have helped directly or indirectly.

References

Crew, P.R., 1964. Some general engineering principles of trawl gear design. In: *Modern Fishing Gear of the World*. Fishing News (Book) Ltd., London, 2: 165–181.

Jaya Naik, Hanumanthappa, B., Prabhu, R.M and Varadaraju, S., 2004. Conducted survey studies on the effect of bridle lengths to find an effective category. In: *Env. and Ecol.*, 22(4): 787–790.

Mathai, T.J., Abbas, M.S. and Mahalathkar, H.N., 1984. Towards optimization of bridle lengths in bottom trawls. *Fish. Technol.*, 21(2): 106–108.

Paine and Gruver, 1996. By catch reduction of the solving by catch workshop September 25–27, 1995, Seattle Washington, USA Alaska Sea Grant College Programme, pp. 87–88.

Rajan, K.C.M., Boopendranath, M.R., Mathai, P.G and Pillai, N., 1990. Studies on the optimization of bridle lengths for demersal trawls. *Fish. Technol.*, 272: 87–91.

Sreekrishna, Y., 1995. Factors affecting fish behaviour. In: *Observation and Ongoing Research on Fish Behaviour and Fish Relation to Fishing Gear and Stimulii, India*. CIFE (Books) Ltd., pp. 11.

Chapter 31

Plant-lore with Reference to Manipuri Proverbs in Association with Various Human Affairs of Manipur State

M.M. Ahmed and P.K. Singh*

Ethnobotany and Plant Physiology Laboratory, Department of Life Sciences, Manipur University, Canchipur, Imphal – 795 003, Manipur (India)

ABSTRACT

This article deals with 21 indigenous plants under 18 families which are commonly referred in 20 Manipuri proverbs of Manipur state. The collection of research data was based on a survey programme conducted during February 2005 to September 2005 by using questionnaire and oral contact of 520 persons of Manipuri community. The present paper aims at the conservation steps of indigenous plants by bringing the clue of plant lore with human affairs. Negligence of the plants by the people of Manipur is known from the fact that the relevance of proverbs use with plants to this present day has been decreased.

Keywords: Proverbs, Manipuri, Plants, Conservation.

Introduction

Inhabitants of Manipur, a small state in the North-east India, speak Manipuri language (a Tibeto-Burman language). But majority of tribal in the hill districts has their own language seeks to speak

* Corresponding Author: E-mail: mustaque.ahmed@rediffmail.com.

Manipuri as their *lingua franca. Pourou* is known for proverbs in Manipuri. In Manipur, there is a way of speaking, talking with proverbs having reference to plants. The study of proverbs helps us to understand the mind of the people. The recitation of a proverb by the eldest member of a family is intended to warn his or her subordinates (Singh, 1993). Both in the valley and hill, people equally use proverb with a great taste that they started signifying plants with human nature, human way of life etc. The study of proverbs with reference to plants includes both the cultivated plants and the wild edible plants. It is indicative of the fact that people have utilized the plants at their surrounding and their usefulness to them is highly recorded. They never forget the plants helping them in different ways especially for food and shelter. Use of proverbs in this language has been transferred from generation to generation. To this, Rao and Hajra (1995) noted, "folklores specially folksongs, proverbs and tales have references to certain interesting properties on aspects of plants". Thus plants have a significant place in the spoken Manipuri language, which is shown in this report for the first time from Manipur. A similar kind of Bio-folklore research was reported by Singh *et al.* (1996), Singh and Singh (1996).

Methods

Use of proverbs along with the plants in the spoken Manipuri language in Manipur is taken as a special case study in the present research. For this purpose, the proverbs enlisted under the enumeration have been collected from a literature survey (Kuber, 1998; Singh, 1995). For personal and oral contact, a list of questionnaire was prepared based on the questions of whether the proverbs having reference to plants, distribution of plants, uses of plants etc. are relevant in day to day life. The test was conducted during a field study from February 2005 to September 2005. The present method of study is adopted by the methods of Rao and Hajra (1992) and Mohanty (2004). Among the informants of both the sexes totaling 520 persons include 180 old aged (age above 45), 201 middle aged (age below 45) and 139 youths (age from 15–25). Of the nine districts of Manipur, one hill district (Chandel) and three valley districts (Imphal east, Imphal west, and Thoubal) were randomly taken for the study. During the survey, plants have been collected and they were properly identified, classified with the help of Lawrence (1974), Pandey (1997) and Sinha (1986). Plants were deposited in the herbarium of Manipur University.

Enumeration

Twenty folk proverbs relating to plants in the comparison of human nature, character and activities are presented in spoken Manipuri language with their transliteration followed by word meaning, scientific name of plants, family, vernacular name and explanation.

1. **Ahing amma tumdaraga *Heitup* masung amatasu phang-ee**
   (Ahing ama-one night, tumdaraga-not sleeping, *Heitup-edible* fruit, masung-piece, amatasu-even if one, phang-ee-receive or achieve)
   ***Grewia microcos* Linn., Fam-Tiliaceae, Vern-*Heitup*.**

A man receives a piece of *Heitup* for not sleeping a night. Big or small there is a good result for a laborious person. *Heitup* is regarded a precious plant because it appears once in rice harvesting time. The smallest reward of a laborious person compares with a piece of a *Heitup.* Human labour does not go waste. *Heitup* is wild edible fruit and it is found in the hills. Usually it is eaten with salt and chilly powder.

2. Angao *Phou* shu chara nongkannei

(Angao-seemingly mad, *Phou* shu-grinding paddy grain, chara-one meal, nongkhannei-feels stomach)

Oryza sativa* Linn.; Fam-Poaceae; Vern-*Phou

Grinding of paddy grain in a seemingly mad way yields a days meal. A work of an unexperienced person may be useful. It compares with a child cooking and feeling the stomach of elders. A man is borne to carry out responsibilities. Any result of work may be bad, even though, always there is a reward. In every Manipur houses wooden mortar and wooden pestle were used for separation of grains from chaff. The process takes time in getting rice grains. Paddy is cultivated throughout the state.

3. *Ee* nana yatpa khayat-tasu shangom mapanga chiniga chai laina yatpa khayatpu chara henba hounabra

(*Ee* nana-with reed leaf, yatpa-carved or marked, khayattashu-to such a mouth, shangom mapanga-with cream milk, chiniga-also with sugar, chai-eat, laina-by God, yatpa-marked, khayatpu-also such mouth, chara-one meal, henba-starving, hounade-impossible)

Imperata cylindrica* Linn.; Fam-Poaceae; Vern-*Ee.

The theme of this proverb is similar to "A mouth marked by a reed leaf eats cream milk and sugar, how could a mouth created by God remain starved". If a man is not lazy, he should not starve. Leaf of reed plant is used in the roofing of houses. This plant is found both in valley and in hills.

4. Ekai khangdagi *Kang* kotli

(Ekai-blushing face, khangdagi-to bear with, *Kang*-free floating plant, kotli-to pick up)

Riccia natans* Corda.; Fam-Riciacea; Vern-*Kang

In a pond, a lady was taking bath with her torn clothes. By then, some men arrived near the pond. Unable to bear with this blushful incident the lady pretends herself to do something else as she splashed water on her face. It can be compared with the lady in trying to pick up *Kang* from inside the pond. *Kang* is used as duck food.

5. Erujadana *Heibi* teiba

(Erujadana-without bathing, *Heibi*-edible fruit plant, teiba-smearing or rubbing)

Vanguiria spinosa* Hook. f.; Fam-Flacouritaecae; Vern-*Heibi

We often listen to elders speaking, "You do not take bath. Bad smell is coming out from your hairs. Even though you put oil on your hair". Also, it compares with a student taking exam without preparation. In the ancient time, before coming of modern lotion into the market, people in Manipur used extract of *Heibi* to control dry skin. The fruit is edible and the plant is found both in valley and in hills.

6. Eshing lui lude *Tharo* marida khang-ee

(Eshig-water, lui-deep, lude-shallow, *Tharo*-water lily, marida-to petiole, khang-ee-understand)

Nelumbo pubescens* Willd.; Fam-Nymphaceae; Vern-*Tharo

A man is boastful of being prestigious, benevolent etc. But others disagree with him. It can be compared with, "A man is known by his deeds or companions". Many people in the valley earn their living around the lake. Deep or shallow of a lake is indicated by the length of *Tharo* petiole. It gives a prior warning to a fisherman working in the lake. Usually, *Tharo* grows in the lakes. Leaves, stem and roots are eaten.

7. *Heinoujom* yumbal telanga yahip
(*Heinoujom-carambola* fruit, yumbal-way of living, telanga-kite, yahip-way of sleeping)
Averrhoea carambola* Linn.; Fam-Averrhoaceae; *Vern-Heinoujom.

A house not in the order is compared with this proverb. Because a fruit of carambola is very delicate that it gets rotten very soon. Most of the time, the fruit is not properly harvested. Falling on the ground wastes such fruits. This is compared with careless human way of living. The fruit of carambola is very sour in taste. Mostly the plant is found wild in the hills and it is grown near the houses in valley.

8. *Kwa* matap amana chin pei
(*Kwa* matap-areca nut and betel leaf with lime over the leaf, amana-one, chin-mouth, pei-curly or bent)
***Areca catechu* Linn.; Fam-Arecaceae; Vern.-*Kwa* maru**
***Piper betle* Linn. Var-Siriboa; Fam-Piperaceae; Vern-*Kwa* mana**

Some one is completely satisfied even by 'giving' a *Kwa-matap* (a piece of Areca nut and a Betel leaf with lime smearing over leaf). The word 'giving' signifies friendly or nearness attitudes shown to each other in the personal relationship. Young and old alike use to eat this composition with great delight. *Areca catechu* is cultivated in the valley area of Manipur.

9. *Laphoi* charingeida khangdraga *Thambou* charakpada khang-ee
(*Laphoi*-banana, charingeida-while eating, khangdraga-not knowing, *Thambou*-rhizome of water lily, charakpada-on eatng, khang-ee-come to public)
Musa paradisiaca* Linn.; Fam-Musaceae; Vern-*Laphu

Secret deed at the beginning is not observed. Later on it comes to know by its consequences. For example, in the sacred, a man and woman have established physical relationship. No one knew until pregnancy was confirmed by relatives. It is compared to eating of plantain fruit producing no sounds but there is much sound while eating rhizome of water lily. *Laphu* is a cultivated plant.

10. Lawai *Langthrei* khongnembi thoudok khuding yaoganbi
(lawai-village, *Langthrei*-a plant of cultural importance, thoudok-occassions, khuding-each, yaoganbi-much wanted)
Eupatorium birmanicum* DC.; Fam-Asteraceae; Vern-*Langthrei

Like a kind of man who is involved / wanted in any matter of Leikai (a suburb of 30–40 houses), *Langthrei* is involved in many cultural feast of Manipur. The plant is found in valley.

11. Manana *Heimang* shaba
(Manana-with leaf, *Heimang*-edible fruits of varnish tree, shaba-substitutes)
Rhus simialata* Murray.; Fam-Anacardaceae; Vern-*Heimang.

In the proverb, the two things differ in taste are compared. Leaf of *Heimang* is bitter in taste while fruit is sour in taste. Leaves pass for the *Heimang* fruit in the cultural occasions. The proverb is used in a sense of mock. It is like a stepmother taking place of a real mother. *Heimang* is wild edible fruit found in the hills.

12. Mangda *Mangra* tanba
(Mangda-in dreams, *Mangra*-sweet potato, tanba-to dig)
Ipomoea batatas* Linn.; Fam-Covolvulaceae; Vern-*Mangra

Harvesting of sweet potato is possible only in the reality but not in the dreams. Sweet potato is compared to someone's goal. Chasing after this goal in the dream is impossible. It is just like a blind man hoping to climb Mount Everest. *Mangra* is a cultivated plant common in the hills.

13. *Morok* metpada jat taba
(*Morok-chilly*, metpada-to prepare a local chatani, jat taba-becoming low caste)
***Capsicum indicum* Linn.; Fam-Solanaceae; Vern-*Morok*.**

Someone maintaining frugality in the preparation of *Morok metpa* deems to make his family standard down. *Morok metpa* is a delicious local dish prepared from chilly, fermented fish and other vegetable spices. It is prepared everyday in every Manipur houses in one or the other meal. Chilly is a common cultivated plant.

14. Okchinda *Pan* thaba
(Okchinda-to a mouth of pig, *Pan*-giant taro, thaba-cultivation)
***Alocasia indica* Linn.; Fam-Araceae; Vern-*Pan*.**

The mouth of a Pig is improper place for growing *Alocasia*. It can be compared with a lady growing cauliflower near grazing field. Growing plant in an improper place is similarly expressed to a condition of growing *Alocasia* in the mouth of pig. *Pan* is a commonly cultivated plant in the hills and valley.

15. Ü-leitaba lamda *Kege* na yumbi öi.
(Ü-tree, leitaba-not available, lamda-to a country, *Kege* na-with castor plant, yumbi-house post, öi-substitute)
Ricinus communis* Linn.; Fam-Euphorbiaceae, Vern-*Kege

This can be compared with the proverb, "In the land of the blind one eyed is king. In a treeless country the castor plant passes for a tree". Castor plant can't use in building a house, but it is found to be useful in a tree less country. People started cultivating this plant.

16. Shagol mamang punjao pubi nungshit mairam *Lashing* kappi
(sagol-horse, mamang-in front, punjao-big earthen pot, pubi-she carries, nungshit-wind, maram-way, *Lashing*-cotton, kappi-processing of cotton fibers)
***Gossypium arboreum* Linn; Fam-Malvaceae; Vern-*Lashing*.**

Doing any work in the improper time is expressed in the proverb. A horse might kick breaking the water pot carrying in its front, while cotton being light might flew by the wind if it is prepared against the wind. A girl, in a rainy day, walks out to fetch a pot of water from courtyard pond. She was so hurry that she broke her pot. This example entails the literal meaning of the proverb. Cotton is a cultivated plant. Earlier the plant was found plenty in the hills.

17. Samu maya thindorakpa *Mange* thenguna yeishinba yade
(Shamu-elephant, maya-trunk, thindorakpa-stretching, Mange-tamarind tree, thenguna-with hammer, yeishinba-recoil, yade-not allowed)
Tamarindus indica* Linn.; Fam-Caesalpinaceae; Vern-*Mange

A hammer of tamarind tree is very hard. Even though, it is impossible to block elephant trunk at its inception by using *Mange* hammer. This can be compared with, "The hammer has nothing to do with the stretching of the elephant's trunk". Likewise, a person of merit can't be easily suppressed. *Mange* is wild edible fruit found only in the hills.

18. Sanagi mahut *Yaingang* na shili
(Sanagi-of gold, mahut-in place of, *Yaingang* na-with turmeric, shilli-replaces)
***Curcuma domestica* Valeton; Fam-Zingiberaceae; Vern-*Yaingang*.**

Gold is replaced by turmeric. For example, fragrant fruits like lemon is required for offering to diety. It is not the season of lemon. In such a' situation, leaves substitute the fruit. Turmeric is gold in colour, although their value is different. Turmeric is cultivated both in the valley and hills.

19. Yongna *Chu* konbagum konba
(Yong-na-by a Monkey, *Chu*-Sugar cane, konbagum-if very fondly) holding, konba-embrace)
Saccharum officinarum* Linn., Fam-Poaceae; Vern-*Chu

If sugarcane is once hold by a monkey, it is not likely to leave it. A monkey is very fond of sugarcane like a child does to a doll. Sugarcane is a cultivated plant. A kind of sweet syrup called *chuhi* is prepared from sugarcane juice.

20. *Waton* na wanglaga kwakna phamdek-ee
(*Waton* na-top of bamboo, wanglaga-being tall, kwakna-by a crow, phamdekee-lowers by sitting)
***Bambusa* sp.; Fam-Poaceae; Vern-*Wa*.**

If a bamboo is too tall, a crow sits on it. For a man of too much pride, there always comes an able man to defeat him. Rabon of Lanka once thought that he was the ablest person on earth. Later on, his Lanka was turned to ashes. By sitting a crow on the top of a bamboo, the height of the latter is decreased. Bamboo is a common plant in Manipur. Bamboo shoot after fermentation is eaten.

Discussion

"Many plant names appear in similies and metaphors in Hindi literature, *e.g.*, a girl breaks away from her family after marriage like the tender stem of castor", as is noted by Agrawal (1997). This came to know from the present study that the plant names appear in Manipuri proverbs. The proverbs enumerated above are still prevailed in the spoken language of Manipur.

Of all the persons interviewed in different age group categories, 130 out of 180 old persons, 108 out of 201 middle aged persons, and 25 out 139 youths were found to know all the proverbs enumerated above. The result shows a decreasing trend in the use of such proverbs in the spoken Manipuri language. Crockett *et al.* (1977) noted of a popular Herbal *An Elizabethan Herbal* by Culpeper (1649) wherein it is written that "lovers could do conversation in the language of flowers". In the same vein, from the present study it is interesting to note that person maintains much frugality is known by the way his family prepares *Morok Metpa* (a common delicious preparation of chilly in every house, otherwise known as chatani). In the ancient time *Heibi* (*Vanguiria spinosa*) extract was used as skin lotion. In the occasional personal contact, 'giving and eating' of *Kwa mana* and *Kwa* (Betel leaf and *Areca* nut) symbolises friendship and to this act personal satisfaction is shown. *Heitup* (*Grewia microcos*) is regarded a precious fruit as a piece of this fruit compares to 'a reward of human labour'. Leave aside the side effects to a feverish man, always there is temptation to eat fermented bamboo shoot as it can be seen in another proverb" Arum apumba *soijin* ahaoba" which means persistent fever tasty *soijin* (fermented shoot of Bamboo). *Langthrei* (*Eupatorium birmanicum*) has been found in many a cultural feast in Manipur, therefore, a person of many activities is compared to this plant.

To give a special attention to some plants such as *Heitup*, *Mange* (*Tamarindus indica*), *Heimang* (*Rhus simialata*), they are wild edible fruits and found only in the hills. These plants along with turmeric (*Curcuma domestica*), gaint taro (*Alocasia indica*), bamboo (*Bambusa* sp.), chilly (*Capsicum indicum*) cotton (*Gossypium arboreum*) etc. have potential economic value in the markets. In today's world, people started neglecting useful plants at our surrounding. This has been inferred with the fact that the relevance of plants use in the proverbs has been decreased. It is right " time that the people of different fields wake up to this reality and take up some awareness programmes on the importance of plants and their conservation. In the future, people will not believe to the truth that our ancestors after their long experiments with the plants selectively and purposefully used to speak plants in their spoken language by indicating various human affairs.

Acknowledgement

The authors are thankful to the Head, Department of Life Sciences, Manipur University for facilities and to research scholars working in the Ethnobotany and plant physiology laboratory for their help.

References

Agrawal, S.R., 1997. Trees, flowers and fruits in Indian folksongs, folk proverbs and folk tales. In: *Indian Ethnobotany*, 3rd edn. (Ed) S.K. Jain. Scientific Publishers, Jodhpur (India).

Crockett, J.U. and Tanner, O., 1997. *Herbs*. Time Life Books, Alexandria, Virginia, pp. 16–17.

Kuber, C., 1998. *Meitei pourou amasung chatnabi londa*. Imphal.

Lawrence, H.M., 1974. *Taxonomy of Vascular Plants*. Oxford & IBH Publishing Co., New Delhi.

Mohanty, R.B., 2004. Traditional wisdom on livestock selection and Management in folk proverbs of Orissa. *Indian Journal of Traditional Knowledge*, 3(1): 92–95.

Pandey, B.P., 1997. *Taxonomy of Angiosperms*. S. Chand and Company Limited, New Delhi.

Rao, R.R. and Hajra, P.K., 1995. Methods of Research in Ethnobotany. In: *A Manual of Ethnobotany*, (Ed.) S.K. Jain. Scientific Publishers, Jodhpur, India, p. 28–34.

Singh, H.B., Singh, P.K., Singh, S.S. and Elangbam, B., 1996. Indigenous bio-folklores and practices: Its role in biodiversity conservation in Manipur. *J. of Hill Res.*, 9(2): 359–362.

Singh, M.K., 1993. *Folk Culture of Manipur*. Manas Publication, Delhi, pp. 87.

Singh, P.K. and Singh, H.B.K., 1996. Superstition in botanical folklore with reference to Meitei culture. In: *Contribution to Indian Ethnobotany*, (Ed) S.K. Jain. Scientific Publishers, Jodhpur, India, p. 367–372.

Singh, S.A., 1985. *Ahallaman pourou praman panthei*. Imphal.

Sinha, S.C., 1986. Ethnobotanical study of Manipur. *Ph.D. Thesis*, Manipur University.

Chapter 32

Microbial Changes During the Fermentation of Sun Dried *Puntius sophore*

Ch. Sarojnalini and T. Suchitra

Department of Life Sciences, Manipur University, Canchipur – 795 003, Manipur, India

ABSTRACT

Variations in the microbial count during the fermentation of sun dried *Puntius sophore* at an interval of 30 days were analysed. Total bacterial and fungal count reached upto 10^5 and 10^3 cfu/g respectively. *Bacillus* and *Micrococcus* spp were found to be dominant amongs the bacteria identified. *Bacillus cereus* was detected only in unfermented fishes but absent in fermented fishes. *E. coli* and *Salmonella* were not detected in all the samples analysed. *Aspergillus, Penicillium, Gliocladium* and *Humicola* were the fungus isolated during the fermentation. The results shows that several microflora with varying percentage were found to be associated with the fermentation of sun dried *P. sophore* and they might also play an important role during fermentation.

Keywords: Sun dried Puntius sophore, Fermentation, Bacteria, Fungus.

Introduction

Fermented foods stuff and condiments remain a key constituents of diets throughout many parts of the World. Many fermented fish products are prepared in different parts of the world and the methods of processing depend upon various factors (Aljedah, 2002). In Manipur, the fermentation of sun dried *P. sophore* by microorganisms naturally to produce a fermented fish product 'Ngari' is very popular. Another fermented fish product includes 'Hentak' which is prepared using powdered sun dried *Esomus danricus* along with aroid plants to form a paste (Sarojnalini and Vishwanath, 1988).

The process of fermentation are often influence by the co-existence of different kinds of microorganisms. Such co-operation of microorganisms are also apparent in the process of production and ripening of a large variety of local and unique fermented food products in South east Asia. Microorganisms involves in fermentation include filamentous fungi, yeasts and bacteria in succession or in combination (Djien, 1982; Steinkraus, 1983). The kinds of microorganisms used in traditional fermentation are natural microflora (Sandhu and Soni, 1989) and are restricted to a relatively few genera including *Aspergillus, Rhizopus, Mucor, Actinobacter, Bacillus* and *Lactobacillus* (Hesseltine and Wang, 1960).

The quality of fish products is judged by its microbiological characteristics. Microorganisms, which occur invariably in nature, are responsible for spoilage of food materials. Growths of unwanted microbes make food unfit for human consumption and add undesirable flavors, appearance and finally the food value is destroyed. Microbial contamination in processed foods is a cause of concern to the health of the consumers (Ramachandra Rao, 1980). The importance of microorganisms in food poisoning outbreak cannot be ignored. Brown *et al.*, 1958, reports, *Bacillus cereus* as a significant source of both human and animal infection. This spore forming bacteria is also a source of food poisoning (Taylor and Gilbert, 1975). Barber and Deibel (1973) reported the incidence of Staphylococcal food poisoning associated with fermented food causing gastro enteritis in human. Determination of total plate count (TPC) and the *Staphylococci* count, are widely accepted parameters in inspection of fish foods (Anon, 1964). Determination of faecal contamination indicator organisms is also important in assessing the quality of fishery products. Their presence indicated the presence of enteric pathogens (Shewan 1962; Frazier and Westhoff, 1983).

The present study was aimed at finding out the different types of microorganisms associated with the fermented fish 'Ngari' prepared in the laboratory to study the microbial count changes during the period of six months of fermentation.

Materials and Methods

Samples of unfermented fishes (sun dried *Puntius sophore*) were procured from the Imphal Market of Manipur.

Preparation of Ngari

Ngari was prepared as according to Sarojnalini and Vishwanath (1988). This indigenous fermented fish is prepared from small and uneconomic sun dried fishes such as *Puntius sophore* (Ham) and *P. ticto* (Ham), allowing to fermentation without salt for five to six months. The process involves a brief washing of the fishes and left dry for 24–48 hrs. The head and bones were pressed hard using stone roller. Mustard oil was applied to the inner wall of the earthened pots known as 'kharung' to check porosity. The fishes were mechanically pressed hard inside the pots using wooden stick. The pots were then sealed airtight and stored at room temperature for five to six month. In laboratory, six small pots in triplicate were prepared. Observations were made at an interval for every 30 days.

Methods

Enumeration of total viable counts of bacteria and fungi were determined by the dilution plate method of APHA (1976) using Trypton glucose agar medium and Potato dextrose agar respectively. The colonies present were counted and expressed as colony forming unit per gram of samples. The isolated fungi were stained with lactophenol cotton blue and the species were identified referring to Gilman (1957) and Ellis (1971, 1976).

Most probable count for coliform and detection of bacteria *viz., Escheria coli, Staphylococcus* and *Salmonella* and faecal *Streptococci* and *Bacillus cereus* were also tested as per the method of APHA (1976).

Coliform counts were determined by incubating diluted sample in brilliant green lactose broth using Durham's tubes. Positive tubes were counted and coliform percentage was expressed as most probable number (MPN) per gram of sample. *E. coli* was determined using brilliant green lactose broth Eosine Methylene blue. Culture from coliforms tubes were streaks across the surface of Eosine methylene blue (EMB) agar and incubated for 48 hrs at 38°C. The colonies on EMB medium were tested using the methods of APHA (1976) and Kiss (1984). For determination of *Staphylococci*, samples was inoculated in Baired Parker Agar medium containing egg yolk solution and 1.0 per cent potassium tellurite and incubated 35–38°C for 48 hrs to obtain an elevated small black colonies with surrounding white zone. *Salmonella* was determined by using Bismuth sulphite agar medium after enriching the samples with selenite cystine broth. Faecal *Streptococci* was determined using KF agar enriched with trichlorotetrazolium chloride (TTC). *Bacillus cereus* was estimated using Polymixin Pyruvate Egg Yolk Mannitol Bromothymol Blue Agar (PEMBA) medium. Bacterial colonies were isolated and identified based on Buchanan and Gibbons (1954) and APHA (1976) and finally confirmed by the Institute of Microbial Technology (IMTECH) Chandigarh and Bioscience U.K centre, Egham.

Results and Discussion

The total plate count of unfermented *Puntius* was ranged from 10^4–10^6 cfu/g as shown in Table 31.1. During aging, the total population of microorganisms increased during the early stage of fermentation and then decreased. Even though the processed fishes such as SW1 dried, salted, fermented and smoked fish owe their long history to the antibacterial effect, bacterial count was still high. The total plate counts of bacteria at 30 days was 8.2×10^5 cfu/g. Maximum bacterial count was observed at 60 days with a value of 4.38×10^6 cfu/g and slightly declined from 90 days and remain constant at the final stage of fermentation. Increase in TPC was reported in various kinds of fermented fish. This observation was in general, in agreement with those reported by Yang and Chung (1995), Ijong and Ohta (1996), Syaefudin *et al.* (1992) in various fermented fish. This finding was also similar to the finding of Aljedah *et al.* (1999) in which a value of 1.5×10^6 cfu/ml was declined to 1.9×10^5 at six months of fermentation in mehiawah, a fish sauce in the Gulf countries Chou *et al.*, 1988 noticed that during the aging period of fermented condiment, the yeast count gradually increased as aging progressed.

Normally food that are produced, ripened or fermented by the action of bacteria will yield high total count (Hall *et al.*, 1967). *Staphylococci* count was maximum at 90 days with a value of 1.56×10^6 and declined slowly to 2.50×10^5 at the final stage. Red blood colonies of faecal *Streptococci* count reached upto 1.23×10^6 at 60 days and reduced to 2.50×10^5 at the end. The *Staphylococcal* count lies within the acceptable limit, as its value was less than 10^8/g, which is considered as unfit for food as according to Almas (1981). The microorganisms in the latter stage of fermentation degrades fish protein leading to the production of volatile compounds from amino acids and small peptides (Lopetchart and Park, 2002). In fermented fish, the growth of bacteria assists in the breakdown of fish muscles, which is necessary for the development of characteristics odours, flavors and textures (Beddows, 1985). The absence of large number of microorganisms associated with the absence of characteristics organoleptic attributes to the products (Sillikar *et al.*, 1963). Hence the increased in microorganisms might be responsible for the development of characteristics taste.

Table 32.1: Microbial Count Changes During the Fermentation of Sun Dried *Puntius sophore*

No. of Days		Unfermented	30 Days	60 Days	90 Days	120 Days	150 Days	180 Days
TPC (Bacteria)	Range	3.2×10^4–5.6×10^7	7.0×10^5–8.6×10^6	1.9×10^6–8.0×10^6	7.0×10^5–8.2×10^6	1.8×10^6–2.9×10^6	4.1×10^5–9.2×10^5	8.5×10^5–9.4×10^5
	Mean	2.53×10^6	8.2×10^5	4.38×10^6	3.30×10^6	2.31×10^6	7.68×10^5	8.92×10^5
TPC (Fungi)	Range	2.0×10^3–9.0×10^4	2.0×10^4–9.1×10^4	3.5×10^4–6.6×10^4	2.0×10^4–3.6×10^4	1.0×10^3–6.0×10^3	1.30×10^2–5.1×10^3	5.0×10^3–1.7×10^3
	Mean	3.44×10^4	2.86×10^3	5.63×10^4	2.66×10^4	3.65×10^3	1.17×10^3	1.10×10^3
Staphylococci count	Range	5.0×10^4–3.00×10^5	2.57×1.0^4–4.7×10^4	1.0×10^6–14×10^6	1.4×10^6–1.8×10^6	1.4×10^5–2.1×10^5	2.80×10^5–3.7×10^5	2.0×10^5–2.78×10^5
	Mean	1.67×10^5	2.57×10^4	1.16×10^6	1.56×10^6	1.48×10^5	2.01×10^5	2.50×10^5
Faecal *Streptococci*	Range	1.01×10^5–4.10×10^5	1.90×10^6–4.20×10^6	1.0×10^6–1.5×10^6	2.0×10^5–7.3×10^5	3.8×10^5–4.2×10^5	2.1×10^5–3.3×10^5	1.20×10^5–2.3×10^5
	Mean	2.13×10^5	2.69×10^6	1.23×10^6	5.27×10^5	4.0×10^5	233×10^5	1.60×10^5

Table 32.2: Percentage of Bacterial Flora Isolated from Fermented *Puntius sophore* During Fermentation Period

Bacteria	*Unfermented (Before Fermentation)*	*No. of Days*		
		60	*120*	*180*
Bacillus cereus	0.00–16.00	ND	ND	ND
Bacillus subtilis	2.00–9.09	25.00–33.33	14.28–26.78	15.49–28.08
Bacillus pumilis	3.44–5.26	25.00–37.50	19.04–35.70	12.76–19.71
Bacillus panthothenticus	9.09–22.72	8.12–18.33	7.14–9.52	15.59–18.30
Bacillus coagulans	4.00–5.26	8.33–18.75	8.92–14.28	11.26–17.02
Staphylococci	2.40–8.00	0.00–12.50	9.52–12.5	8.45–11.70
Micrococcus	15.78–36.00	16.66–25.00	8.92–34.61	13.82–28.16
Unidentified	0.00–18.00	0.00–9.52	2.81–11.53	10.63–12.67

Table 32.3: Percentage of Fungal Flora Isolated During the Fermentation of *R.sophore*

Microflora	*Unfermented (Before Fermentation)*	*No. of Days*		
		60	*120*	*180*
Aspergillus fumigatus	0.00–14.28	15.00–30.00	7.69–42.85	12.00–27.27
A. candidus	ND	5.26–22.22	ND	ND
A. sydowi	25.00–28.57	50.00–66.66	0.00–20.00	0.00–14.28
A. niger	0.00–22.22	ND	ND	ND
Penicillium citrinum	11.11–28.57	0.00–33.33	14.28–50.00	0.00–29.41
P. fellutanum	10.00–33.33	14.28–20.00	33.33–62.50	11.76–45.45
P. regulosum	ND	0.00–20.00	4.28–50.00	9.00–50.0
Gliocladium penicilloides	12.50–16.66	13.33–20.00	14.28–50.00	0.00–12.5
Humicola sp.	ND	0.00–10.00	ND	ND

The results of the present investigation indicate that gram-positive rod, *Bacillus* and gram-positive cocci, *Micrococcus* spp. predominate the bacterial flora during fermentation. The *Bacillus* species isolated consists of *B. cereus, B. coagulans, B. pumilis, B. subtilis* and *B. Pantothenticus.* The percentage of *Bacillus subtilis* was range from 25.00–33.33. The value slightly decreased at the end of fermentation. The average percentage of *Staphylococcus* and *Micrococcus* was ranged from 8.45–11.70 and 13.82–28.16 at 180 days of fermentation.

The occurrence of large percentage of *Bacillus* species during initial stage suggested that spore forming *Bacilli* might play an important active role during fermentation. These spores forming *Bacillus* spp. in the completely fermented products may reflect the resistant nature of these organisms (Crisan and Sand, 1975). According to Oettero *et al.* (2003) decrease in *Staphylococcus* sp. could be attributed to the lack of specific nutrients and probably presence of substances that could inhibit its growth or to competition with other microorganisms. *E. coli* and *Salmonella* were not recovered during fermentation. Total fungal count was ranged from 10^3–10^4 cfu/g in unfermented fish while a value of 10^2–10^3 cfu/g was noticed at the end of fermentation. According to Abraham *et al.*, 1993, fungal count of cured fishes ranged from 10^3–10^4/g.

Many fungal floras were found associated with the fermented fishes. Amongst these *Penicillium* and *Aspergillus* constituted the dominant genus. These two genera were very common in various fermented fishes including fish sauces and fish paste (Crisan and Sands, 1978). *A. niger* which was observed before fermentation, was not detected during fermentation. *Cladosporium, Cunninghamella, Curvularia, Gliocladium* and *Humicola* were also isolated from the unfermented samples. The percentage of fungal flora decreased due to the reduction of total plate count of fungus. The incidence of fungus in the unfermented samples might be due to unhygienic drying, handling, transportation and storage during processing. According to Refai *et al.*, 1993, the surrounding especially the air, is the main source of fungal contamination. Essuman (1992) reported that moulds are able to grow under dry conditions better than bacteria. The authors added that spores of moulds that are often present in the air and soil contaminate fish during processing. According to Sarojnalini and Vishwanath, (1987), many varieties of mould, including *Penicillium, Cladosporium, Geotrichium* sp. *and Myrothecium* sp. were isolated from fish paste 'Hentak' of Manipur and the fermentation was reported to be caused by enzymes released by the ferments on the microorganisms.

It can be concluded from the study that the occurrence of microorganisms is important for fish fermentation to be proceed. However, the occurrence of faecal *Streptococci* and *Staphylococci* in 'Ngari' is of concern to the health and hygienic of consumers.

References

Abraham, J.J., Sukumar, S., Shanmugam, S.A. and Jeyachandra, P., 1993. Microbial stability of certain cured Fishery products. *Fish Technol.*, 30: 134–138.

Aljedah, J.K., Ali, M.Z. and Robinson, R.K., 1999. Chemical and microbiological properties of Mehiawah: A popular fish sauce in the Gulf. *J. Food Sci. Technol.*, 36(6): 561–564.

Almas, A.K., 1981. *Chemistry and Microbiology of Fish and Fish Processing.* Department of Biochemistry. Norwegian Institute of Technology, University of Trondheim, Norway.

Anon., 1964. An evaluation of public health hazards from microbiological contamination of food. Publication 1195, National Academy of sciences, National Research Council, Washington, D.C.

APHA, 1976. *Compendium of Methods for Microbiological Examination of Foods*, (Ed.) M.L. Speak. American Public Health Assn., Washington, pp. 701.

Barber, L.E. and Deibel, R.H., 1973. Effect of pH and oxygen tension on staphylococcal growth and enterotoxin formation in fermented sausage. *Applied Microbial.*, 24(6): 891–898.

Brown, R.F. and Gibbons, N.E., 1974. *Bergey's Manual of Determinative Bacteriology*. William and Willins, Baltimore, pp. 1146.

Chou, C.C., Hwang, G.R. and Frank, M.H., 1988. Changes of microbial flora and enzymes activity during the aging of Tou-Pan-Chang: A Chinese fermented condiment. *J. Ferment. Technol.*, 66(4): 472–478.

Crisan, E.V. and Sands, A., 1975. Microflora of four fermented fish sauce. *Applied Microbial.*, 29: 106–108.

Djein, K.S., 1982. Indigeneous fermented foods. In: *Fermented Foods, Economic Microbiology,* (Ed.) A.H. Rose. Academic Press Inc., London, pp. 17–15.

Eliss, M.B., 1976. More *Dermaticeous Hypomycetes*, CMI, Kew, Surrey England, 507.

Essuman, K.M., 1992. *Fermented Fish in Africa: A Study on Processing, Marketing and Consumption.* FAO Fisheries Technical Paper, Rome, FAO, 329: 80.

Frazier, W.C. and Westhoff, D.D., 1983. *Fundamentals of Microbiology*. McGraw-Hill, New York.

Eliss, M.B., 1976. *Dermaticeous Hypomycetes.* CMI, Kew, Surrey, England, pp. 608.

Gilman, J.C., 1957. *A Manual of Soil Fungi.* The Iowa State University Press, Iowa, USA.

Hall, H.E., David, F.B. and Keith, H.L., 1967. Examination of market foods for coliforms organisms. *Applied Microbial.*, 15(5): 1062–1069.

Hesseltine, C.W. and Wang, H.L., 1969. Traditional fermented food. In: *Symposium on Flavoring Compounds by Fermentation*, Presented at the 152nd Annual Meeting of the American Chemical Society, New York.

Ijong, F.G. and Ohta, Y., 1996. Physico-chemical and microbiological changes associated with bakasang processing: A traditional Indonesian fermented fish sauce. *J. Sc. Fd. Agric.*, 71(1): 69–74.

Kiss, I., 1984. *Testing Methods in Food Microbiology*. Elsevier, Amsterdam, Oxford, New York, Tokyo, pp. 437.

Lopechart, K. and Park, J.W., 2002. Characteristics of fish sauce made from Pacific Whiting and surimi by products during fermentation stage. *J. Food. Sci.*, 67: 511–516.

Oetterer, M., Sergio, D.P., Claudio, R.G, Lia Ferraz de Arruda, Ricardo, B. and Ana Maria, Paschoal, Da Cruz, 2003. Monitoring the sardine (*Sardinella brasiliensis*) fermentation process to obtained Anchovies. *Scientia Agricola.*, 60(3): 511–517.

Ramachandra Rao, T.N., 1980. Developments in industrial and food microbiology. *J. Fd. Sci. Technol.*, 17: 83–88.

Refai, M., Mansour, N., Nagar, A.E.I. and Abdel, A.A., 1993. The fungal flora in modern Egyptian slaughter houses. *Fleischertchaft*, 73(2): 191–193.

Sandhu, D.K. and Soni, S.K., 1989. Microflora associated with Indian Punjabi wari fermentation. *J. Fd. Sci. Technol.*, 26(1): 21–25.

Sarojnalini, Ch. and Vishwanath, W., 1988. Biochemical composition and fungal flora of fermented fish paste "Hentak" of Manipur. *Int. J. Acad. Ichthyol.*, 8(1): 9–12.

Shewan, J.M., 1962. Food poisoning caused by fish and fishery products. In: *Fish as Food*, Vol. 2 (Ed.) G. Borgstrom. Academic Press, New York, pp. 443–466.

Sillikar, J.H., 1963. Total count as index of food quality. In: *Microbial Quality of Foods*, (Eds.) L.W. Stanetz, C.O. Chichester, A.R. Ganfin and Z.I. Ordal. Academic Press, New York, London, pp. 273.

Steinkraus, K.H., 1983. *Handbook of Indigenous Fermented Foods*. Marcel Dekkar, New York, pp. 671.

Syaefudin, K., Budianto and Sunarya, S., 1992. Changes in microbiological and chemical composition during fermentation of fermented fish (peda). *FAO Fisheries Report*, Rome, pp. 136–141.

Taylor, A.J. and Gilbert, R.J., 1975. *Bacillus cereus* food poisoning: A–provisional–serotyping scheme. *J. Med. Microbial.*, 8: 543.

Yang, H.C. and Chung, H.J., 1995. Changes of microbial and chemical component in salt-fermented youbsak during the fermentation. *Korean J. Fd. Sci. Technol.*, 27(1): 185–192.

Chapter 33

Study on Haemogram of Yak (*Poephagus grunniens* L.) while Carrying Load in Cross Country Mode

***B.C. Das*[1]**, M. Sarkar*[2]*, D.N. Das*[3]*, D. Gogoi*[2]*, D.B. Mondal*[4]*, A. Basu*[2]*, M. Mazumder*[2]*, P. Bora*[5] *and M. Ahmed*[6]**

[1]*Department of Veterinary Physiology, WBUA&FS, 37, K.B. Sarani, Kolkata – 37*

[2]*National Research Centre on Yak (ICAR), Dirang, West Kameng, Arunachal Pradesh – 790 101*

[3]*Scientist, NDRI, SRS, Adugodi, Bangalore*

[4]*Senior Scientist, Veterinary Medicine Division, IVRI, Izatnagar, Bareilly – 243 122, U.P.*

[5]*NRC Pig (ICAR), Six Mile, Guwahati – 22*

[6]*C/O CORPF, 56 APO*

ABSTRACT

Yaks were subjected to load 15 per cent of their body weight and allowed to walk in cross country mode on ups and downs of the hilly terrain almost at around 10,000 ft altitude. Haemogram in respect of total erythrocyte count (RBC), Haemoglobin (Hb), Packed-cell volume (PCV), total leukocyte count (WBC) in seven adult male yak were studied subjecting them to load. Among haematological parameters only RBC and WBC changed significantly ($p < 0.5$) between treatment, like before, in-between and after carrying load. However changes in ESR, Hb and PCV were found non-significant, which reflects better adaptability of yak for load carrying.

Keywords: Yak, Load and Haemogram.

* Corresponding Author.

Introduction

Yak is a multipurpose bovid of high altitude and mostly found at 6000–10,000 ft and even as high as upto 30,000 ft above msl. Surprisingly at such altitude apart from producing meat and milk, it is used as pack animals by yak herders, where partial pressure of oxygen is very less. It is a sure-footed animal and carries the load in hilly terrain without much difficulty. It is the only means of transportation for hilly people as because of difficult topography and landscape. Changes in haematological parameters are often used as indices of strain due to work (Thomas, 1969). Studies on pack ability on yak and effects on haemogram are very meagre. In the present investigation efforts have been made to study haematological parameters.

Materials and Methods

The present study has been conducted at Nuykmadung yak farm of National Research Centre on yak. The altitude of farm is around 8500 ft from msl. Meteorological attributes during experimental period was moderately cold humid. Temperature in morning and afternoon was 18±0.4 and 18.2±0.3°C respectively. Relative humidity morning to evening ranges from 74.2±5.2 to 73.7±2.6 per cent.

Apparently healthy and free from respiratory diseases, eight adult male yaks, above 2–3 years of age selected from the herd of NRC Yak, Dirang, Arunachal Pradesh. These animals were of similar age and body weight(s). All the experimental animals were maintained under identical conditions of feeding and management. Animals were fed as per schedules adopted at NRC Yak and feeds and fodder were fed depending upon the availability from time to time. A total of 7 male yak selected from farm taken daily in the morning for 30 min walk and this training continued for a period of 20 days. All the animals were subjected to, weight 15 per cent of their body weight and allowed to move for about 11 km in downslide through approachable kachha road. Loads were placed on the back of yak with the help of a wooden shaddle locally made. Three times blood samples were collected, before carrying load, after walking around 11 km and at last when returned to yak farm. So a total of 22 km journey made during this study. Blood samples were measured (Jain, 1986; Benjamin, 1974) within few hours of collection.

Results and Discussion

Average RBC, WBC, ESR, Hb and PCV values presented in Table 33.1 and ANOVA for the said parameters in Table 33.2. Erythrocyte count between load carrying increased but it was not significant. However, after completing the load carrying the value increased significantly ($p < 0.05$). Total leukocyte count had a significant difference between treatment ($p < 0.05$). However, within treatment there was no significant difference. There was non significant change in ESR, PCV and Hb level, though Hb increased to a higher level between carrying load. Unlike cattle (Georgie *et al.*, 1970, Rana *et al.*, 1977) wherein Haemoglobin and PCV did not decrease in case of yak, might be due to very minimum haemodilution and destruction of blood cells (Thomas and Razdan 1973) and not much change in PCV. In case of cattle PCV increased significantly due to work (Singh *et al.*, 1982). There was increased level of RBC and WBC and might be due compensation mechanism in hypoxic condition in addition to load carrying and work done (Singh *et al.*, 1980).

Haematological indices are the most important parameters to be considered for an animal used for pack and work purposes. From the present study in yak it is revealed that most of the parameters changed non-significantly except RBC and WBC. Erythrocytes and leukocytes as increased to compensate hypoxic situation and yak could effectively be used for carrying load. However, further studies are required for other factors associated with work stress and physiological system.

Table 33.1: Hematological Parameters Cross Country

Treatment	Observation	RBC	WBC	ESR	Hb	PCV
Before	7	3.936±0.15[a]	10744.05±408.35[a]	3.71±0.26	12.65±0.17[a]	35.62±1.03
Between	7	4.189±0.15[a]	11641.43±408.35[ab]	3.64±0.26	13.22±0.17[b]	36.33±1.03
After	7	4.627±0.15[b]	12646.43±408.35[bc]	4.12±0.26	12.99±0.17[ab]	37.14±1.03

Different superscripts indicate significant difference at 5 per cent level.

Table 33.2: ANOVA of Hematological Parameters Cross Country

Source	df	RBC MSS	WBC MSS	ESR MSS	Hb MSS	PCV MSS
Between Treatment	2	0.857*	6340093.26*	0.465[NS]	0.578[NS]	4.072[NS]
Within Treatment	18	0.150	1167272.71	0.490	0.202	7.397

*: Indicates significant difference at 5 per cent level.

NS: Indicates non-significant.

References

Benjamin, M.M., 1974. *Outline of Veterinary Clinical Pathology*. The Iowa State University Press, Iowa, USA, pp. 56–64.

Georgie, G.C., Sastry, N.S.R. and Razdan, M.N., 1970. Studies on the work performance of crossbred cattle. 2. Some physiological and chemical responses of blood. *Indian Journal of Animal Production*, 1: 115–119.

Jain, N.C., 1986. *Schalm's Veterinary Hematology*, 4th edn. K.M. Varghese Co., Bombay.

Rana, R.D., Singh, N., Nangia, O.P. and Ahmad, A., 1977. Effect of exercise on some haematological parameters in buffalo calves. *Haryana Agricultural University Journal of Research*, 7(4): 237–241.

Singh, N., Nangia, O.P. and Dwarkanath, P.K., 1980. Effect of exercise on biochemical constituents of blood in entire; castrated and vasectomised buffalo males. *Indian Journal of Dairy Science*, 33(3): 299–303.

Singh, N., Nangia, O.P., Garg, S.K. and Dwarkanath, P.K., 1982. Draught capacity in male buffaloes: Autonomic and haematological responses in untrained entire castrated and vasectomised animals. *Indian Journal of Animal Sciences*, 52(4): 216–221.

Thomas, C.K., 1969. Studies on adaptability of Sahiwal-Brown Swiss crossbred cattle to tropical condition. *Ph.D. Thesis*, Punjab University, Chandigarh.

Thomas, C.K. and Razdan, M.N., 1973. Adaptability of Sahiwal-Brown Swiss cattle to subtropical conditions. *Indian Journal of Animal Sciences*, 53(12): 335–337.

Chapter 34

Seed Germination and Seedling Growth Response of Some Crop Plants to Solid Waste of a Chlor-Alkali Industry of Orissa

B. Padhy[1], P.K. Gantayet[2] and S.K. Padhy[1]

[1]*Division of Physiology and Biochemistry, Department of Botany, Berhampur University, Berhampur – 760 007, Orissa*

[2]*Department of Botany, Rayagada College, Rayagada – 765 001*

ABSTRACT

The toxic effect of solid waste of chlor-alkali industry (Jayashree Chemicals of Ganjam District, Orissa) on seed germination and seedling growth of one cultivar from each group of crops *i.e.* cereal (*Oryza sativa* L.cv. IR-36), pulses (*Phaseolus aureus* cv PDM-54) and oil-yielding crop (*Brassica campestris* cv Pusabold) were tested. All the concentrations of solid waste extract (SWE) ranging from 10 to 100 per cent considerably inhibited seed germination and seedling growth parameters (shoot and root lengths, fresh and dry weights of both shoot and roots and photosynthetic pigments (chl-a, chl-b, and total chlorophyll). The synthesis of total carbohydrate and protein in the leaves of the seedlings was also greatly affected by SWE. Hence, it is suggested that the solid waste should not be dumped or used as land refilling nearby agricultural fields, which may cause directly or indirectly harm to the vegetation of that locality.

Keywords: *Solid waste, Chlor-alkali industry, Seed germination, Seedling growth, Oryza, Phaseolus, Brassica.*

Introduction

The toxicity of mercury (Hg^{++}) obtained from solid waste and effluents of chlor-alkali industries, world over, is a challenging topic. A lot of reports on mercury toxicity by the chlor-alkali industries on.

various plants have been reported in India and abroad. Reports are available on the effects of heavy metals like Ni, Cu, Zn, Cd and Hg on uptake and translocation of nutrients in barley plants (Oberlander and Rath, 1978). Reports are also available on accumulation of Hg^{++} in cereal crops (Dudas and Pawluk, 1977), soyabean and potato (MacLean, 1974), algae (Hannan *et al.*, 1973), water and soil (Bull *et al.*, 1977). Literature on effect of solid waste on physiology of crop plants are scanty. The present work is aimed at studying the changes in seed germination, seedling growth, chlorophyll, total carbohydrate and protein content influenced by solid waste from the chlor-alkali factory installed at Ganjam town of Orissa, where mercury is used as a cathode in the electrolytic process and solid wastes are dumped on the fields around the industry area whose pollutants are leached out and enter into the environment.

Materials and Methods

Local varieties of one cultivar from each group of crops *i.e.* cereal (*Oryza sativa* L cv IR-36), pulses (*Phaseolus aureus* cv PDM-54) and oil yielding crop (*Brassica campestris* cv Pusabold) were used as target crop, Solid wastes were collected from the dump site of chlor-alkali industry dried in sun light and powdered, soaked in distilled water at the rate of 1 kg/l, stirred well, allowed to leach for 24 hours at room temperature and then filtered through cheese cloth followed by filter paper. The filtrate was considered as 100 per cent concentration (stock solution) and referred hereafter as solid waste extract (SWE). The different dilutions of 10, 25, 50, 75 and 100 per cent solutions were prepared from the stock solutions of SWE using distilled water.

Pure line seeds of rice (*Oryza sativa* L cv IR-36), green gram (*Phaseolus aureus* cv PDM-54) and mustard (*Brassica campestris* cv Pusabold) were procured from Regional Agricultural Research Station of Orissa University of Agriculture and Technology (OUAT) located at Berhampur, Visually selected seeds for uniform size, shape and colour were surface sterilized with 0.03 per cent formalin solution for half an hour and then allowed to germinate in petridishes (5 cm in diameter) @ 20 seeds per petridish lined with blotting paper soaked in different concentrations of SWE. Each set has 5 replicates for each treatments. Seeds soaked in distilled water served as control. The petridishes were kept in a seed germination chamber maintained at 30±1°C. Emergence of coleoptile/radicle from seeds was considered as the criteria for germination. The per cent of germination was recorded at an interval of 24 hours from the time of seed soaking till more than 95 per cent germination observed in respective control sets.

After germination, the petridishes were transferred to seedling growth chamber maintained at 30±1°C and provided with continuous illumination from two fluorescent electric tube lights supplying 2±0.5 K lux intensity of light. During the period of observation, the petridishes were supplied with equal volume of respective concentrations of SWE or distilled water periodically to maintain wetness. Seedling growth parameters (shoot and root lengths, fresh and dry weights of both shoots and roots), changes in chlorophyll, total carbohydrate and protein content of first leaves of seedlings of control and treated sets of all test crops in relation to control were recorded at an interval of 2 days till 14 days after soaking.

Results and Discussion

All the concentrations of solid waste extract ranging from 10 to 100 per cent considerably inhibited seed germination process in all test cultivars. Higher the concentration of extract more was the inhibition effect. The shoot and root lengths, fresh and dry weights of both shoot and root of the seedlings of the test cultivars were also considerably checked by all the concentrations of SWE tried. A progressive decline in chlorophyll, total carbohydrate and protein content in first leaves was noticed with increase of the concentrations of leachate throughout the period of observation.

Table 34.1: Effect of Different Concentrations of Solid Waste Extract (SWE) of Chlor-alkali Industry on Seed Germination (%) of Rice (*Oryza sativa*), Green Gram (*Phaseolus aureus*) and Mustard (*Brassica campestris*)

Name of the Crop	Hours After Soaking (HAS)	Control	Concentration of SWE (%)				
			10	25	50	75	100
Rice	48	52.3	48.1	40.2	36.2	20.3	12.5
	72	68.4	60.5	55.5	50.4	42.1	28.4
	96	96.2	78.0	62.4	59.6	53.4	43.4
Green gram	24	78.4	75.5	66.8	63.1	51.6	21.2
	48	88.4	85.6	76.7	71.4	62.5	34.5
	72	100	95.8	85.8	80.4	69.4	46.2
Mustard	24	77.5	74.4	69.3	61.4	51.6	42.5
	48	90.5	87.2	81.2	73.6	64.4	57.8
	72	98.4	94.3	89.5	79.2	71.6	64.5

Data are mean of 10 seedlings.

Table 34.2: Effect of Different Concentrations of Solid Waste Extract (SWE) of Chlor-alkali Industry on Seedling Growth of Rice (*Oryza sativa* L cv IR-36)

Concentrations of SWE (%)	Days After Soaking (DAS)	Shoot Length (in cm)	Root Length (in cm)	Shoot Fresh Weight (in mg)	Root Fresh Weight (in mg)	Shoot Dry Weight (in mg)	Root Dry Weight (in mg)
Control	8	6.55	4.54	3.43	2.18	0.64	0.61
	10	6.88	4.74	3.52	2.41	0.72	0.64
	12	7.05	5.21	4.61	2.61	1.53	0.95
	14	7.24	5.71	4.78	3.71	1.75	0.93
10	8	6.18	3.48	3.36	2.14	0.55	0.51
	10	6.42	3.78	3.39	2.32	0.59	0.56
	12	7.05	4.95	4.13	2.62	0.81	0.83
	14	7.16	2.27	4.24	3.65	0.85	0.85
25	8	5.46	2.54	2.18	1.67	0.41	0.41
	10	5.62	2.63	2.95	1.84	0.46	0.46
	12	6.84	2.86	3.54	2.53	0.49	0.49
	14	6.98	2.97	3.81	2.72	0.54	0.53
50	8	3.21	1.03	0.45	0.34	0.71	0.32
	10	3.33	1.07	0.51	0.37	0.84	0.41
	12	4.48	2.01	0.57	0.39	0.88	0.52
	14	4.58	2.09	0.66	0.44	0.49	0.59
75	8	2.37	0.93	0.21	0.12	0.54	0.19
	10	1.95	0.97	0.26	0.15	0.58	0.28
	12	2.42	1.18	0.31	0.18	0.68	0.38
	14	2.87	1.23	0.33	0.21	0.71	0.46
100	8	0.28	0.05	0.11	0.05	0.34	0.16
	10	0.38	0.07	0.13	0.08	0.42	0.19
	12	0.46	0.08	0.18	0.11	0.46	0.24
	14	0.56	0.12	0.22	0.16	0.55	0.31

Data are mean of 10 seedlings.

Table 34.3: Effect of Different Concentrations of Solid Waste Extract (SWE) of Chlor-alkali Industry on Seedling Growth of Green Gram (*Phaseolus aureus* cv PDM-54)

Concentrations of SWE (%)	*Days After Soaking (DAS)*	*Shoot Length (in cm)*	*Root Length (in cm)*	*Shoot Fresh Weight (in mg)*	*Root Fresh Weight (in mg)*	*Shoot Dry Weight (in mg)*	*Root Dry Weight (in mg)*
Control	8	9.8	8.2	4.32	3.21	10.9	0.84
	10	10.9	9.4	5.09	3.74	1.78	0.98
	12	12.6	10.5	6.21	3.95	2.23	1.06
	14	13.4	11.4	7.25	4.09	2.89	1.15
10	8	9.1	7.7	4.21	3.12	0.95	0.74
	10	10.3	8.9	4.91	3.49	1.54	0.79
	12	11.9	9.5	5.69	3.67	1.95	0.95
	14	12.3	10.3	6.57	3.84	2.05	1.01
25	8	8.1	7.1	3.89	2.89	0.65	0.54
	10	8.9	7.7	3.91	2.94	0.74	0.56
	12	10.3	8.8	4.23	3.12	0.85	0.63
	14	10.8	9.2	4.46	3.34	0.98	0.68
50	8	6.1	5.9	2.95	2.15	0.54	0.34
	10	7.2	6.9	3.09	2.36	.0.69	0.38
	12	8.5	8.1	3.34	2.67	0.84	0.41
	14	8.9	8.7	3.84	2.71	0.89	0.42
75	8	4.3	4.1	2.05	1.68	0.41	0.25
	10	4.8	5.4	2.11	1.96	0.49	0.29
	12	5.7	6.2	2.71	2.23	0.54	0.34
	14	6.6	7.1	3.17	2.52.	0.71	0.45
100	8	3.3	2.6	1.89	1.17	0.34	0.19
	10	3.9	3.5	1.95	1.51	0.38	0.27
	12	4.4	4.3	2.13	1.67	0.42	0.32
	14	4.9	4.7	2.26	1.88	0.49	0.33

Data are mean of 10 seedlings.

The inhibitory effect on seed germination, seedlings growth and chlorophyll, total carbohydrate and protein content in first leaves exhibited positive correlation with increase of concentrations of leachate. Padhi (1990) reported that, mercury contaminated medium inhibit seed germination by arresting or inhibiting the vital functions in other cultivars of green gram seeds during germination process. Mishra and Choudhury (1996) reported the membrane damage in seeds during germination in some cultivars of pulses influenced by mercury contaminated effluent of a chlor-alkali industry. The present results corroborate with the reports of Misra (1984), Misra and Misra (1984), Patnaik (1990), Mishra (1990) and Radha (1993) on seed germination of some of rice cultivars. Mishra and Choudhury (1997) reported that the inhibition of germination induced by mercury ions (Hg^{++}) was not only due to non-availability of sugars from starch hydrolysis by arresting α-amylase activity but also injuring the embryos as a result the germination is delayed and/or inhibited.

Table 34.4: Effect of Different Concentrations of Chlor-alkali Industry Solid Waste Extract (SWE) on Seedling Growth of Mustard (*Brassica campestris* cv Pusabold)

Concentrations of SWE (%)	*Days After Soaking (DAS)*	*Shoot Length (in cm)*	*Root Length (in cm)*	*Shoot Fresh Weight (in mg)*	*Root Fresh Weight (in mg)*	*Shoot Dry Weight (in mg)*	*Root Dry Weight (in mg)*
Control	8	6.85	6.78	0.79	0.48	0.27	0.15
	10	7.89	7.57	0.87	0.56	0.29	0.17
	12	8.35	8.55	0.93	0.61	0.32	0.19
	14	9.42	9.02	0.98	0.65	0.35	0.22
10	8	6.31	6.53	0.74	0.45	0.24	0.13
	10	7.65	7.31	0.79	0.52	0.27	0.15
	12	8.25	7.89	0.87	0.57	0.29	0.17
	14	8.84	8.17	0.91	0.61	0.32	0.19
25	8	5.95	6.18	0.67	0.39	0.21	0.11
	10	7.23	6.89	0.72	0.45	0.23	0.13
	12	8.09	7.42	0.78	0.49	0.26	0.15
	14	8.21	7.98	0.86	0.54	0.29	0.17
50	8	5.21	5.64	0.58	0.32	0.18	0.08
	10	6.52	6.35	0.66	0.39	0.21	0.11
	12	7.36	6.69	0.71	0.43	0.23	0.13
	14	7.84	7.12	0.79	0.49	0.25	0.15
75	8	4.87	4.67	0.44	0.23	0.14	0.06
	10	5.78	5.62	0.53	0.27	0.16	0.08
	12	6.52	6.01	0.57	0.32	0.18	0.11
	14	6.91	6.43	0.61	0.38	0.21	0.13
100	8	3.92	3.87	0.38	0.14	0.07	0.03
	10	4.97	4.98	0.45	0.18	0.09	0.06
	12	5.66	5.26	0.51	0.22	0.12	0.07
	14	6.02	5.58	0.54	0.26	0.15	0.08

Data are mean of 10 seedlings.

Regarding seedling growth the pollutants present in SWE might have checked the synthesis of chlorophyll pigments, plant growth regulators and other bio-macromolecules responsible for growth and development of seedlings. Misra (1984) reported that mercury ions present in the solid waste and/or effluents inhibit the ATP synthesis and ATPase activity. Further Tsuzuki *et al.* (1979) reported the binding of mercury with pyrimidine nucleotide of electron transport system. Lapina and Povov (1970), opined that chlorophyll protein synthesis is regulated by various salts and ions, hence mercury ions and salts present in SWE at higher concentration might have inhibited the protein and carbohydrate synthesis. Lee *et al.* (1973), reported that bioaccumulation of mercury in different plant tissues alter and/or inhibit the Hill reaction. Sahu *et al.* (1988) and Shaw *et al.* (1988) reported that mercury at very

Table 34.5: Effect of Different Concentrations of Solid Waste Extract (SWE) of Chlor-alkali Industry on Changes in Chlorophyll Content in the First Leaves of Rice (*Oryza sativa* Lcv IR-36), Green Gram (*Phaseolus aureus cv* PDM-54) and Mustard (*Brassica campestris* cv Pusabold) During Seedling Growth

Concentration of SWE (%)	*Name of the Cultivar*	*Chl-a Days After Soaking*				*Chl-b Days After Soaking*				*Total Chlorophyll Days After Soaking*			
		8	*10*	*12*	*14*	*8*	*10*	*12*	*14*	*8*	*10*	*12*	*14*
Control	Rice	0.36	0.51	0.46	0.39	0.16	0.23	0.19	0.17	0.52	0.74	0.65	0.56
	Green gram	0.31	0.46	0.38	0.32	0.14	0.21	0.17	0.15	0.45	0.67	0.55	0.47
	Mustard	0.29	0.42	0.36	0.31	0.12	0.19	0.16	0.13	0.41	0.61	0.52	0.44
10	Rice	0.29	0.48	0.39	0.32	0.12	0.17	0.14	0.12	0.41	0.65	0.53	0.44
	Green gram	0.24	0.42	0.36	0.26	0.09	0.15	0.13	0.11	0.33	0.57	0.49	0.37
	Mustard	0.21	0.38	0.31	0.23	0.07	0.13	0.11	0.09	0.28	0.51	0.42	0.31
25	Rice	0.24	0.38	0.31	0.26	0.08	0.15	0.12	0.09	0.32	0.53	0.46	0.35
	Green gram	0.21	0.36	0.29	0.24	0.07	0.14	0.13	0.08	0.28	0.4	0.42	0.32
	Mustard	0.18	0.29	0.21	0.19	0.05	0.11	0.09	0.06	0.23	0.4	0.29	0.25
50	Rice	0.21	0.32	0.26	0.23	0.07	0.14	0.11	0.08	0.28	0.46	0.37	0.31
	Green gram	0.19	0.29	0.23	0.21	0.06	0.11	0.09	0.07	0.25	0.4	0.32	0.29
	Mustard	0.15	0.25	0.19	0.16	0.04	0.09	0.07	0.05	0.19	0.34	0.26	0.21
75	Rice	0.18	0.28	0.24	0.21	0.06	0.12	0.09	0.07	0.24	0.4	0.34	0.29
	Green gram	0.16	0.25	0.21	0.18	0.05	0.09	0.07	0.06	0.21	0.34	0.29	0.24
	Mustard	0.11	0.19	0.16	0.12	0.03	0.08	0.05	0.04	0.14	0.27	0.21	0.16
100	Rice	0.11	0.23	0.19	0.12	0.04	0.11	0.08	0.05	0.15	0.24	0.27	0.17
	Green gram	0.09	0.19	0.15	0.11	0.03	0.09	0.07	0.04	0.12	0.28	0.22	0.15
	Mustard	0.07	0.12	0.09	0.08	0.02	0.07	0.06	0.03	0.09	0.19	0.15	0.11

Data are mean of 10 seedlings expressed in mg/g fresh wt.

Table 34.6: Effect of Different Concentrations of Solid Waste Extract (SWE) of Chlor-alkali Industry on Total Carbohydrate Content in the First Leaves of Rice (*Oryza sativa* LCv IR-36), Green Gram (*Phaseolus aureus* cv PDM-54) and Mustard (*Brassica campestris* cv Pusabold) at Different Days After Soaking (DAS)

Name of the Crop	*Concentration of SWE (%)*	*Total Carbohydrate*			
		8 DAS	*10 DAS*	*12 DAS*	*14 DAS*
Rice	Control	0.36	0.46	0.42	0.39
	10	0.33	0.42	0.38	0.34
	25	0.29	0.38	0.35	0.31
	50	0.26	0.35	0.31	0.28
	75	0.22	0.31	0.28	0.23
	100	0.18	0.29	0.25	0.19

Contd...

Table 34.6–Contd...

Name of the Crop	Concentration of SWE (%)	Total Carbohydrate			
		8 DAS	10 DAS	12 DAS	14 DAS
Green gram	Control	0.33	0.41	0.39	0.35
	10	0.29	0.38	0.32	0.31
	25	0.27	0.36	0.31	0.28
	50	0.25	0.34	0.29	0.26
	75	0.21	0.28	0.25	0.22
	100	0.16	0.25	0.21	0.17
Mustard	Control	0.29	0.39	034	0.31
	10	0.27	0.33	0.31	0.28
	25	0.24	0.31	0.28	0.25
	50	0.21	0.28	0.25	0.22
	75	0.18	0.26	0.21	0.19
	100	0.16	0.23	0.19	17

Data are mean of 10 seedlings expressed in mg/g fresh wt.

Table 34.7: Effect of Different Concentrations of Solid Waste Extract (SWE) of Chlor-alkali Industry on Protein Contents in the First Leaves of Rice (*Oryza sativa* L cv IR-36), Green Gram (*Phaseolus aureus* cv PDM-54) and Mustard (*Brassica campestris* cv Pusabold) at Different Days After Soaking (DAS)

Name of the Crop	Concentration of SWE (%)	Protein			
		8 DAS	10 DAS	12 DAS	14 DAS
Rice	Control	14.34	19.27	18.52	15.23
	10	12.14	17.32	15.34	13.21
	25	10.68	16.12	14.21	11.51
	50	9.84	13.73	12.11	10.24
	75	5.39	11.26	9.43	6.21
	100	4.11	8.23	6.21	5.06
Green gram	Control	16.23	21.32	19.11	17.14
	10	14.79	19.15	16.21	15.32
	25	13.17	17.29	15.42	14.11
	50	11.33	16.12	13.37	12.57
	75	9.75	13.42	12.14	10.37
	100	7.66	11.27	10.57	8.24
Mustard	Control	17.67	22.39	19.78	18.29
	10	15.29	21.18	20.15	17.13
	25	14.25	19.69	17.65	15.59
	50	13.21	17.51	15.38	14.29
	75	11.15	16.79	14.32	12.24
	100	9.37	14.28	13.11	9.81

Data are mean of 10 seedlings expressed in mg/g fresh wt.

low concentration act as a growth regulator where as higher concentration of the same acts as growth inhibitor. Hemalath *et al.* (1997) reported that heavy metals effect on certain biochemical constituents and nitrate reductase activities in *Oryza sativa* L. seedlings. Radha *et al.* (2002) reported that the solid waste of Chlor-alkali industry alter chlorophyll, carotenoid, pheophytin, photosynthetic rate and respiratory rate in other rice cultivars.

As a whole, the toxic chemical like mercury present in the SWE might have checked the germination process partially or fully by inhibiting various metabolic processes through several essential enzyme activities.

Conclusion

Hence, such crop should not be raised near the chemical factory or in the other words such wastes should not be allowed to dispose off near or around the crop fields as land refilling. General awareness should be created among the public and farmers regarding the harmful effect of the solid waste on crop plants which are discharged from the factory through various social organizations.

References

Bull, K.R., R.D. Roberts, M.J. Inskip and Goodmann, G.T., 1977. Mercury concentration in soil, grasses, earthworms and small animals near an industrial emission source. *Environ. Pollution*, 12(3): 135–140.

Dudas, M.J. and Pawluk, S., 1997. Heavy metals in cultivated soils and cereal. *Can. J. Soil Sci.*, 57(3): 329–339.

Hannan, P.J., P.E. Wilkinss, C. Potouillet and Corr, R.A., 1973. Measurements of mercury absorption by algae. *N.R.I. Report*, pp. 7628.

Hemalath, S., A. Anburas and K. Francis, 1997. Effect of heavy metals on certain biochemical constituents and nitrate reductase activity in *Oryza sativa* L. seedlings. *J. Environ. Biol.*, 18(3): 313–319.

Lapina, L.P. and Povov, B.A. 1970. Effect of NaCl on the photosynthetic apparatus of tomato. *Sov. Plant Physiol.*, 17: 477.

Lee, S.S., R.D. Wauchope, R. Haque and Fang, S.C., 1973. Binding of mercury compounds to the chloroplast in relation to inhibition of Hill reaction. *Pestc. Biochem Physiol.*, 3: 225–229.

MacLean, A.J., 1974. Mercury in plants and retention of mercury by soil in relation to properties and added sulphur. *Can. J. Soil Sci.*, 54(3): 287–292.

Misra, S.R., 1984. Analysis of the effect of solid waste extract of a caustic chlorine factory on growth and physiology of rice seedlings. *Ph.D. Thesis*, Berhampur University, Berhampur, Orissa, India.

Mishra, A. and Choudhury, M.A., 1997. Differential effect of Pb^{++} and Hg^{2+} on inhibition of germination of seeds of two rice cultivars. *Indian J. Plant Physiol.*, 2(1): 41–44.

Mishra, Jyostna, 1990. Analysis of factors concerning yield in rice. I. Pollen and spikelet sterility. *M. Phil. Dissertation*, Berhampur University, Berhampur, Orissa, India.

Mishra, A. and Choudhury, M.A., 1996. Possible implication of heavy metals (Pb^{++} and Hg^{++}) in free radical mediated membrane damage in two cultivars of rice. *Indian J. Plant Physiol.*, New series, 1: 43–47

Misra, S.R. and Misra, B.N., 1984. Studies on the solid waste extract from a chlor-alkali factory. I. Morphological behavior of rice seedlings grown in the waste extract. *J. Environ. Pollut.*, 37: 17–28.

Oberlander, H.E. and Rath, K., 1978. Effect of heavy metals chromium, nickel, copper, zinc, cadmium, mercury and lead on uptake and translocation of K and P by young barley plants. *Z. Pflazenernahr, Boden Kd.*, 141(1): 107–116.

Padhi, N.P., 1990. Ecophysiological response of germinating rice seedlings to an organomercurial fungicide, MEMC. *Ph.D. Thesis,* Berhampur University Berhampur, Orissa, India.

Patnaik, D.K., 1990. Toxicological effect of mercury contained solid waste on crop plant. *M.Phil. Dissertation*, Berhampur University, Berhampur, Orissa, India.

Radha, S., 1993. Toxicological effects of a mercury contained solid toxicant on a crop plant. *Ph.D. Thesis*, Berhampur University, Berhampur, Orissa, India.

Radha, S., C.V. Raju and Panigrahi, A.K., 2002. Toxic effect of solid waste from chlor-alkali factory on pigments and photosynthetic rate, respiration rate in rice seedlings *Poll. Res.*, 21(3): 315–318.

Sahu, A., B.P. Shaw, A.K. Panigrahi and Misra, B.N., 1988. Effect of PMA on oxygen evolution in a nitrogen fixing blue-green alga. *Microbios Letters*, 37: 119–123.

Shaw, B.P., A. Sahu and Panigrahi, A.K., 1988. Effects of the effluent from a chlor-alkali factory on a blue-green alga, changes in oxygen evolution rate. *Microbios. Letters*, 37: 89–96.

Tsuzuki, Y., Y. Takamichi and Koji, I., 1979. The relation between pyrimidine nucleotides and mercury compounds in artificial intra-cellular and extra-cellular fluids. *J. Toxicol. Sci.*, 4: 417–426.

Chapter 35

Study of Fluctuation of Groundwater Level in Somni Stream Watershed, Patan Block, Durg District, Chhattisgarh

Prashant Shrivastava[1] *and Anupama Asthana*[2]

[1]*Department of Geology,* [2]*Department of Chemistry*
Government Arts and Science College, Durg (C.G.)

ABSTRACT

The groundwater level fluctuation mainly depends on the climatic conditions of the area along with other factors like rainwater harvesting, artificial recharge etc. The Somni stream watershed receives average annual rainfall 1010.9 mm. The climate alternates between hot and dry summer and cold winter. The maximum day temperature during summer shoot up as high as 47°C and minimum falls to 35°C while during winter temperature rises up to 37°C and minimum 8°C.

The main groundwater structures present in the area are dug well, dug cum bore well and tube wells. The pre monsoon and post monsoon static water levels (SWL) have been recorded from these wells. On the basis of these two season data it is found that the post monsoon static water level in the water shed area is generally lowered during pre monsoon by 4 to 10 meters.

Introduction

Somni stream water shed is situated in the northern part of Patan Block, Durg district, Chhattisgarh. It can be easily approached both by rail and road. The national highway no. 6 and Mumbai-Howrah rail route (via Nagpur) runs along northern boundary of the watershed area. The Somni stream water shed falls in Survey of India Toposheet no. 64G/8 and G/12. It is bounded by latitudes 21 5′–21 14′N and longitudes 81 23′–81 34′ E.

Materials and Methods

The groundwater structures present in the area are dug wells, dug cum bore wells and tube wells. According to Survey report of PHE Department (1989) there are 274 dug wells and 284 tube wells are present in the study area. Most of the dug wells are private and vary in depth between 5 to 18 meters bgl. The PHED bore wells normally range in depth in between 60 and 100m bgl, 38 dug wells distributed over the entire watershed area were selected as observation wells. The pre monsoon and post monsoon static water levels have been recorded from these wells. The well inventory data is tabulated in Table 35.1.

Post Monsoon Water Table

The post monsoon static water level of the Somni stream watershed area range in between 271.85 and 308.90m above mean sea level. The highest post monsoon SWL is recorded from the village Parsada. The SWL values are plotted on the watershed map and the water table contours are constructed (Figure 35.1). It is clear from the Figure 35.1 that, the general flow of groundwater during

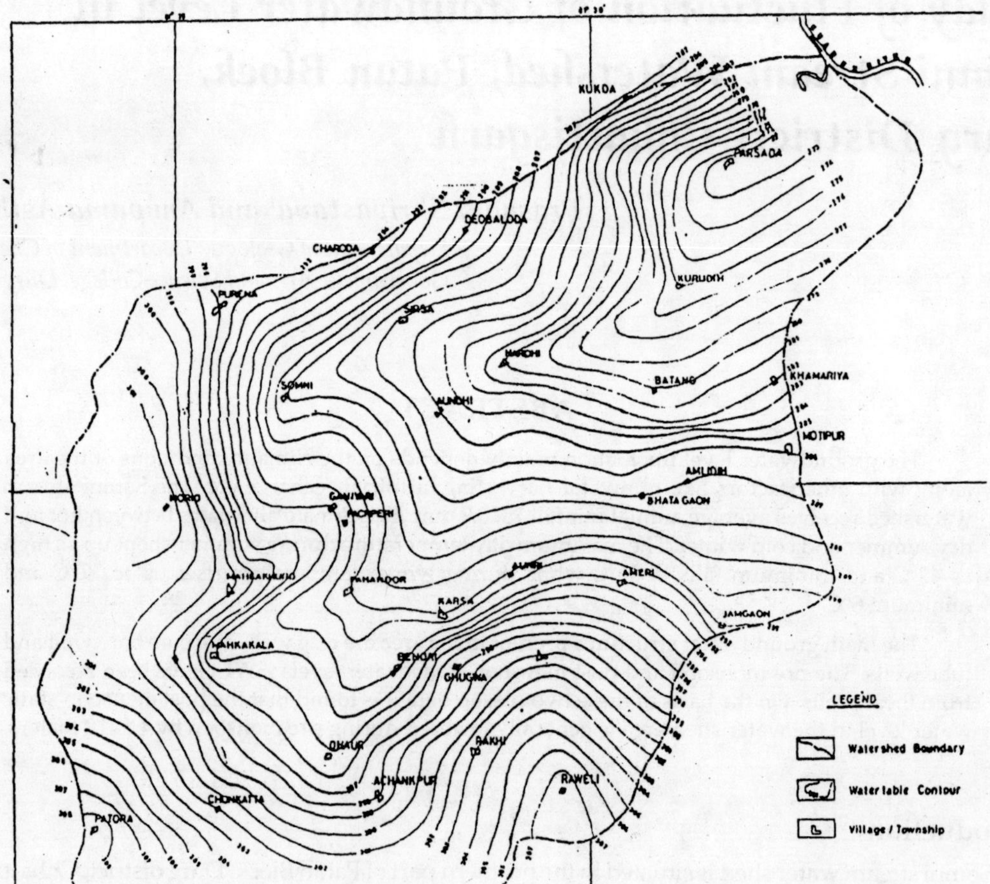

Figure 35.1: Post Monsoon Water Table Contour Map of Somni Nala Basin

Table 35.1: Well Inventory Data of Pre Monsoon and Post Monsoon Season, Somni Nala Basin

Sl.No.	Name of Village	Name of Owner	Diameter of Well in Meters	Height of the Parapet in Meters	Total Depth in Meters	Depth of Water Table (in meter)		R.L. of Water Table (in meter)		Water Table Fluctuation (in meter)	Remarks Lined/Unlined and Location
						Post Monsoon	Pre Monsoon	Post Monsoon	Pre Monsoon		
1.	Urla	Mr. Marar	3.00	G.L.	15.10	3.70	12.95	284.740	275.490	9.25	Lined/Near Tank
2.	Kukda	Govt. well	2.50	1.0	13.90	4.30	11.80	282.690	275.190	7.50	Lined/Near Road
3.	Charoda	Public well	1.90	0.50	7.20	3.10	5.65	290.990	287.340	2.55	Lined/Near Jyoti School
4.	Devbaloda	Jagdish	2.90	0.50	11.80	4.50	9.65	285.670	279.020	5.15	Lined/Near Temple
5.	Sirsa	Kharag Singh	1.80	0.90	15.00	4.85	13.20	–	–	8.35	Lined/Near Road
6.	Purena	Public well	2.40	0.30	7.90	3.60	5.85	–	–	2.25	Lined/Near Village
7.	Somni	Public well	2.10	–	11.70	4.40	10.45	–	–	6.05	Lined/Near Market
8.	Nardhi	Bisahu Shah	2.10	0.40	14.65	4.20	12.75	278.990	270.140	8.55	Lined/Near Bari
9.	Aundhi	Public well	2.90	0.60	15.60	4.80	11.85	283.150	276.100	7.05	Lined/Near Village
10.	Bhatagaon	Public well	3.20	0.40	13.40	4.10	11.75	286.366	278.610	7.65	Lined/Near Village
11.	Pahanda	Govt. well	2.50	0.40	11.50	4.80	10.35	277.920	272.370	5.55	Lined/Near Village
12.	Khamaria	Public well	2.80	0.60	14.50	3.20	11.75	283.170	274.620	8.55	Lined/Near Village
13.	Motipur	Gram Panchayat	3.60	0.60	9.50	4.15	8.55	286.625	282.225	4.40	Lined/Near Village
14.	Batang	Public well	2.50	G.L.	10.50	4.50	10.10	282.270	276.640	5.60	Lined/Near Village
15.	Morid	Rajiya Bai	3.60	0.70	9.50	3.40	7.40	303.690	299.690	4.00	Lined/Near Canal
16.	Amlidih	Khaman	3.00	1.00	12.50	3.10	11.00	285.090	277.190	7.90	Lined/in Village
17.	Mohka Khurd	Harlal Dau	2.85	G.L.	13.10	4.90	8.35	298.075	295.590	3.45	Lined/in Village
18.	Mohka Kalan	Public well	2.50	0.20	9.00	2.60	7.95	296.410	291.060	5.35	Lined/Near Temple
19.	Ganiyari	Leela Satnami	3.20	0.65	13.75	2.75	10.60	293.160	283.510	7.85	Lined/Near Temple
20.	Pachperi	Hulsa Satnami	3.00	0.90	14.90	2.30	11.75	291.520	282.070	9.45	Lined/in Village

Contd...

Table 35.1–Contd...

Sl.No.	Name of Village	Name of Owner	Diameter of Well in Meters	Height of the Parapet in Meters	Total Depth in Meters	Depth of Water Table (in meter)		R.L. of Water Table (in meter)		Water Table Fluctuation (in meter)	Remarks Lined/Unlined and Location
						Post Monsoon	Pre Monsoon	Post Monsoon	Pre Monsoon		
21.	Pahandor	Govt. well	3.00	0.90	9.60	2.50	6.75	286.620	282.370	4.25	Lined/in school
22.	Aunri (B)	Public well	2.80	1.00	14.00	3.90	12.95	286.940	277.890	9.05	Lined/in Village
23.	Kurudih	Ashwani	2.50	0.60	14.50	2.50	12.85	279.135	267.285	10.35	Lined/in Village
24.	Parsada	Bughas Ram	2.50	G.L.	12.00	3.15	11.45	274.215	265.915	8.30	Lined/in Field
25.	Ameri	Public well	3.20	1.00	10.15	2.80	9.00	293.160	286.760	6.20	Lined/in Village
26.	Karsa	Public well	1.40	G.L.	12.75	3.95	12.30	284.695	276.345	8.35	Lined/in Village
27.	Parewadih	Chamariya	1.30	G.L.	6.35	2.00	5.95	298:835	294.085	3.95	Lined/in Village
28.	Jamgaon	Public well	2.50	0.50	13.30	4.40	10.85	294.305	287.855	6.45	Lined/in Village
29.	Gabnra	Public well	2.70	0.60	11.80	4.15	9.55	298.525	293.125	5.40	Lined/in Village
30.	Ghugwa	Manrakhan	2.10	0.70	17.90	3.10	15.60	292.405	279.895	12.50	Lined/in Field
31.	Patora	Bhuneswar	3.60	–	13.30	3.10	12.90	310.630	300.830	9.80	Lined/Near canal
32.	Chunkatta	Chanu	1.05	–	5.19	1.11	5.00	305.290	301.400	3.89	Lined/on Road
33.	Murpar	Vishal	0.95	–	7.10	2.00	6.00	301.410	296.310	4.00	Lined/on Road
34.	Dhaur	Bishram Singh	3.65	–	10.64	3.41	9.90	290.670	284.780	6.49	Lined/in Field
35.	Bendri	Kanhaiya	1.85	–	10.10	1.45	9.00	289.100	280.650	7.55	Lined/in Field
36.	Rakhi	Govt. well	2.95	0.90	14.00	2.00	12.12	304.630	254.510	10.12	Lined/on Field
37.	Achanakpur	Bhagelan	3.14	0.87	11.13	2.02	9.20	295.040	287.160	7.18	Lined/Near Tank
38.	Raweli	Public well	3.11	8.82	14.21	2.86	13.20	304.535	295.955	10.34	Lined/at Garden

post monsoon is towards NE. The close contour spacing in the areas between Kukda, Parsada, Ghughwa and Karsa village indicate the groundwater gradient is steep in these portion of the watershed area.

Pre Monsoon Water Table

The pre monsoon highest SWL 299.1 m above msl and lowest 263.5m above msl are recorded from the villages Patora and Parsada respectively. The pre monsoon water table contour map is presented in Figure 35.2. The general flow of groundwater is in NE direction. The close spacing in water table contours in between villages Purena-Somni-Sirsa-Charoda and near Ghughwa, Aunri indicate the steep gradient of groundwater around these villages.

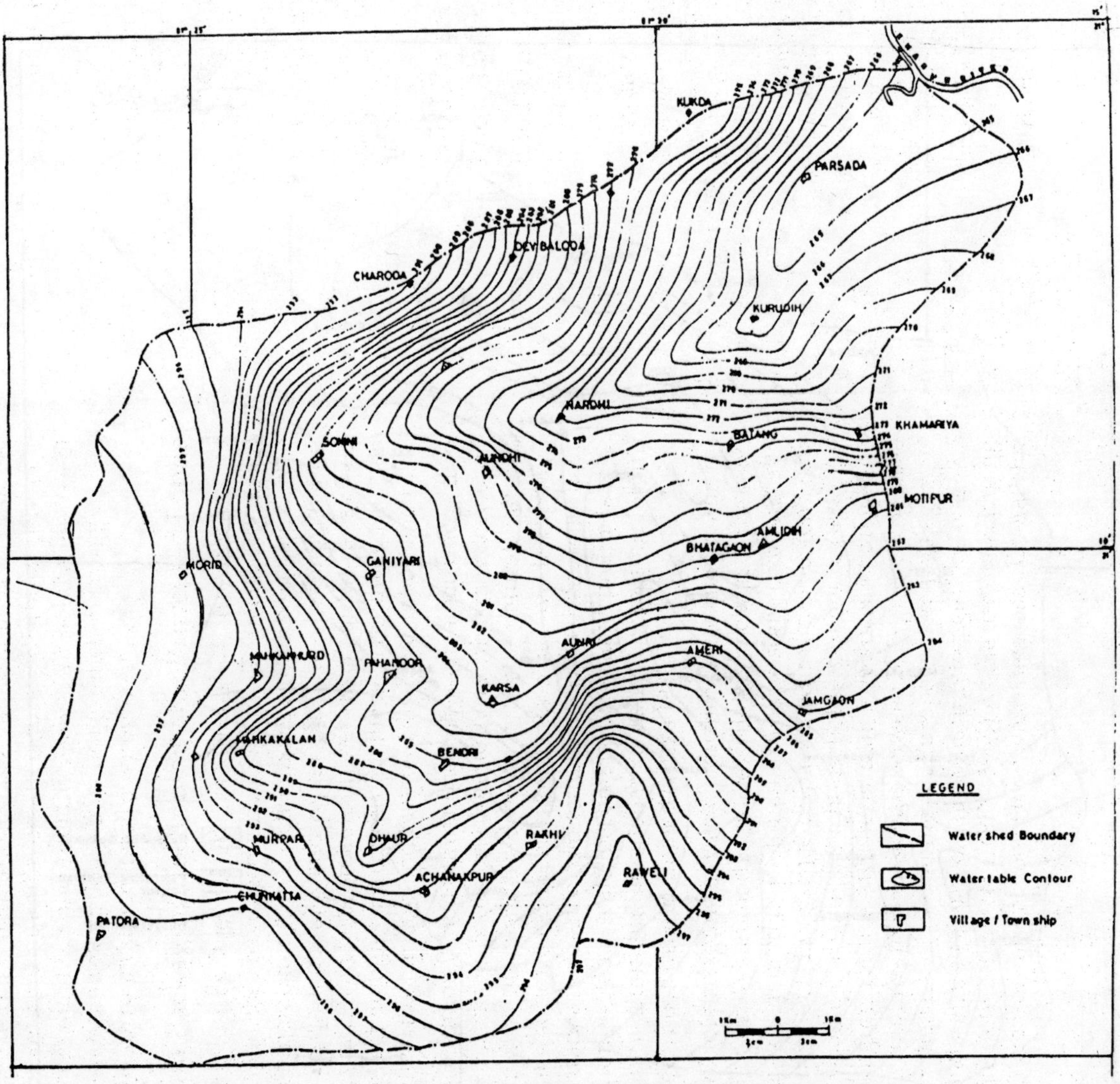

Figure 35.2: Pre Monsoon Water Table Contour Map of Somni Nala Basin

Results and Discussion

Groundwater Fluctuation

The post monsoon SWL in the watershed area is generally lowered during pre monsoon by 4–10 m. The water table contours constructed on the basis of the difference in the pre monsoon and post monsoon SWL are presented in Figure 35.3. It is clear from the Figure 35.3 that the groundwater level fluctuation is greater in wells of Ghughwa (max. 11 meters), Rakhi, Raweli, Patora as compared to the fluctuations observed in wells of NW portion of the area between Mohkakalan-Morid-Purena-Charoda and in the wells of SE part of the area near Gabnra-Ameri-Jamgaon and Motipur. The SWL fluctuation in this area is low up to 5 meters.

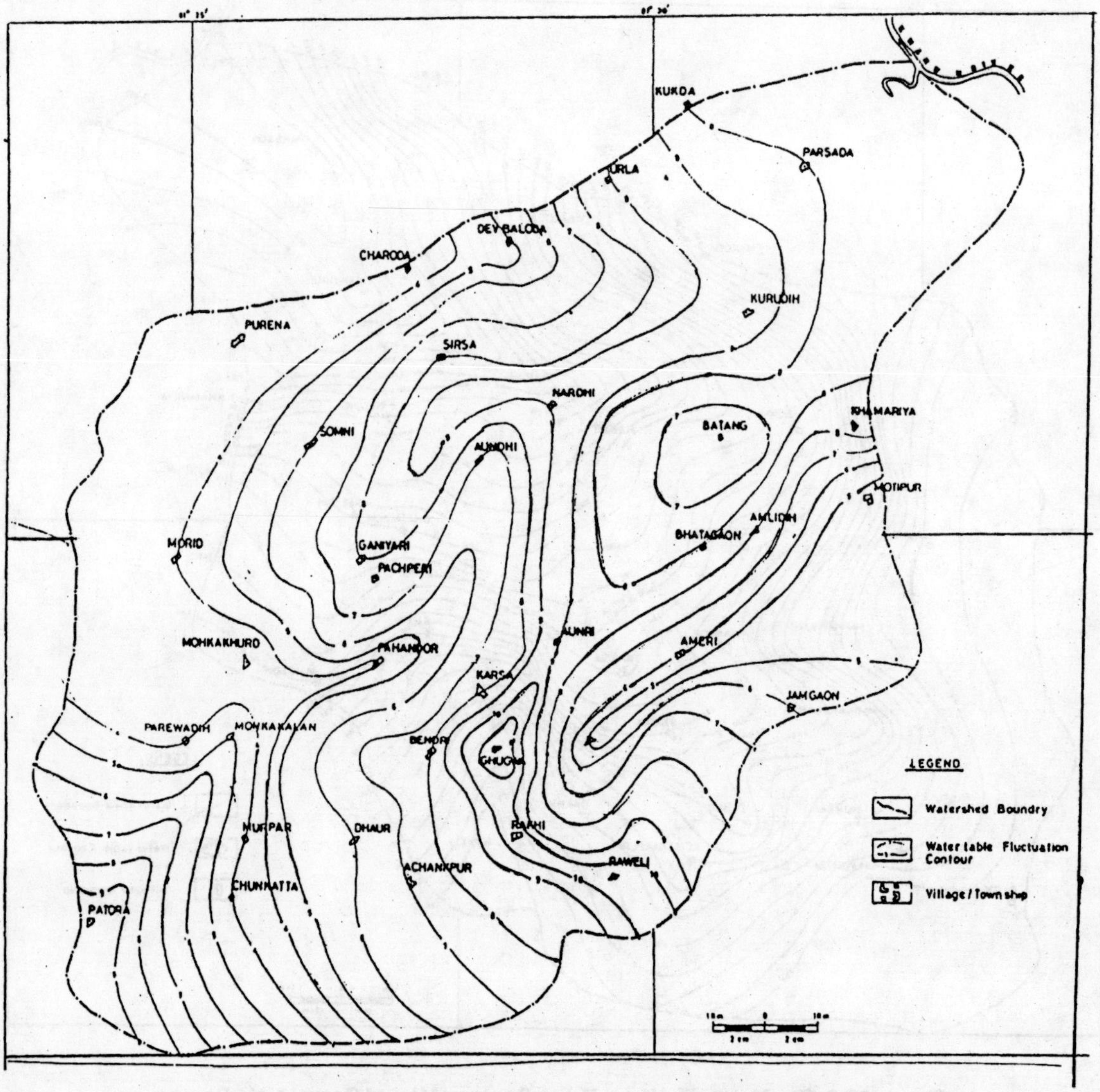

Figure 35.3: Water Table Fluctuation Map of Somni Nala Basin

Acknowledgement

The authors are highly thankful to Dr. S.K. Pande, Reader, School of Studies of Geology and WRM, Pt. Ravishankar Shukla University, Raipur (C.G.) for providing valuable guidance in connection with this work. Authors are also thankful to UGC for providing financial support under minor research project.

References

Shrivastava, P.K., 1998. Impact of Bhilai Steel Plant Effluent on Quality of Groundwater of Somni Nala Basin, District Durg, Madhya Pradesh, India, pp. 66–98.

State Groundwater Board of M.P., 1979. Geohydrological Survey Report of the Patan and Gurur Block, Durg District, Madhya Pradesh, pp. 1–81.

Chapter 36

Emetine an Antioxidant from *Melothria purpusilla* (Blume) Cogn: A Well Known Home Remedy Herbal for Humankind

S.R. Singh and M. Neshwari Devi

Postgraduate Studies Centre, HRDRI, Canchipur – 795 003, Imphal

ABSTRACT

Melothria, one of the best home remedy herbal used therapeutically in Jaundice in Manipur, since immemorial time have been investigated for exploratory chemical examination on alkaloids by using TLC Chromatography so as to supportive to the traditional home remedial knowledge of associativeness Emetine, an antioxidant absorber of free radicals both in cells and tissues have been identified from the test herbal.

Keywords: *Melothria purpusilla, Jaundice, Antioxidant, Emetine, Chloroform extract, Home remedy herbal.*

Introduction

Melothria (Fam. Cucurbitaceae) a large genus of herbs with 105 species (Kirtikar and Basu, 1935) found in warmer region of the world, is slender scandent or prostrate annual or with a perennial root monoecious or very rarely dioecious. *Melothria purpusilla* (Blume) Cogn a species out of the 105 species is a high valued medicinal significance. The plant is bitter and all aerial parts possess a potent of antihepatoxic and miraculous healing capacity from dreadful disease of Jaundice, the roots recommends as laxative antihelmintic, diuretic and in constipation (Sinha, 1990,1996). The seed in decoction are sudonitic, inflatulence, seeds in crushed used an aching bodies specially as sprained backs and when masticated relieves tooth ache, the tendrils shoots and tender leaves are used as a gentle aperlent and present bed in vertigo and biliousness. A review of literature did not reveal any information on alkaloids contains and antioxidants study of the plants.

An antioxidant is a chemical that presents the oxidation of other chemicals. In biological system the normal processes of oxidation (plants a minor contribution from ionizing radiation) produce highly reactive free radicals. These can readily react infection and damage other molecules in some cases the body uses this to enhance infection. In other cases the damage may be to the body's own cells. The presence at extremely easily oxidisable compounds in the system can make up free radicals before the damage other essential molecules. Based on all the fact and findings it is a worthwhile to assess the antioxidant alkaloid present in the indigenous herb *Melothria purpusilla* (Blume) Cogn for supportive and associative in efficiency towards the health human, health care system and understanding life saving drug plants.

Moreover, investigation of chemical constituents of *Melothria purpusilla* the authors identified, cucurbitacin glucosides, cucurbitacin aglycones, Tocopherol etc. The present paper reports the identification of emetine in the widely used semi processed novel natural products of *Melothria purpusilla*.

Materials and Methods

All the aerial parts of fresh *Melothria purpusilla* were collected from the campus of the M.U. Campus Canchipur and authenticated at the Botanical Survey of India (BSI), Shillong. A voucher specimen of the sample has been deposited at the H.R.D.R.I., Canchipur. The plants were washed thoroughly with tap water and then with distilled water. Then they were dried under shade and all the dried aerial parts were cut and ground in a mixture to obtain the plant powder. The plant powder was divided into three equal portions and each portion was subjected to extraction in Soxhlet separately with $CHCl_3$, MeOH and H_2O extraction execute with 10 gm dried powdered with 98 per cent chloroform at 10/100 ml v/v for 52 hrs, 10/100 ml v/v in MeOH for 58 hrs, and with distilled water at 10/40 ml v/v for 36 hrs. 5 ml of concentrated was carried on using TLC on Silica Gel G chromatoplate and $I/CHCl_3$ (5 per cent) was used as spraying agent.

For general qualitative test of emetine, the plant powder extract was shaking with 5 times the amount of cold HCI acid (Sp. gr. 1.19), filter and sprinkle some chlorinated lime upon the liquid and checking the distinct characteristic colour of fire red.

Results and Discussion

In general qualitative test the characteristic of fire red colour was detected showing the presence of alkaloid emetine.

The $CHCl_3$ concentrate on TLC (Silica Gel) analysis showed the spot as the standard with R_f 0.41, consequently the compound is verified to be Emetine. The structure of compound is shown in Figure 35.1.

Pure emetine forms a white non-crystallizable powder, which turns brown by exposes to light and air. It is very slightly soluble in water, the solution taste bitter and is alkaline to litmus paper. It dissolves readily in dilute acid as well as $CHCl_3$, alcohol, warm benzene, and ether is also soluble in fixed oil and benzol, but insoluble in caustic alkali and in essential oil. With acids, emetine forms neutral soluble bitter acrid and for the most part uncrystallizable salts.

Emetine ($C_{29}H_{40}N_2O_4$) M 480

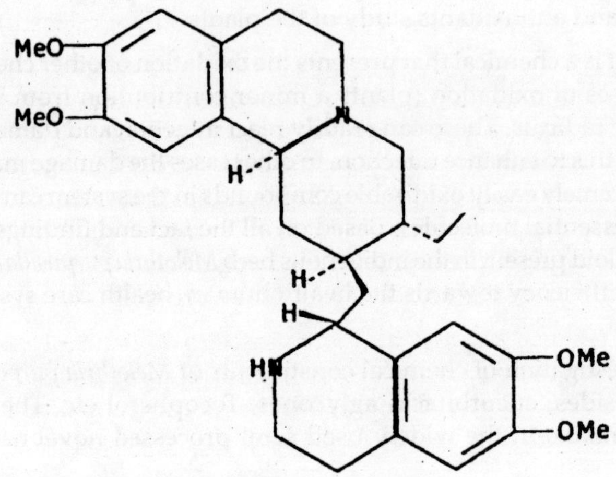

Figure 36.1: Structure of Emetine

The result of present study show that in chloroform extract of aerial part of Melothria possess definite characteristic spot with hR_f value 41, with yellow white colour compound emetine. In TLC the hR_f are to be regarded only as guide value for the migration distance, when all experiment data are accurately measured. Certain factors like layer thickness, chamber saturation, air humidity, separation effects of solvent mixture etc. which are all difficult to reproduce, can exert a marked influence to identified (Ganshirt, 1996, Stahl, 1969) compound emetine a prerequisite antioxidant.

Stahl (1964) detected the emetine and cephaeline and distinguishing between *Rio* and *Cartagena ipecacaunha* by using TLC and determined hR_f values.

In testing the stability of alkaloids Machovicova and Parrak (1964) showed TLC as a tool is far better than paper chromatography for determining the purity of emetine, a very essential antioxidant.

The similar R_f of ~0.4 or hR_f 40 with yellow white colour spot on TLC was reported by Wagner and Bladt (1996) from ipecacaunhae radix as emetine and confirmed the alkaloid emetine an antioxidant. Antioxidant being compound that protect cells against the damaging effect of oxygen species such as singlet oxygen, superoxide, physiological system. The resultant, overall pharmacological effect on the organism could be synergistic or antagonistic action (Izaddoost and Robinson, 2002). Because of the complementary and opposing physiological effects of the compounds, the expected pharmacological effects following administration of opium are complicated. The numerous interactions indicate an expected effect can result from the use of plant exudates.

During the last few decades, successions of so called "Wander Drugs" (*e.g.* reserpine, quinine, ephedrine, cocain, emetine, colchicine, digoxin, D-tubocurarine, artimisinine and gugulipid) have been discovered from plants with rich herbal home remedial role in tribal societies (Maheshwari, 1996).

The natives of Madagascar valued the rosy periwinkle (*Catharanthus roseus*) as an oral hypoglycemic (reduction of the sugar content of the blood) agent; it yielded two powerful drugs, vinblastine and vincristine, effective against Hodgkin's disease and childhood leukaemia

(Maheshwari, 1996). However larger number of folk medicines have remained endemic to certain regions or tribes. Even today, some miraculous medicines are known to the tribals and much acquired knowledge through experience of ages is usually passed on by oral traditions as a guarded secret of certain families.

Of the 120 active compounds currently isolated from higher plants and used in medicine, 74 per cent show a positive correlation between their modern therapeutic use and the traditional use of the plant from which they derived. The ancient medical knowledge of various tribes and folklore systems of medicine, sometimes referred as ethnotherapeutics, has therefore provided a powerful and more effective strategy for the discovery of clinically useful compounds. Some of the tribal medicines have been peroxyl radicals, hydroxyl, peroxy nitrile, etc. had caused oxidative stressing and cellular damage when imbalance exist, between antioxidants, reactive oxygen species, oxidative stress.

Oxidative stress being linked to ageing antherosclerosis, ischemic injury inflammation and new to degenerating diseases (Parkinson's and Alzheimer's) (Halliwell and Gulleridge 1989; Buhler, 2003) a sincere emphasized to challenge to cope with them is an antioxidant. Further, Emetine inhibits multiplication of viruses (Boyd and Knight, 1964) and exhibits an anti-inflammatory effect (10 mg/kg) (Koltai and Minkey, 1969). Growth of mouse tumors can be inhibited with emetine (Hartwell *et al.*, 1946) and the compound causes a regular reduction of stroke volume of the heart in laboratory animals (Bluthges and Marquardt, 1950).

Even though the compound alkaloid emetine was coined by Pelletier and Magendie cited in Richard 1.898 to an isolate from true ipecaec root in 1891 Kunz found the formula $C_{30}H_{40}N_2O_5$ in 1887 and established, the dyed nature of the alkaloid emetine in its saturation power with acids which in 1890 was confirmed by Blunt and Simonson.

Pharmacological effects from administration of whole plant extract can be quite different from the effects of administering on single active principle.

A recognized fact of drug treatments is that "the presence of inert substances may modify or prevent the absorbability or potency of the active constituents" (Claus and Tyler 1965). Moreover, more than one active constituent may be present in drug plant extracts. At the molecular level 2 or more active constituents could compile for the same site of action or act at different sites that are both connected with a larger incorporated in the organized systems of medicine (Chinese, Ayurveda, Unani-Tibb, Siddha, Homeopathy, Tibetan etc.). Further, therapeutically, in general Melothria are very usefull to major indication of liver enlargement, infantile jaundice, hepatopethia constipation with symptom of burning in chest, heaviness of abdomen without hungry feeling, enlargement of liver with jaundice sign, a high colour of urine and sometimes thick and yellow colour.

The action of emetine as constituent of home remedial herbal *Melothria purpusilla* interaction with other constituents like Cucurbitacin glycoside, Cucurbitacin aglycone, Saponin, Ephedrine etc. may have synergistic effect in action. Consequently, emphasized to source of emetine, an antioxidant from a home remedial herbal, *Melothria purpusilla,* and it highlightened the unlimited resource by conserving nature, and spontaneously patenting the report.

Acknowledgements

The Authors are thankful to the Director Botanical Survey of India, Shillong for information of confirming the identity of test plant and are grateful to Dr. P.N. Singh, Senior Physician for clinical and therapeutic-discussion and suggestions.

References

Blunt, 1890. *Pharm. Jour. Trans.*, 20: 809.

Boyd, E.M. and Knight, L.M., 1964. The expectorant action of cephaline, emetine and 2-dehydro emetine. *J. Pharm. Pharmacol.*, 12: 118–124.

Boericke, W., 1972. *Materia Medica*. B. Jain Publishers, San Francisco.

Buhler, 2003. *Antioxidant Activities of Flavonoids*. www.google.com.File: / A:\aza.htm.

Bluthgen, U. and Marquardt, P., 1950. Therapeutic use and pharmacology of emetine. *Pharmazie*, 5: 415–420.

Claus, E.P. and V.E. Tyler Jr., V.E., 1965. *Pharmacognosy*, 5th Edition. Lea and Febiger, Philadelphia, pp. 45.

Ganshirt, H., 1996. *Documentation of Thin Layer Chromatography in Thin layer Chromatography*. Springer International Students Edition, Japan.

Halliwell, B. and Gutterdge, J.M.C., 1989. *Free Radicals in Biology and Medicine*. Clarendon Press, Oxford.

Hartwell Shear, J.L.M.J., Johnson, J.M. and Kornberg, S.R.L., 1946. Selection and J.L.M.J. synthesis of organic compounds (for action against tumors in mice). *Cancer Res.*, 6: 489.

Izaddoost, M. and Robinson, T., 2002. Synergism and antagonism in the pharmacology of alkaloidal plants. *Herbs, Spices and Medicinal Plants*, Vol. 2.

Kirtikar, K.R. and Basu, B.D., 1973. *The Indian Medicinal Plant*, Vol. 1. Lalit Mohon Basu, Allahabad.

Koltai, M. and Minkey, E., 1969. Effect of various nucleic acids and protein synthesis inhibitors an anaphylactoid inflammation in rats. *J. Pharmacol.*, 6: 175–182.

Kunz, H., 1887. *Jahresb. der Pharm.*, pp. 416.

Machovicova, F. and Parrak, F., 1964. *Ceskoslov. Farm.*, 13: 200.

Maheshwari, J.K., 1996. New vistas in ethnobotany (Editorial). *J. Econ. Taxon. Bot.*, Additional Series, 12(1–11).

Richard, A., 1898. Ipecacuanha (USP). *Ipecac.*

Semonson, W., 1890. *Proc. Amer. Pharm. Assoc.*, pp. 188.

Sinha, S.C., 1996. *Medicinal Plant of Manipur*. Mass and Sinha, Imphal.

Sinha, S.C., 1990. Notes on ethnomedicinal plants of Manipur (New report). *Curr. Pharm. Letters*, 1(1): 3–7.

Stahl, E., 1969. *Thin Layer Chromatography*. Springer International Students Edition, New York, Japan, pp. 104.

Stahl, E., 1964 (in English). *An Gew.* Chem. Intern. Ed., 3: 784.

Wagner, H. and Bladt, S., 1996. *Plant Drug Analysis*. Springer-Verlag, Berlin, Heidelberg, pp. 384.

Willmar, S., 1991. *Homeopathic Repititorium*. German Edition.

Chapter 37

Growth Analysis of Cowpea [*Vigna unguiculata* (L.) Walp] as Influenced by Phosphorus, Bioinoculants, Zinc and Sulphur

Charanjit Singh Kahlon and Sharanappa
Department of Agronomy, GKVK, UAS, Bangalore – 65

ABSTRACT

A field experiment was conducted during rainy season of 2003–04 to study the effect of phosphorus, bioinoculants, zinc and sulphur on growth parameter. Application of 50 kg P_2O_5, 20 kg S, 10 kg $ZnSO_4$ ha^{-1} along with Phosphorus solubilizing Bacteria (PSB) and Vesicular Arbuscular mycorrhiza (VAM) resulted in higher Net Assimilation Rate (14.30 g $m^{-2}day^{-1}$, Crop Growth Rate (11.96 g $m^{-2}day^{-1}$ whereas, Relative Growth Rate was higher with 75 kg P_2O_5, 20 kg S, 10 kg $ZnSO_4$ ha^{-1} along with PSB and VAM (0.040 g $g^{-1}day^{-1}$) and Absolute Growth Rate (g $plant^{-1}$ day^{-1}) with 50 kg P_2O_5, 20 kg S ha^{-1} along with PSB and VAM (0.74 g $plant^{-1}$ day^{-1}).

Keywords: Cowpea, Phosphorus solubilizing bacteria, Vesicular arbuscular mycorrhiza, Phosphorus levels.

Introduction

Pulses are always given secondary importance compared to cereals, their prime value rests with their high nutritive value in the diet. Cowpea contains 23.4 per cent protein, 60.3 per cent carbohydrates and 1.8 per cent fat besides being a good source of vitamins and phosphorus.

Phosphorus is one of the important macronutrients of plant, involved virtually in all major metabolic process in plant growth and development, but its availability to crops seldom exceeds 20 per cent as it gets fixed in the soil and is unavailable to plants Goswamy *et al.* (1983). This reduces the applied phosphorus use efficiency leading to improper P nutrition. Further escalating costs of P fertilizers resulted in lower P fertilizer application by farmers resulting in nutrient imbalance of NPK. Thus phosphorus management through the biological sources that are cheap and ecofriendly play an important role in enhancing the efficiency of applied phosphorus and crop productivity. Use of P solubilizing and mobilizing microorganisms has been reported to reduce P fixation and increase the available phosphorus content in soil both from soluble and insoluble phosphatic sources Shinde and Patil (1985) by solubilizing the unavailable and fixed soil phosphorus Sperber (1958). Sulphur was a neglected plant nutrients and results in sulphur deficiency and has been receiving increasing attention world wide during the past few years. Zinc is recognized as key element in protein synthesis and also involved in nitrogen fixation in leguminous crops. Hence, present investigations on cowpea were conducted to analyze crop growth and yield as influenced by Phosphorus, Bioinoculants, Zinc and Sulphur.

Materials and Methods

The field experiment was conducted at the Zonal Research Station, Gandhi Krishi Vignana Kendra, University of Agricultural Sciences, Bangalore during rainy season 2003–2004. The soil was red sandy clay loam with pH 6.5 and organic carbon content of 0.46 per cent. Available nitrogen, phosphorus, sulphur and zinc content of soil was 205, 22.68, 14.0 and 1.27 kg ha^{-1} respectively. The experiment was conducted in randomized block design replicated thrice keeping 12 treatments as detailed in Table 37.1. The uniform dose of 25 kg nitrogen and potassium were applied for all the treatments. Whereas, phosphorus was applied at the rate of 0, 25, 50 and 75 kg ha^{-1} as per treatments. Sulphur was applied at 20 kg ha^{-1} through gypsum and zinc was supplied @ 10 kg $ZnSO_4$ ha^{-1}. Phosphorus solubilizing culture was used *Bacillus megaterium* var. *phosphaticum* while mycorrhiza culture used was *Glomus mosseae.* The fertilizer and microbial culture were used as basal dose. Cowpea variety TVX-944-02C was sown in rows spaced at 45 cm with an intra row spacing of 15 cm. Thinning was done to retain one seedling per hill. Gap filling was attended with 10 days after sowing by dibbling the seeds to maintain the optimum plant population. Periodical observations on various parameters were recorded during the crop season. The computation for Leaf area per plant, AGR West *et al.* (1920), RGR Fisher (1921), NAR Gregory (1926), COR Watson (1952) were done from the required data taken in field.

Results and Discussion

Leaf Area (cm^2)

Leaf area as influenced by phosphorus, bioinoculants, sulphur and zinc recorded at 30 and 60 DAS and at harvest is presented in Table 37.1. Leaf area has increased progressively upto 60 DAS and then declined at harvest. It did not differ significantly due to various treatments at all growth stages. At 30 DAS it was highest with the application of 75 kg P_2O_5, 20 kg S ha^{-1} along with PSB and VAM (495 cm^2) where as, it was lowest with no phosphorus, 20 kg S, 10 kg $ZnSO_4$ ha^{-1} along with PSB and VAM (313.7 cm^2). At 60 DAS it was recorded highest with the application of 25 kg P_2O_5, 20 kg S ha^{-1} along with PSB and VAM (900.4 cm^2) and lowest with 25 P_2O_5 per ha (497 cm^2). At harvest it was highest with 75 kg P_2O_5, 20 kg S and 10 kg $ZnSO_4$ ha^{-1} along with PSB and VAM (487.3 cm^2) and lowest with no phosphorus, 20 kg S and 10 kg $ZnSO_4$ ha^{-1} along with PSB and VAM (286.3 cm^2). Lower leaf area over control attribute due to the antagonistic effect of zinc on phosphorus and results in lower availability of phosphorus.

Table 37.1: Leaf Area (cm^2) and Grain Yield (kg ha^{-1}) as Influenced by Phosphorus Bioinoculants, Sulphur and Zinc in Cowpea

Treatments	*30 DAS*	*60 DAS*	*At Harvest*	*Grain Yield kg ha^{-1}*
T_1: Zero P_2O_5 ha^{-1} (control)	369.6	645.5	309.8	1541
T_2: 25 kg P_2O_5 ha^{-1}	380.7	479.0	389.3	1680
T_3: 50 kg P_2O_5 ha^{-1}	353.0	631.6	361.9	1900
T_4: 75 kg P_2O_5 ha^{-1}	434.5	804.6	405.3	1925
T_5: Zero P_2O_5 + 20 kg S ha^{-1} + PSB + VAM	444.7	612.3	371.6	1794
T_6: 25 kg P_2O_5 + 20 kg S ha^{-1} + PSB + VAM	417.1	900.4	442.2	2050
T_7: 50 kg P_2O_5 + 20 kg S ha^{-1} + PSB + VAM	406.7	766.4	383.0	2124
T_8: 75 kg P_2O_5 + 20 kg S ha^{-1} + PSB + VAM	495.4	811.2	356.1	2130
T_9: Zero P_2O_5 + 20 kg S + 10 kg $ZnSO_4$ ha^{-1} + PSB + VAM	313.7	608.6	286.3	1780
T_{10}: 25 kg P_2O_5 + 20 kg S + 10 kg $ZnSO_4$ ha^{-1} + PSB + VAM	404.6	679.6	401.0	2192
T_{11}: 50 kg P_2O_5 + 20 kg S + 10 kg $ZnSO_4$ ha^{-1} + PSB + VAM	362.7	647.0	392.6	2208
T_{12}: 75 P_2O_5 + 20 kg S + 10 kg $ZnSO_4$ ha^{-1} + PSB + VAM	386.5	654.0	487.3	2032
CD (P = 0.05)	NS	NS	NS	220

DAS: Days After Sowing; NS: Non Significant.

Table 37.2: Net Assimilation Rate (g $m^{-2}day^{-1}$) and Crop Growth Rate (g $m^{-2}day^{-1}$) as Influenced by Phosphorus Bioinoculants, Sulphur and Zinc in Cowpea

Treatments	*Net Assimilation*		*Crop Growth Rate*	
	30–60 DAS	*60 DAS–Harvest*	*30–60 DAS*	*60 DAS–Harvest*
T_1: Zero P_2O_5 ha^{-1} (control)	5.30	5.74	3.93	4.01
T_2: 25 kg P_2O_5 ha^{-1}	5.22	4.10	3.26	2.60
T_3: 50 kg P_2O_5 ha^{-1}	6.23	9.91	4.50	7.20
T_4: 75 kg P_2O_5 ha^{-1}	4.50	7.05	4.10	6.30
T_5: Zero P_2O_5 + 20 kg S ha^{-1} + PSB + VAM	4.20	4.16	3.30	2.99
T_6: 25 kg P_2O_5 + 20 kg S ha^{-1} + PSB + VAM	4.03	8.83	3.90	8.74
T_7: 50 kg P_2O_5 + 20 kg S ha^{-1} + PSB + VAM	4.59	13.49	3.97	11.39
T_8: 75 kg P_2O_5 + 20 kg S ha^{-1} + PSB + VAM	5.09	9.54	4.90	8.20
T_9: Zero P_2O_5 + 20 kg S + 10 kg $ZnSO_4$ ha^{-1} + PSB + VAM	4.70	7.12	3.19	4.69
T_{10}: 25 kg P_2O_5 + 20 kg S + 10 kg $ZnSO_4$ ha^{-1} + PSB + VAM	4.80	12.61	3.83	10.02
T_{11}: 50 kg P_2O_5 + 20 kg S + 10 kg $ZnSO_4$ ha^{-1} + PSB + VAM	4.95	14.30	3.66	11.96
T_{12}: 75 P_2O_5 + 20 kg S + 10 kg $ZnSO_4$ ha^{-1} + PSB + VAM	3.71	3.37	2.83	2.83
CD (P = 0.05)	NS	2.53	NS	2.08

Relative Growth Rate (g $g^{-1}day^{-1}$)

Relative growth rate (RGR) differed significantly during 60 DAS to harvest while it did not differ significantly from 30 to 60 DAS. It was recorded significantly higher with 75 kg P_2O_5, 20 kg S, 10 kg

$ZnSO_4$ ha^{-1} along with PSB and VAM (0.040 g $g^{-1}day^{-1}$) from 60 DAS to harvest (Table 37.3) and on par with 25 kg P_2O_5, 20 kg S, 10 kg $ZnSO_4$ ha^{-1} along with PSB and VAM (0.37 g $g^{-1}day^{-1}$) and 50 kg P_2O_5, 20 kg S, 10 kg $ZnSO_4$ ha^{-1} along with PSB and VAM (0.033 g $g^{-1}day^{-1}$) whereas it was lowest with control (0.018 g $g^{-1}day^{-1}$). Highest could be attributed due to more availability of phosphorus and less fixation as PSB and VAM also contributed their effect on availability of phosphorus.

Table 37.3: Relative Growth Rate (g $g^{-1}day^{-1}$) and Absolute Growth Rate (g $plant^{-1}day^{-1}$) as Influenced by Phosphorus Bioinoculants, Sulphur and Zinc in Cowpea

Treatments	*Relative Growth Rate*		*Absolute Growth Eate*	
	30–60 DAS	*60 DAS–Harvest*	*30–60 DAS*	*60 DAS–Harvest*
T_1: Zero P_2O_5 ha^{-1} (control)	0.041	0.018	0.26	0.27
T_2: 25 kg P_2O_5 ha^{-1}	0.039	0.022	0.22	0.34
T_3: 50 kg P_2O_5 ha^{-1}	0.042	0.024	0.32	0.49
T_4: 75 kg P_2O_5 ha^{-1}	0.040	0.028	0.28	0.54
T_5: Zero P_2O_5 + 20 kg S ha^{-1} + PSB + VAM	0.033	0.021	0.23	0.31
T_6: 25 kg P_2O_5 + 20 kg S ha^{-1} + PSB + VAM	0.037	0.028	0.27	0.55
T_7: 50 kg P_2O_5 + 20 kg S ha^{-1} + PSB + VAM	0.037	0.035	0.26	0.74
T_8: 75 kg P_2O_5 + 20 kg S ha^{-1} + PSB + VAM	0.042	0.025	0.33	0.53
T_9: Zero P_2O_5 + 20 kg S + 10 kg $ZnSO_4$ ha^{-1} + PSB + VAM	0.042	0.021	0.24	0.30
T_{10}: 25 kg P_2O_5 + 20 kg S + 10 kg $ZnSO_4$ ha^{-1} + PSB + VAM	0.034	0.037	0.20	0.65
T_{11}: 50 kg P_2O_5 + 20 kg S + 10 kg $ZnSO_4$ ha^{-1} + PSB + VAM	0.038	0.033	0.22	0.54
T_{12}: 75 P_2O_5 + 20 kg S + 10 kg $ZnSO_4$ ha^{-1} + PSB + VAM	0.034	0.040	0.19	0.71
CD (P = 0.05)	NS	0.010	NS	0.20

Absolute Growth Rate (g $plant^{-1}day^{-1}$)

Absolute growth rate differed significantly from 60 DAS to harvest while it did not differ significantly from 30 to 60 DAS whereas, it was highest with 50 kg P_2O_5, 20 kg S ha^{-1} along with PSB and VAM (0.74 g $plant^{-1}day^{-1}$) (Table 37.3) and on par with 25 kg P_2O_5, 20 kg S, 10 kg $ZnSO_4$ ha^{-1} along with PSB and VAM (0.65 g $plant^{-1}day^{-1}$), 25 kg P_2O_5, 20 kg S ha^{-1} along with PSB and VAM (0.55 g $plant^{-1}day^{-1}$) and 50 kg P_2O_5, 20 kg S, 10 kg $ZnSO_4$ ha^{-1} along with PSB and VAM (0.54 g $plant^{-1}day^{-1}$) while it was lowest in control (0.27 g $plant^{-1}day^{-1}$). This may be attributed to better dry matter accumulation from 60 days after sowing to harvest and more availability of phosphorus which being a major nutrient promotes growth, development and assimilation of photosynthates Abbas *et al.* (1994) and Zarong and Munns (1980).

Net Assimilation Rate (NAR)

Net assimilation rate and crop growth rate as influenced by phosphorus, bioinoculants, sulphur and zinc is presented in Table 37.2. Net assimilation rate did not differ significantly from 30 to 60 DAS. It differed significantly from 60 DAS to harvest while it whereas, it was recorded highest with 50 kg P_2O_5, 20 kg S, 10 kg $ZnSO_4$ ha^{-1} along with PSB and VAM from 60 DAS to harvest (14.3 g $m^{-2}day^{-1}$) which was on par with 50 kg P_2O_5, 20 kg S per ha along with PSB and VAM (13.498 g $m^{-2}day^{-1}$) and 25 kg P_2O_5, 20 kg S, 10 kg $ZnSO_4$ ha^{-1} along with PSB and VAM (12.61 g $m^{-2}day^{-1}$) while it was lowest with 75 kg P_2O_5, 20 kg S, 10 kg $ZnSO_4$ ha^{-1} along with PSB and VAM (3.37 g $m^{-2}day^{-1}$) which was on par with control (5.74 g $m^{-2}day^{-1}$). This increased NAR could be attributed due to better dry matter accumulation during this period in the plants and resulted in better photosynthesis and utilization of solar energy by crop canopy. The better uptake of zinc was facilitated by VAM may also be ascribed to increasing photosynthetic efficiency of cowpea which lead to higher assimilate production. The results of this study are in agreement with the findings of Edward and Mohammed (1973), Shorti *et al.* (1983) and Agrawal *et al.* (1996) that zinc helps in increasing the photosynthetic efficiency.

Crop Growth Rate (CGR)

Crop growth rate did not differ significantly from 30–60 DAS, however it was differed significantly from 60 DAS to harvest. The highest CGR was recorded with 50 kg P_2O_5, 20 kg S, 10 kg $ZnSO_4$ ha^{-1} along with PSB and VAM (11.96 g $m^{-2}day^{-1}$) from 60 DAS to harvest which was on par with 50 kg P_2O_5, 20 kg S ha^{-1} along with PSB and VAM (11.39 g $m^{-2}day^{-1}$) (Table 37.2) and 25 kg P_2O_5, 20 kg S, 10 kg $ZnSO_4$ ha^{-1} along with PSB and VAM (10.02 g $m^{-2}day^{-1}$) while it was lowest with 50 kg P_2O_5 (2.60 g $m^{-2}day^{-1}$) which was statistically on par with control (4.01 g $m^{-2}day^{-1}$). This higher CGR could be attributed due to better leaf area index and net assimilation rate.

Grain Yield (kg ha^{-1})

The grain yield of cowpea differed significantly due to treatments (Table 37.1). It was significantly highest with the application 50 kg P_2O_5, 20 kg S, 10 kg $ZnSO_4$ ha^{-1} along with PSB and VAM (2208 kg ha^{-1}) and was on par with application of 25 kg P_2O_5, 20 kg S, 10 kg $ZnSO_4$ ha^{-1} along with PSB and VAM (2192 kg ha^{-1}), 75 kg P_2O_5, 20 kg S ha^{-1} along with PSB and VAM (21.30 kg ha^{-1}), 50 kg P_2O_5, 20 kg S ha^{-1} along with PSB and VAM (2124 kg ha^{-1}) and 25 kg P_2O_5, 20 kg S ha^{-1} along with PSB and VAM (2050 kg ha^{-1}) and significantly superior to rest of treatments, whereas control recorded significantly lower grain yield (1541 kg ha^{-1}) and was statistically on par with 25 kg P_2O_5 ha^{-1} (1680 kg ha^{-1}) while the other treatments were intermediate in grain yield. The higher seed yield may be attributed to favourable effect of phosphorus, sulphur and zinc on effective number of nodules, dry matter accumulation, photosynthetic efficiency and nitrogen fixation coupled with more release of other nutrients and uptake by PSB and VAM. Similar effect of phosphorus was also reported by Kumar *et al.* (1993) and Sepatnekar, (1994).

References

Abbas, M., Singh, M.P., Nigam, K.B. and Kandalkar, V.S., 1994. Effect of phosphorus, plant densities and plant types on different growth and physiological parameters of soybean (*Glycine max*). *Indian J. Agron.*, 39: 246–248.

Agrawal, V.K., Dwivedi, Shrivastava, S., Patel, R.S., and Nigam, P.K., 1996. Influence of phosphorus and zinc application on physiological determinants of growth and productivity of soybean (*Glycine max* L.) Merr. *Crop Res.*, 12(2): 192–195.

Edward, G.C. and Mohammad, A.K., 1973. Reduction in carbonic anhydrse activity in zinc deficient leaves of *Phaseolus vulgaris* L. *Crop Sci.*, 13: 351–354.

Fisher, 1921. Some remarks on the methods formulated in a recent article on "The quantitative analysis of plant growth". *Ann. Appl. Soil.*, 7: 367–372.

Goswamy, N.N., Mahapatra, I.C. and Komath, M.B., 1983. Strategy for P and K fertilizer recommendations in multiple cropping systems with particular reference to their use efficiency pp 73– 78. In: *Fertilizer Use Under Multiple Cropping Systems*, FAO Fertilizer and Plant Nutrition Bulletin, 5 FAO, Rome.

Gregory, P.G., 1926. The effect of climate conditions on the growth of barley. *Ann. Bot.*, 40: 1–26.

Kumar, A., Sinha, M.N. and Raj, P.K., 1993. Comparative performance of chickpea, lentil and kla at various P levels and inoculation. *Annals of Agricultural Research*, 14(3): 345–347.

Sepatnekar, H.G., Patil, P.Z. and Rasal, P.H., 1994. Efficient utilization of phosphate fertilizers through solubilizers in wheat. In: *34th Annual Conference of AMI*, PAU, Ludhiana, February, 9–11.

Shinde, B.N. and Patil, A.J., 1985. Use of phosphate solubilizing cultures and rock phosphate on phosphate availability to wheat in black calcareous soil. *Curr Res. Report*, 1: 105–106.

Shorti, C.K., Mohanti, P., Rathore, V.S. and Tewari, M.N., 1983. Zinc deficiency limits the photosynthetic enzyme activities in *Zea mays*. *Biochem. Physiol. Planz.*, 178: 213–217.

Sperber, J.I., 1958. The incidence of apetite solubilizing organisms in the rhizosphere and soil. *Australian J. Agric. Res.*, 9: 778.

Watson, D.J., 1952. The physiological basis of variation in yield. *Adv. Agron.*, 4: 101–144.

West, C., Brigass, G.B. and Kidd, F., 1920. Methods and significant relations in a quantitative analysis of plant growth. *New Phytol.*, 19: 200–207.

Zarong, M.G. and Munns, D.N., 1980. Effect of phosphorus and sulphur nutrition on soluble sugars and growth in *Clitroria ternatea* L. *Plant and Soil*, 55: 243–250.

Chapter 38

Effect of Isopod Parasite, *Cymothoa indica* on Pearl Spot, *Etroplus suratensis* (Bloch) from Parangipettai Coastal Waters (Southeast Coast of India)

M. Rajkumar, P. Perumal, P. Santhanam and N. Veerappan*

Centre of Advanced Study in Marine Biology, Annamalai University, Parangipettai – 608 502, Tamil Nadu, India

ABSTRACT

The present study reported for the first time the effect of isopod parasite, *Cymothoa indica* infestation on *Etroplus suratensis*, a culturable and commercially important fish from Parangipettai, South East Coast of India. The weight loss in host fishes (male 33.23 per cent; female 49.57 per cent) was observed. The weight of uninfected female fish was found to be higher than that of infested one. The calus like thickening developed on the gill arch and gill filaments of host fish due to the constant irritation caused by the appendages of the parasite. The reduction of gill surface area was due to the attachment of the parasites. The observed maximum reduction in the first gill arch was mainly due to the heavy pressure exerted by the parasite. Details of gross lesions observed in the branchial chamber, buccal cavity and body surface were enumerated. Heavy infestations of parasitic juveniles have the potential to kill small fingerlings. The swimming capacity of the fish was also found to be affected.

Keywords: Isopod fish parasite, Loss of weight, Gross lesions, Gill surface reduction.

---

* Corresponding Author: E-mail: arunachalashivamdr@yahoo.com.

Introduction

Several facultative and obligatory parasitic members of the order Isopoda are deleterious to fishes, capable of exhibiting tremendous destructive activity (Trilles, 1979; Maxwell, 1982). Parasites interfere in the biology of fishes as they affect their behaviour, health and distribution (Rohde, 1993). Isopods associated with fish constitute a significant population of crustacean ectoparasites, yet little is known about their biology (Bunkley-Williams and Williams, 1998).

To date, about 450 isopod species have been identified as parasites in freshwater and marine teleosts, as well as elasmobranchs (Kabata, 1985; Moiler and Andress, 1986). Most are cymothoids flabelliferans, a group of isopods with a short free-living planktonic phase. Cymothoids are exclusively parasitic and the presence of a few adults can cause damage to hosts. They are protandric hermaphrodites, living on the skin, in the gill chambers, or in the mouth of the fish the position is thus highly specific (Baer, 1951 and Trilles, 1969). They feed on host blood, plasma in wounds and fish tissue, sometimes resulting in massive tissue damage (Bunkley-Williams and Williams, 1998).

Diseases are a most serious limiting factor in the aquaculture production and knowledge of the fish diseases and parasites is highly essential for a successful aquaculture, particularly in a country like India, which has long and highly productive coastal waters. But, a glance of the relevant literature shows that the tropical coastal and brackish-waters, which likely to contain many unknown parasites and pathogens, are poorly documented.

Isopoda absorb their nourishment directly from the host's body and thus depend upon their hosts for feeding (Pillai, 1958). Segal (1987) observed the death of host fish, *Meniida beryllina* died due to the attack of isopod parasite, *Nerocila acuminate* which lived the physical damage produced by the isopod parasite upon the host has been studied by a few workers. Mini Nair (1987) however, found that the feeding intensity of fishes was unaffected due to the infestation by isopods.

In spite of all these reports, works pertaining to pathological effects of isopod parasites on the physiology of host fishes are scanty particularly from Parangipettai environments. Hence the present attempt was made to study the effect of isopod infestation on (host) fish, with reference to reduction of the respiratory surface area and loss of weight in the pearl spot, *Etroplus suratensis*.

Materials and Methods

The samples of fish, *E. suratensis* were collected from Vellar estuary (Lat. 11°29′N, Long. 79°96′E) [(Parangipettai coast (Figure 38.1)] and examined thoroughly for the presence of isopod parasites. *E. suratensis* is an Asian cichlid (Family: Cichlidae) endemic to Southern India and the Island of Sri Lanka and they are popularly called as Pearl spot (Green chromide) (Keshava *et al.*, 1988). It is an oval-shaped fish, laterally compressed fish with a pointed head. The body colour is ereamy olive green to greenish brown. The body is marked with six to eight slightly diagonal transverse bars which may at times, be indistinct. Each scale has a golden spot and the fins are the same colour as the body. The anal fin may have some blue iridescence. The back of the dorsal fin is a reddish-orange colour. Their face lacks the spangling of the body and is instead a metallic apple green, which is very pronounced on the forehead and radiating into their eye. The dorsal fin is long and the base extends to the caudal peduncle. The soft rays are more elongated than the hard front rays of the dorsal fin. The total length of the fish was ranged from 400–460 mm. It is a common species occurring throughout the year (Mills and Lambert, 2004). The site of attachment, orientation of parasites (on the host) and the number of parasites in each location were recorded.

Figure 38.1: *Cymothoa indica* Parasite Attach on Mouth of the *Etroplus suratensis*

To study the effect of infestation on the pearl spot fish, (*E. suratensis*), the length and weight data were analysed. The fishes were categorised as infested and uninfested. The average length and weight of infested and uninfested fishes were determined. The loss of weight and percentage loss of weight of male and female fishes were calculated. Respiratory surface area of gill arch was measured to find out the influence of parasites on the gill surface area. The gill arches of uninfested and infested fishes were carefully dissected out and blotted to remove the moisture. Then the total surface area of gill arch of both infested and uninfested fish was compared, and the differences were noted and considered as the reduction of respiratory area due to infestation.

Results

Effect on Weight of Host Fish

The effect of *C. indica* infestation on the weight of *E. suratensis* is presented in (Tables 38.1 and 38.2) (Figures 38.2 and 38.3). It is evident from the results that the infestation affected the weight was 6.7 g when compared to that of parasitized fish. Further the weight loss was more pronounced in female (49.57 per cent) than in males (33.23 per cent).

Table 38.1: Effect of *Cymothoa indica* in Relation to the Body Weight (g) of *Etroplus suratensis*

Nature of Infestation	*Mean Body Weight*	*Reduction of Weight*	*Percentage Reduction of Weight*
Uninfected	87.97	18.24	6.72
Infected	66.64	10.40	

Table 38.2: Percentage Weight (g) Variation in Male and Female of *Etroplus suratensis*

	Sex of Host	*Uninfected*	*Infected*
Mean Weight	Male	54.72	30.48
	Female	41.60	35.56
Loss of weight	Male	–	10.13
	Female	–	17.63
Percentage loss of weight	Male	–	33.23
	Female	–	49.57

Table 38.3: Infestation Rate of *Cymothoa indica* on *Etroplus suratensis* During Different Months (July 2003–June 2004), from Parangipettai Coast

Months	*No. of Fishes Examined*	*No. of Fishes Infested (% Prevalence)*	*No. of Parasites Collected (Mean Intensity)*
July 2003	41	18 (43.9)	23 (1.2)
August	44	13 (29.5)	16 (1.2)
September	43	16 (37.2)	22 (1.3)
October	58	14 (24.1)	25 (1.7)
November	22	10 (45.4)	12 (1.2)
December	18	7 (38.8)	13 (1.8)
January 2004	21	5 (23.8)	10 (2.0)
February	32	8 (25.0)	10 (1.2)
March	37	11 (29.7)	12 (1.0)
April	20	6 (30.0)	8 (1.3)
May	26	3 (11.5)	5 (1.6)
June	35	9 (25.7)	10 (1.1)
Total	**397**	**120 (30.2)**	**166 (1.3)**

Effect on Length and Weight Relationship

The effects on length and weight relationship of *E. suratensis* by the infestation of *C. indica* in both male and female fishes were compared and the results are as follows:

Length and Weight Relationship in Male Fish

The rate of growth is almost similar in uninfested and infested fishes. It can be inferred that the rate of growth by weight and average size was not affected much by infestation.

Length and Weight Relationship in Female Fish

The rate of increased growth in terms by weight in uninfested female fish was found to be higher than that of the infested. It can be inferred that the weight gain is more in uninfested fish than in the infested fish.

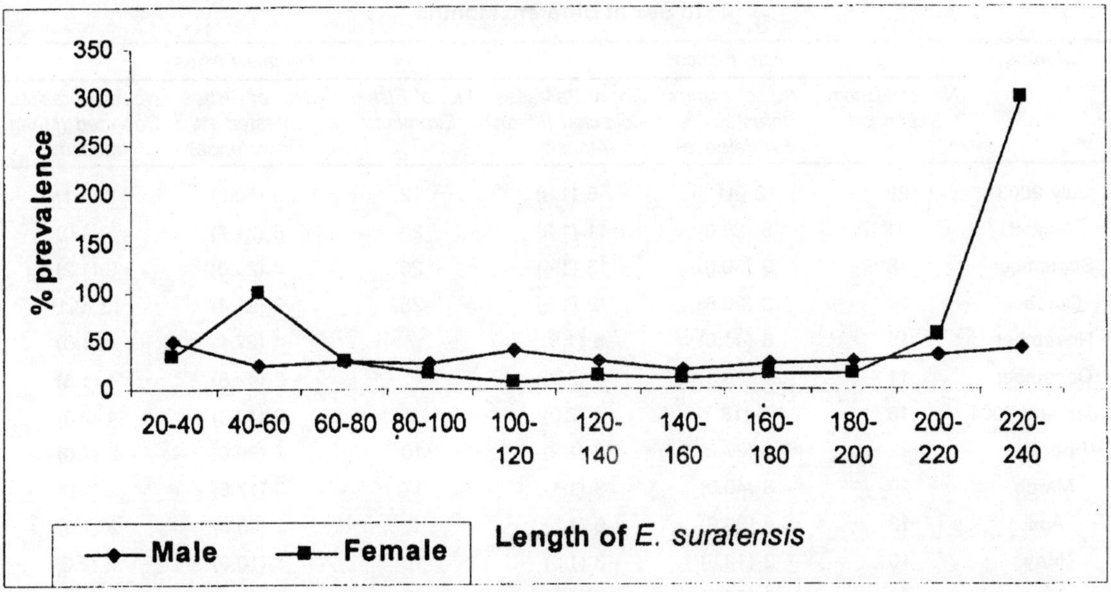

Figure 38.2: The Percentage Prevalence in Relation to Size and Sex of *Etroplus suratensis*

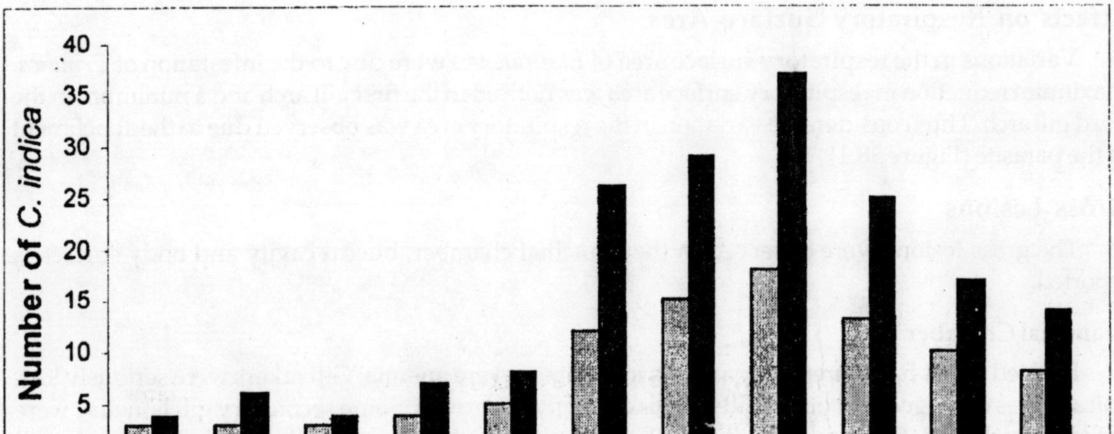

Figure 38.3: The Number of Male and Female *C. indica* in Relation to Length of *Etroplus suratensis*

Table 38.4: Infestation of *Cymothoa indica* on *Etroplus suratensis* in Relation to Sex at Different Months

Months	*Male Fishes*			*Female Fishes*		
	No. of Fishes Examined	*No. of Fishes Infested (% Prevalence)*	*No. of Parasites Collected (Mean Intensity)*	*No. of Fishes Examined*	*No. of Fishes Infested (% Prevalence)*	*No. of Parasites Collected (Mean Intensity)*
July 2003	29	12 (41.3)	16 (1.3)	12	6 (50.0)	7 (1.1)
August	21	8 (38.0)	11 (1.3)	23	5 (21.7)	5 (1.0)
September	18	9 (50.0)	13 (1.4)	25	7 (28.0)	9 (1.2)
October	30	8 (26.6)	12 (1.5)	28	6 (21.4)	13 (2.1)
November	15	6 (40.0)	8 (1.3)	7	4 (57.1)	4 (1.0)
December	11	4 (36.3)	9 (2.2)	7	3 (42.8)	4 (1.3)
January 2004	16	3 (18.7)	6 (2.0)	5	2 (40.0)	4 (2.0)
February	22	6 (27.2)	8 (1.3)	10	2 (20.0)	2 (1.0)
March	20	8 (40.0)	9 (1.1)	17	3 (17.6)	3 (1.0)
April	12	4 (33.3)	6 (1.5)	8	2 (25.0)	2 (1.0)
May	16	2 (12.5)	3 (1.5)	10	1 (10.0)	2 (2.0
June	23	6 (26.0)	7 (1.1)	12	3 (25.0)	3 (1.0)
Total	**233**	**76 (32.6)**	**108 (1.4)**	**164**	**44 (26.8)**	**58 (1.3)**

Effects on Respiratory Surface Area

Variations in the respiratory surface area of *E. suratensis* were due to the infestation of *C. indica*. Maximum reduction in respiratory surface area was noticed in the first gill arch and a minimum in the third gill arch. Thus considerable variation in the respiratory area was observed due to the attachment of the parasite (Figure 38.1).

Gross Lesions

The gross lesions were observed in the branchial chamber, buccal cavity and body surface is reported.

Branchial Chamber

Infested fishes had extremely pale gills indicating severe anemia. Gill rakers were seriously lost, apical edges damaged and out of gill lamellae heavily destroyed. Some secondary gill lamellae were fused or thickened. Between the gill lamellae calus-like thickening were observed. Further, the gill lamellae of the first and second gill arches were found eroded towards posterior position due to *C. indica.* Due to the lodging of parasites at the gill clefts, the gill arches showed lesions with a wide depression. The parasites further stimulated over secretion of the mucus production.

Buccal Cavity

Gross lesions observed in the buccal cavity of infested fishes showed small pin-holes in the cartilage support of the gill arches (developed), through which dactyls of peropods of isopod were penetrated claws are dig into the host tissues. The result is usually a localized destruction of the epidermis and inflammatory response around the attachment area was noticed.

Body Surface

Isopods make frequent shifts in position on the host causing a serious wound. They often move about as they feed. The activities also stimulate mucus secretion, epidermal proliferation and dilation of dermal capillaries. Mucus cells increase in number in the epidermis peripheral to the wound. Due to increased mucus secretion, there are signs of inflammation in the dermis beneath the wound. The digestive secretion apparently causes significant damage to the host's tissues.

Discussion

E. suratensis is commonly distributed and occurring throughout the year in the Parangipettai coastal waters. They are commercially important and contribute to sizeable market landings from the brackish waters of the southeast coast of India. *E. suratensis* being the largest have many desirable features which make them ideal fishes for aquaculture. Its wide salinity tolerance, ability to breed in confined waters, fast rate of growth, good body weight, tasty flesh, highly adaptable feeding habits, robust and sturdy body and good market price are some of the favourable characteristics for selection of this fish as a candidate species for brackish water aquaculture. In some parts of South India, *E. suratensis* is considered a delicacy and priced as much as Rs. 30/- per Kg which is comparable to the market value of other high quality table fishes (Keshava *et al.*, 1988).

Infestation causes serious problems to host animals either directly or indirectly affecting the physiological status of host. Loss of weight has been probably the most common effects on crustacean infestation. The present study also revealed that there was significant weight loss in the infested pearl spot fish, *E. suratensis* thereby proving the fact that infestation causes weight loss in fishes.

Kabata (1984) expressed that the general effects as loss of weight can be attributed to more than one cause and the most obvious one could be the loss of food reserve drawn from the various depots and other tissues to help in coping with the ranges of infestation. Low weight of infested fish might result from failure to grow normally. However, the present study on the length and weight relationship proved that the infestation alter the growth of the fish. It shows that the increase in length of fish is not in accordance with the increase in weight of fish due to infestation. Trilles (1979) found that several species of European cymothoids were capable of slowing down the growth in their hosts, although they did not affect weight-size ratios of the fish. However, Maxwell (1982) reported that the significant difference in the length and weight relationship of infested and uninfested fishes of Jack Mackeral, *Trachurus decliuis* when infested by *Cerotothea impricatus.*

The present study showed significant reduction of respiratory surface area because of the infestation of cymothoids in the branchial chamber was noticed. The dorsal surface of the parasite was always in close contact with the first gill arch, causing more atrophy of the first gill filament. The pressure exerted towards second to fourth gill arches were comparatively lesser than the first gill arch and the damage was also less.

Kroger and Guthrie (1972) stated that male, *Olencira praegustor* damaged the gill of juvenile Atlantic Menhaden, *Brevoorita tyrannus.* Stephensen (1976) stated that the gill damage was owing to buccal parasites. Bowmen (1960) found that the gill filaments of Hawaiian Moray eel, *Gymnothorax eurostus* were missing in the anterior or posterior most regions of gill arches due to the attachment of *Lirneca puhi.* Overstreet (1978) reported that the similar males and juveniles of isopod, *Olencira praegustator* parasitic on Gulf Menhaden cause change on the gill filaments through feeding.

In the present observation the parasites (*C. indica*) mostly attach with buccal region and the branchial chamber of the (host-fish) first gill arch. *C. indica* first described from Bangkok, Thailand has been recorded by Chilton (1924) from Chilka Lake and is stated to infest the mouth of *Gobius giuris*, causing deformation of the host's tongue. Subsequently, Panikkar and Aiyar (1937) recorded *C. indica* on *Etroplus maculatus*, *E. suratensis* and *Glossogobius giuris* from the Adyar estuary, Chennai (India). Evangeline (1963) reported the occurrence of this parasite, in the Adyar fish farm, from a variety of hosts, *viz.*, *E. suratensis*, *Tilapia mossambica*, *Macrones gulio*, *Gobius giuris*, *Polynemus tetradactylus*, *Pomadasys hasta* and *Sphyraena obtusata*. It is thus evident that *C. indica* exhibits a wide geographic and host distribution. Jayadev Babu and Sanjeeva Raj (1984) noticed the infestation in several regions like the chin, nape and pectoral fin base and the buccal cavity. But, Ravichandran *et al.* (1999) found that the dorsal surface of the parasite facing the first gill arch where the parasites were attached in *Joryma brachysoma*. The damage of gill filaments thus was not only for feeding but also the pressure exerted by the dorsal side of the parasite.

The gross size and shape of parasites can act as physical irritants, which may be responsible for the presently observed damages of the branchial tissues. The reduction in the surface area was thus due to several factors such as mode of attachment, movement, size and duration of stay of the parasites.

Pale gills damaged in gill rakers and erosion of gill lamellae are the severe gross lesions were observed as a consequence of isopod infestation. Pale gills of infested fishes indicate anemia, which may be due to the loss of blood and the construction of branchial circulation by the attachment of parasite and of the haemophageous nature of the branchial cymothoids (Romestand and Trilles, 1977; Romestand, 1979). The callus like thickening observed on the gill arch and gill filaments may be due to the constant irritation caused by the body and appendages of the parasite. Erosion and thickening are the two unique morphological changes noticed owing to the infestation. These changes are mainly due to the heavy pressure exerted by the parasite and also by their feeding nature. Kabata (1985) observed destruction of host tissues as a result of the pressure exerted by the parasite body, when present in the gill cavity. Longer stay of parasite within the gill chamber may also prevent and obstruct the normal growth of the gill arches. This may be the reason for the erosion of gill arch and fusion of gill lamellae. More mucus secreted in the infested gills may be due to the reaction to irritation created by the parasite. Romestand and Trilles (1977) observed that the secretion of mucus on the ligament surface as a reactive response of the host against infestation. As a consequence of the attachment of peropods host tissues were compressed and eroded at the attachment sites, which were surrounded by an inflammed peripheral welt of peropods.

Lesions associated with reproduction of parasite to the host are related to the direct activity of the parasites. A marked increase in the size of the parasite may be seen with the development of marsupium filled with juvenile isopods. This can significantly increase the pressure atrophy caused by the presence of the parasites. These reproduction related activity in the host increases the chance that the parasite will serve as a vector for microbial parasites such as hematozoans (Smith, 1975). Uninfested fishes were more so active than the infested, and were quite still or swam feebly than the infested fishes, resulting in low rate of activity.

Assessment of the general effects of parasites on the condition of their hosts is beset with numerous difficulties. The main difficulty in garging the extent of damage inflicted on the fish lies the normal condition of the fish. Certain reports suggested that the effects of parasitization might become apparent after a latent period of fairly long duration (Reichenbach-Klinke *et al.*, 1968). The parasitic infections also result in the abnormal behaviour of the host. The parasitic infections change the blood picture and also the parasitic condition of many species leads to the secondary infections by bacteria and fungi. Some reports (Venkataraman and Sreenivasan, 1952 and 1954) indicated that the environmental factors may influence the bacterial flora of the skin and gills.

In general, parasitic infection of fishes mainly depends upon a host of factors such as age, size, sex, maturity, stage, behaviour, feeding and breeding, lifecycle, physico-chemical and particularly environment factors. The negative impact of parasites on host's growth and survival has been demonstrated for several parasite-host systems, both in aquaculture and in natural populations (Sindermann, 1987). However, host-parasite relationships are generally very complex and difficult to clarify. With the exception of cases of mass mortalities caused by outbreaks of parasites, assessment of the effects of parasite infection in natural fish populations is particularly difficult because of the presence of predators or scavengers which rapidly remove the moribund or dead fish.

It appears that *C. indica* occurs in a limited range of teleost fishes and that it takes shelter in the host mainly for the purpose of breeding (Misra and Nandi, 1986). In the present study, the pearl spot fish, *E. suratensis* parasitised by the isopod, *C. indica* showed loss of weight, loss of fat content, changes in the water content of various tissues and reduction in respiratory surface area of the host fish.

References

Baer, J.G., 1951. Ecology of the family Anchuridae (Crustacea : Isopoda) with remarks on certain morphological peculiarities. *J. Linn. Soc. (Zool.)*, 36: 109–160.

Bowmen, T.E., 1960. Description and notes on the biology of *Lironeca puhi*, sp. (Isopoda : Cymothoidae). Parasite of the Hawaiian murrey eel, *Gymnothorax eurostus (Abbott)*. *Crustaceana L.*, pp. 82–91.

Bunkley-Williams, L. and E.H. Jr. Williams. 1998. Isopods associated with fishes: A synopsis and corrections. *J. Parasitol.*, 84(5): 893–896.

Chilton, C., 1924. Fauna of the Chilka Lake II. Tanaidacea and Isopoda. *Mem. Indian Mus.*, 5: 877–878.

Evangeline, G., 1963. Occurrence of the isopod fish parasite, *Cymothoa indica* in the Adyar Fish Farm, Madras. *Madras J. Fisheries*, 1.

Jayadev Babu, S. and S. Sanjeevaraj, 1984. Isopod parasites of fish of Pulicat Lake. In: *Proc. Symp. Coastal Aquaculture (Finfish)* 3: 818–823.

Kabata, Z., 1984. Diseases caused by metazoans: Crustaceans in diseases of marine animals. Vol. IV, Part I. *Introduction Pisces Biologische Anstalt, Helgoland* (Ed.) O. Kinne, pp. 321–399.

Kabata, Z., 1985. *Parasites and Disease of Fish Cultured in Tropics: Isopoda*. Taylor and Francis, London pp. 265–271.

Keshava, P. Santha Joseph and M. Mohan Joseph, 1988. Reproduction of Pearl-spot, *Etroplus suratensis* (Bloch) in the Nethravati-Gurpur estuary, Mangalore. In: *The First Indian Fisheries Forum, Proceedings of Asian Fisheries Society*, Indian Branch, Mangalore, (Ed.) M. Mohan Joseph. pp. 237–241.

Kroger, R.L. and J.F. Guthrie, 1972. Incidence of parasitic Isopods, *Olencira praegustator* in juvenile Atlantic Menhaden. *Copeia*, 2: 374–379.

Maxwell, J.G.H., 1982. Infestation of jackmackeral, *Trachurus dacliuis* (Jengus), with cymothoid isopod, *Ceratothoa imbricatus* (Labricus) in Southeastern Australian waters. *J. Fish. Biol.*, 29(3): 341–349.

Mills, Dick and Lambert Derek, 2004. *The Aquarium Fish Handbook: The Complete Reference from Anemone Fish to Zamora Woodcats*, 1st Edition for the United Kingdom.

Mini Nair, 1987. Studies on the systemics and ecology of isopods of the southwest coast of India. *Ph.D. Thesis*, University of Kerala.

Misra, A. and N.C. Nandi, 1986. A new host record of *Cymothoa indica* Schioedte and Meinert (Crustacea : Isopoda) from Sundarbans, West Bengal. *Indian J. Fish.*, 33: 229–231.

Moller, H. and K. Anderss, 1986. *Diseases and Parasites of Marine Fishes*. Moller, Kiel.

Overstreet, R.M., 1978. Marine maladies worms germs and other symbionts from the Northern gulf of Mexico Mississippi. *Alabama Sea Grant Consortium*, pp. 140.

Panikkar, N.K. and R.G. Aiyar, 1937. On a cymothoan parasite on some brackishwater fishes from Madras. *Curr. Sci.*, 5(8): 429–430.

Pillai, N. K., 1958. Tanaidocea and isopods of Travancore. *Ph.D. Thesis*, University of Kerala, pp. 152.

Ravichandran, S., A.J.A. Ranjit Singh, N., Veerappan and T. Kannupandi, 1999. Effect of isopod, parasite *Joryma brachysoma* on *Ilisha melastoma* from Parangipettai coastal waters (South east coast of India). *Ecol. Env. and Cons.*, 5(2): 95–101.

Reichenbach-Klinke, W., F. Braun, H. Held and S. Riedmuller, 1968. Vorlaufige Ergebnisse Vergleichender-physiologischer undersuchungen an coregonen verschiedener obserbayerischer seen (Fettgehatt, Blutbild, Fermentspiegel, Parasitisierung). *Arch Fischereiwiss*, 19: 114–130.

Rohde, K., 1993. *Ecology of Marine Parasites*, 2nd Edn. CAB International, Wallingford, pp. 1–298.

Romestand, B., 1979. Etude ecophysiologique des parasitoses a Cymothoadiens. *Ann. Parasitol. Hum. Comp.*, 54(4): 423–448.

Romestand, B. and J.P. Trilles, 1977. Influence decymothoiddicus (Crustacea, Isopoda, Flabellifera) succrtaines constand hematologiques despoissons hotes. *Zeitschrift fur. Parasitenkunde*, 52: 91–95.

Segal, E., 1987. Behaviours of juvenile *Nerocila acuminata* (Isopoda : Cymothoidae) during attack and attachment and feeding on fish prey. *Bull. Mar. Sci.*, 41(2): 351–360.

Sindermann, C.J., 1987. Effects of parasites on fish populations: Practical considerations. *Int. J. Parasitol.*, 17: 371–382.

Smith, F.G., 1975. *The Pathology of Fishes*. Chap. 6: Crustacean parasites of marine fishes, pp. 189–203.

Trilles, J.P., 1969. Recherches sur les isopodes 'Cymothoidae' des cotes francaises. Apercu general et comparatif sur la bionomie et la sexualite de ces Crustaces. *Bull. Soc. Zool. Fr.*, 94(3): 433–445.

Trilles, J.P., 1979. Les cymothoidae (Isopoda, Flabellifera, parasites de poissons) du Rijksmuseum van Naturlijke Historie de leidern II. Afrique, Amerique et regions indo-ouest-pacifiques. *Zool meded, Leide.*, 54: 245–275.

Venkataraman, R. and A. Sreenivasan, 1952. A preliminary investigation of the bacterial flora of the mackerals of the west coast. *Ind. J. Med. Res.*, 40: 524–533.

Venkataraman, R. and A. Sreenivasan, 1954. Bacteriology of inshore and of mackerals off Telicherry (Malabar). *Proc. Nat. Inst. Sci.*, 20: 651–655.

Chapter 39

Investigation of Artificial Neural Networks and its Applications in Medicine

J. Justin Anand[1], J. Justin Suresh[2] and P. Dhasarathan[3]

[1]*Department of Computer Science, Sri Kaliswari College, Sivakasi – 626 130*

[2]*Department of Computer Science, Adaikalamatha College, Vallam, Thanjavur*

[3]*Microbial Biotech Unit, Department of Biotechnology, Sri Kaliswari College, Sivakasi – 626 130*

ABSTRACT

Artificial neural networks (ANN) are currently a hot research area in medicine and it's believed that they will receive extensive application to bio-medical systems in the next few years. At the moment the research is mostly on modelling parts of the human body and recognising disease from various scans. The neural networks are ideal in recognising diseases using scans since there is no need to provide the specific algorithm on how to identify the disease. Neural networks learn by example so the details of how to recognise the disease are needed. Model neural networks are used experimentally to model the human brain. Building a model of the brain system of an individual and comparing it with the real time physiological measurements taken from the patient can achieve diagnosis. The use of ANN technology is the ability of ANNs to provide sensor fusion, which is the combining value from several different sensors. In medical modelling and diagnosis, this implies that even though each sensor in a set may be sensitive. Only to a specific physiological variable, ANNs are capable of directly complex medical condition by fusing the data from the individual biomedical sensors.

Keywords: Neural network, Artificial neuron, Sensor and Medical model.

Introduction

An Artificial Neural Network, sometimes referred to as simply a Neural Network, is a computer program designed to model the human brain and its ability to learn tasks (Haykin, 1994). An artificial neural network differs from other forms of computer intelligence in that it is not rule-based, as in an expert system. An ANN is trained to recognize and generalize the relationship between a set of inputs and outputs. Early artificial neural networks were inspired by perceptions of how the human brain operates. In recent years the developments in ANN technology have made it more of an applied mathematical technique that has some similarities to the human brain. Artificial neural networks retain as primary features two characteristics of the brain: the ability to 'learn' and to 'generalize' from limited information (Hewitson and Crane, 1994).

Neural Networks, both biological and artificial, employ massive, interconnected simple processing elements, or neurons. In artificial neural networks, the knowledge stored as the strength of the interconnection weights (a numeric parameter) is modified through a process called learning, using a learning algorithm. This algorithmic function, in conjunction with a learning rule, (*i.e.*, back-propagation) is used to modify the weights in the network in an orderly fashion (Fiesler and Beale, 1997). Unlike most computer applications, an *ANN* is not 'programmed', rather it is 'taught' to give an acceptable answer to a particular problem. Input and output values are sent to the ANN, initial weights to the connections in the architecture of the ANN are assigned, and the ANN repeatedly adjusts those interconnection weights until the ANN can successfully produce output values that match the original values. This weighted matrix of interconnections allows the neural network to learn and remember (Obermeier and Barron, 1989).

One of the original aims of artificial neural networks (ANN) was to understand and shape the functional characteristics and computational properties of the brain when it performs cognitive processes such as sensorial perception, concept categorization, concept association and learning.

Materials and Methods

An artificial network performs in two different modes, learning (or training) and testing. During learning, a set of examples is presented to the network. At the beginning of the training process, the network 'guesses' the output for each example. However, as training goes on, the network modifies internally until it reaches a stable stage at which the provided outputs are satisfactory. Learning is simply an adaptive process during which the weights associated to all the interconnected neurons change in order to provide the best possible response to all the observed stimuli. Neural networks can learn in two ways: supervised or unsupervised (Fujita *et al.*, 1992). Supervised-learning network is trained using a set of input-output pairs. The goal is to 'teach' the network to identify the given input with the desired output. Unsupervised learning the network is trained using input signals only. In response, the network organises internally to produce outputs that are consistent with a particular stimulus or group of similar stimuli. Inputs form clusters in the input space, where each cluster represents a set of elements of the real world with some common features.

Neural networks are typically arranged in layers. Each layer in a layered network is an array of processing elements or neurons. Information flows through each element in an input-output manner. In other words, each element receives an input signal, manipulates, it and forwards an output signal to the other connected elements in the adjacent layer. A common example of such a network is the Multilayer Perception (MLP) (Karl, 1997). The hidden units in turn send an output signal towards the neurons in the next layer. This adjacent layer could be either another hidden layer of arranged processing elements or the output layer. The units in the output layer receive the weighted sum of incoming signals and process it using an activation function. Information is propagated forwards until the network produces an output-input Layer Hidden Layer Output Layer Flow of Information.

In the human brain, a typical neuron collects signals from others through a host of fine structures called dendrites. The neuron sends out spikes of electrical activity through a long thin strand known as axon, which splits into electrical effects that inhibit or excite activity from the axon into electrical effects that inhibit or excite activity in the connected neurons. When a neuron receives excitatory input that is sufficiently large compared with its inhibitory input, it sends a spike of electrical activity down its axon. Learning occurs by changing the effectiveness of the synapses so that the influence of one neuron on another changes (Simpson, 1992).

Results and Discussion

The history of artificial neural networks is filled with colourful, creative individuals from many different fields, many of whom struggled for decades to develop concepts. First, one must have a concept, a way of thinking about a topic, some view of it that gives clarity not there before. This may involve a simple idea, or it may be more specific and include a mathematical description. The first practical application of artificial neural networks came in the late 1950s, with the invention of the perception network and associated learning rule by Beale and Jackson (1990).

Two new concepts were most responsible for the rebirth of neural networks. The first was the use of statistical mechanics to explain the operation of a certain class of recurrent network, which could be used as an associative memory. The second key development of the 1980s was the back propagation algorithm for training multi-layer perception networks, which was discovered independently by several different researchers. The most important advances in neural networks almost certainly lie in the future. Although it is difficult to predict the future success of neural networks, the large number and wide variety of applications of this new technology are very encouraging.

The applications are expanding because neural networks are good at solving problems, not just in engineering, science and mathematics, but in medicine, business, finance and literature as well. Their application to a wide variety of problems in many fields makes them very attractive. Also, faster computers and faster algorithms have made it possible to use neural networks to solve complex industrial problems that formerly required too much computation. The brain consists of a large number (approximately 10^{11}) of highly connected elements (approximately 104 connections per element) called neurons. For our purposes these neurons have three principal components: the dendrites, the cell body and the axon. The dendrites are tree-like receptive networks of nerve fibers that carry electrical signals into the cell body. The cell body effectively sums and thresholds these incoming signals.' The axon is a single long fiber that carries the signal from the cell body out to other neurons. The point of contact between an axon of one cell and a dendrite of another cell is called a synapse. It is the arrangement of neurons and the strengths of the individual synapses, determined by a complex chemical process that establishes the function of the neural network (Hertz *et al.*, 1991).

Artificial neural networks do not approach the complexity of the brain. There are, however, two key similarities between biological and artificial neural networks. First, the building blocks of both networks are simple computational devices (although artificial neurons are much simpler than biological neurons) that are highly interconnected. Second, the connections between neurons determine the function of the network.

The first step in utilizing an ANN to solve a problem is to train the ANN to 'learn' the relationship between the input and outputs. This is accomplished by presenting the network with examples of known inputs and outputs, in conjunction with a learning rule. The ANN maps the relationship between the inputs and outputs and then modifies its internal functions to determine the best relationship that can be represented by the ANN (Mancuso and Nicese, 1999).

The inner workings and processing of an ANN are often thought of as a 'black box' with inputs and outputs. One useful analogy that helps in the understanding of the mechanism occurring inside the 'black box' is to consider the neural network as a super-form of multiple regressions. Just as in linear regression, which finds the relationship such that {y} = f {x}, the neural network finds some function f {x} when trained. However, the neural network is not limited to linear functions. It finds its own best function as best as it can, given the complexity used in the network, and without the constraint of linearity (Hewitson and Crane, 1994). Artificial Neural Networks have been proven to be useful in the interpretation of natural resource information.

Acknowledgement

The authors are thankful to Thiru A.P. Selvarajan, Secretary, the Principal and the Management of Sri Kaliswari College for providing facilities to carry out the work.

References

Beale, R. and T. Jackson, 1990. *Neural Computation*. Institute of Physics publishing, Bristol, pp. 22–30.

Fiesler, E. and R. Beale, 1997. *The Handbook of Neural Computation*. Oxford University Press, London, pp. 21–29.

Fujita, H., Katafuzhi, T., Uehara, T. and T. Nishimura, 1992. Neural network approach for the computer aided diagnosis of coronary artery diseases in nuclear medicine. In: *International Joint Conference on Neural Networks*, Baltimore, USA, pp. 215–220.

Haykin, H. and S. Simon, 1994. *Neural Networks: A Comprehensive Foundation*. Macmillan College Publishing Company, New York, pp. 414.

Hertz, J., Krogh, A. and R. Palmer, 1991. *Introduction to the Theory of Neural Computation*. Addison Wesley, Redwood City, California, pp. 56–63.

Hewitson, Bruce C. and Robert G. Crane, 1994. *Neural Nets, Application in Geography*. Kluwer Academic Publishers, Boston, pp. 1–9.

Karl, B., 1997. Elliptic Fourier analysis and classification for a multifunction prosthetic arm. In: *New Trends in Artificial Intelligense and Neural Networks*, (Eds.) T. Ciftciba, M. Karaman and V. Atalay. EMO Scientific Books, Ankara, pp. 105–110.

Mancuso, S. and F.P. Nicese, 1999. Chestnut (*Castanea sartiva* Mill.) genotypes identification: An artificial neural network approach. *J. Hort. Sci. Biotechnol.*, 74: 777–784.

Obermeier, Klaus K. and Janes J. Barron, 1989. "Time to Get Fired Up" in Byte, August, 1981, pp. 217–233.

Simpson, R., 1992. Biological pattern recognition by neural networks. *Marine Ecology Progress Series*, 79: 303–308.

Chapter 40

Investigation on Sub Surface Water Quality of Tarikere Taluk with Special Reference to Physico-Chemical Characteristics

K. Harish Babu and E.T. Puttaiah

Department of P.G. Studies and Research in Environmental Science, Kuvempu University, Shankaraghatta –577 451, Shimoga Distt., Karnataka

ABSTRACT

The research study was carried out in Tarikere taluk, Karnataka (India). This article mainly address the physico-chemical concentration of 56 groundwater sample during September 2004. The results of all the findings are discussed in details which reflect the present status of the groundwater quality of the study area.

Keywords: Groundwater quality analysis, Tarikere taluk.

Introduction

Rural India relies a mainly on groundwater for drinking and agriculture practices. Villages once relied on sources like wells, lakes, ponds and streams for their water needs. Contamination of most surface water sources has rendered them unfit for consumption. Also increase in water demand by an increasing population has necessitated resource to tapping groundwater. Unsustainable withdrawal of groundwater has led to the spectra of depleting the problem of water scarcity.

Every human society, be it rural or urban, industrially or technologically advanced, disposal of waste exceeds the limit of natural scavenging or removal process, they are bound to effect the normal functioning of the ecosystems and consequently they bear an adverse effect on the water resources (Miller, 1984).

Supplying inhabitants with safe and clean drinking water is one of the most common problems in developing countries like India, especially in arid and semi-arid regions. Hazardous pathogenic germs and anthropogenic substances do not only contaminate the available water quality but also geogenic substances is adversely affect the water supply of many regions.

The groundwater of Tarikere taluk had many threats such as anthropogenic activities, quality deterioration by agricultural activities and over exploitation. Persistence of continuous drought condition during 2001–2004 had declined the amount of groundwater in the region. With this background an effort has been made to know the quality of groundwater in this region.

Methodology

Study area

Tarikere taluk is located between 75°45′00″–75°52′30″ E and 13°30′00″–13°47′30″ N and it has a geographical area of 1,222 sq. kms. It comprises 45 Gram Panchayats and 245 villages with a total population of 2,54,093 as per 2001 census. As a whole, the region is a flat plain land with an average elevation of about 590.58m msl (Gazetteer of India, 1981).

Meteorology

The climate of Tarikere taluk is semi arid and enjoys all the three seasons *viz.*, pre-monsoon (February to May), monsoon (June to September) and post-monsoon (October to January). The monthly mean temperature ranges from 8.9 to 38°C. The relative humidity varies between 50 to 84 per cent and it is highest during the month of August and September and lowest humidity observed during the months of January and February.

Precipitation

The precipitation and distribution of rainfall in Tarikere taluk is highly erratic. The annual average rainfall is 980 mm received over 55 rainy days. It varies from as low as 621 mm in the east and as high as 920 mm in west. About 2/3rd of the geological area of Tarikere taluk receives less than 750 mm of annual rainfall. The study area had been receiving rainfall mainly from southwest monsoon and slightly from northeast monsoon.

Geology

Stratigraphically the study area comes under Bababudan belt of Dharwar super group. Bababudan belt is well known for its iron ore (horse shoe shape). Peninsular gneiss, which forms the basement to the younger green stone belts like Shimoga, Chitradurga and Tarikere (Swaminath and Ramakrishnan, 1981). An important gneissic exposure is near Tarikere that transect the structural trend lines of the schist belts while the gneissic foliation is conformable to the schist belt boundary. The tonalitic gneisses occur in Tarikere taluk (Naqvi, 1983).

Pillow breccias and conglomerates are also seen. Metagabbro with titaniferous magnatite is inter-sheeted with in the metabasites at north of the Tarikere dome of basement gneisses. Serpentinites and talcose schist form a minor component of the Meta volcanic association in the Bababudan group (Fareeddudin *et al.*. 1988). Geological status of Tarikere taluk were represented in Figure 40.1.

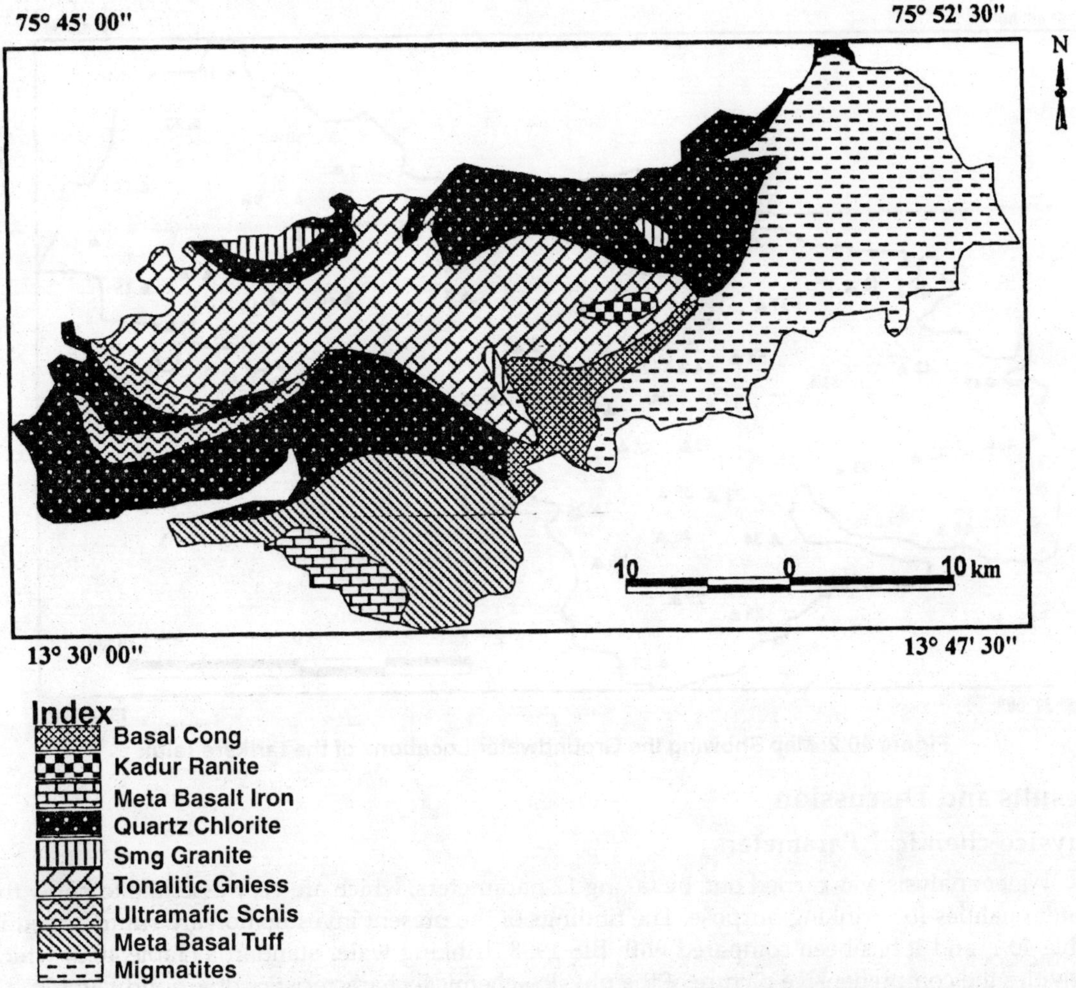

Figure 40.1: The Geological Status of Tarikere Taluk

Analysis of the Samples

The groundwater samples from the 56 sampling sites were collected and analyzed during September 2004 and Sampling locations have been shown in Figure 40.2.

Water samples from the sampling sites were collected from the bore wells. Initially the water was allowed to run for 15 minutes in order to flush out stationary water. Further, the sample bottles were also flushed with water before the samples were collected. The parameters of water such as dissolved oxygen; total dissolved solids, electrical conductivity and pH were analyzed on the spot with the help of water analysis kit (Elico). The remaining parameters were analyzed in the laboratory. Hence, the water was carried to the laboratory in suitable inert bottles. The samples were analyzed using analytical method of APHA (1995) and all analysis was done in triplicate.

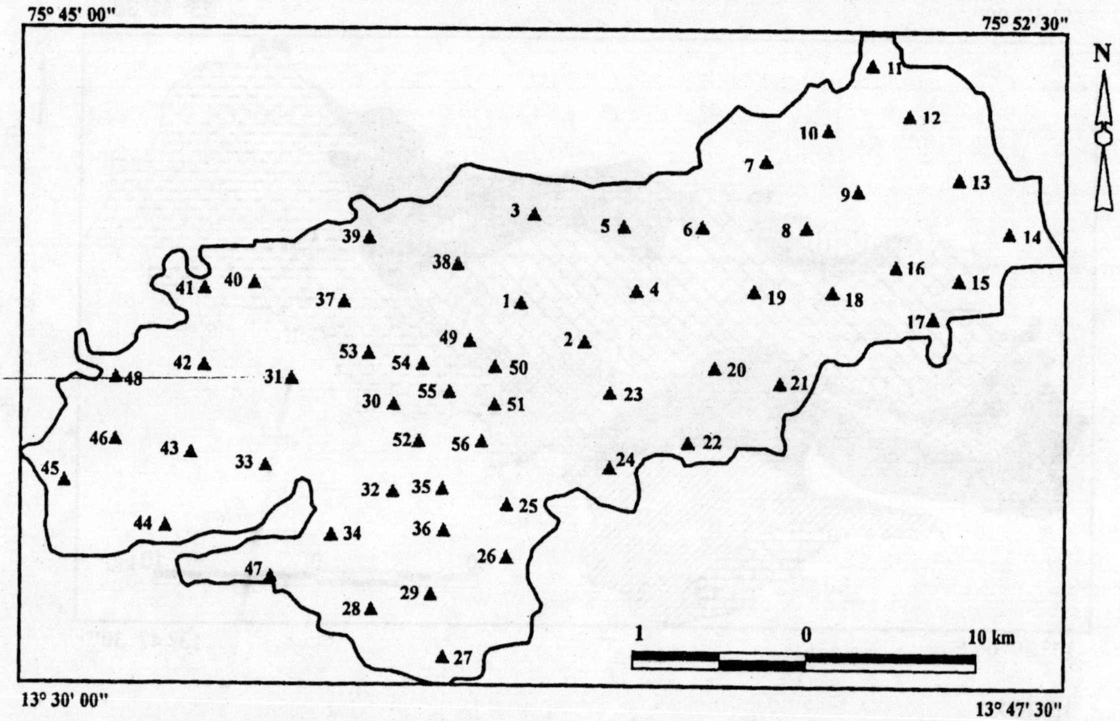

Figure 40.2: Map Showing the Groundwater Locations of the Tarikere Taluk

Results and Discussion

Physico-chemical Parameters

Water analysis was carried out, by taking 12 parameters, which are very essential to know the water qualities for drinking purpose. The findings of the present investigation are summarized in Table 40.1, and it has been compared with BIS 1998 drinking water standards (Table 40.2) which provides the comprehensive picture of the physico-chemical characteristics of groundwater in the study area.

Turbidity (TUR)

In the present study the turbidity values ranged between 0.15 and 86.1 NTU with a mean value of 9.03. The BIS (1998) acceptable limit for turbidity is 25 NTU. It is proved from the present study, 12.5 per cent of total number of samples cross their permissible limit with reference to the BIS (1998) standards. It is observed from the results that these parameters which crossed their permissible limit are unfit for drinking purposes. It causes health problems like gastro-intestinal disorders, headache and also associated with respiratory diseases (Maiti, 1982; Peter, 1974).

pH

In the present investigation, pH values found to be 6.7 to 8.7 with a mean value of 7.54. The recommended value of pH for drinking purposes is between 6.5 to 8.5 (BIS, 1998). In the present study all the water samples analyzed are all well with in the safer limits except in few sampling stations. However, higher values of pH hasten the scale formation in water heaters and reduce the germicidal potential of chlorine (Mohapathra and Purohit, 2000).

Table 40.1: Water Quality Data of Physico-chemical Parameters of the Study Sites During September 2004

Sam. No.	*TUR*	*pH*	*EC*	*TDS*	*DO*	*TH*	Ca^{2+}	Mg^{2+}	Cl^-	*Alk*	SO_4^{2-}	F^-
S1	11.5	8.35	1558	975	3.8	1325	327.5	218	678	815.5	170.5	1.8
S2	3.3	8.2	2338	1440	5.7	1775	364	228.5	700	295	135	1.55
S3	6.75	7.95	897.5	560	5.15	560	177.5	34.5	265	162	27.1	1.5
S4	0.15	7.85	1150	750	4.4	970.5	332.5	171	409	190	90.5	1.1
S5	7.6	7.2	1751	1051	5.4	1225	395	205.5	670.5	188.5	120.5	1.4
S6	1.2	6.85	1289	782	4.1	845	371	186.5	350	14.7	101.3	1.6
S7	0.15	7.15	449	276	4.5	350.5	119.5	102.1	75.35	146.5	104	1.25
S8	2.7	7.15	1125	718	5.3	790	130.5	108	281.5	231	125.5	1.4
S9	76.15	7.9	1600	970	5	947.5	164.5	173	515	156.5	170.6	1.55
S10	1.25	8.45	715	458	4	470	130	23.6	650	188.5	78.75	1.15
S11	1.7	7.4	544.5	352	4.7	508	186	27.6	616	120	21.8	1
S12	1.15	8.15	1136	771	4.15	970	191	170.5	451	537.5	126	1
S13	0.75	7.9	1699	1056	6.05	972.5	210.5	104.6	750	532.5	133	1
S14	0.2	7.3	1100	670	4.15	759.5	165	125.5	445	172.5	171	1.1
S15	36.5	8.15	1315	815	4.05	735	234.9	160	425	260	118.5	1.15
S16	41	8.3	633.5	421	4.55	395	165.5	110	330	184.5	143	0.9
S17	69.5	6.7	595	370	4.15	505	68	39.95	565	194	22.7	0.9
S18	2.15	8.7	1125	705	5	800	275	9.45	426.5	615	215.2	0.6
S19	1.3	8.4	955	606	4	695	206	21	163.5	119.5	176	0.7
S20	1.6	7.65	1469	950	3.8	1175	440.5	25.35	650	913	117	0.7
S21	1.4	6.7	101.8	68	4.7	85	26.5	3.5	25.65	116	174.5	0.3
S22	1.85	7.25	210	121	6.04	177	24.5	16.3	33.1	130	115.5	0.55
S23	0.3	7.6	1047	636	4.4	594	124	63	241.5	110	98.5	0.45
S24	0.2	6.8	227	138	5.6	270	45	37.5	110.1	46	98.95	0.6
S25	1.3	7.3	1278	758	4.6	970	220.5	108	553	130	165.1	0.725
S26	10.9	8.05	1373	830	4.8	986	180.5	115.5	367	96	136.3	0.625
S27	0.6	7.3	242	147	5.15	385	103	57.05	346	87	128	0.7
S28	0.25	7.15	805	482	5.35	625	135.5	102	467	110	171.5	0.85
S29	0.65	8.1	1608	995	3.65	960	137	131	547	132	100	0.975
S30	0.15	7.1	603	362	5.2	456	122.5	66.2	183.1	144	140	1
S31	0.9	7.25	466.5	280	3.9	400.5	105.5	32.85	285.5	110	246.5	1.05
S32	2.4	6.75	486.5	286	4.8	460	132	92.35	69.35	168	149.1	0.95
S33	6	6.7	778	467	4.12	623	305.5	128.7	70.2	86	104.1	1.05
S34	1.15	7.75	923	554	4.3	644	47.72	110.5	251	68	231	0.85
S35	3.15	7.25	1037	622	4.4	610	127.5	100.5	206	91.8	130.1	0.95
S36	0.2	7.45	987.5	593	4.63	570	177	91.55	266.5	180	145.1	1.05

Contd...

Table 40.1–Contd...

Sam. No.	TUR	pH	EC	TDS	DO	TH	Ca^{2+}	Mg^{2+}	Cl^-	Alk	SO_4^{2-}	F^-
S37	0.2	8.4	840	520	3.75	695	139	84.45	177	82.5	249	1.1
S38	3.45	7.85	545	315	4.05	1700	395.5	229.5	670	145.5	231.5	0.8
S39	1.6	7.95	360	260	3.95	595	83.9	63.35	200.5	79	132	1.55
S40	0.25	8.25	252.5	150	5.1	1160	189	170	735	138.5	215	1.075
S41	0.15	7.7	324	195	4.5	370	76.55	23.05	124.5	44.95	112.1	0.975
S42	10.55	8.35	322.5	178	3.85	1065	181.5	142	535	765.5	171	0.85
S43	1.55	7	452	263	4.6	1125	299.5	119.5	406	74	151.5	0.6
S44	0.5	8.15	784.5	478	4.15	580	279	11.1	285	66.5	141.5	0.95
S45	0.5	7.1	475.5	294	5.1	283	54.45	35.5	156.5	172	161.2	0.8
S46	11.65	7.05	1193	705	4.9	672	234.5	101.5	93	81	139.5	0.9
S47	0.35	6.85	693	425	4	397	108.1	56.7	118	92.1	265.5	1
S48	10.7	7.5	1550	925	3.55	882	212.5	186.2	86.95	75	262	1.05
S49	0.2	7	957	577	5.6	625	78.85	92.5	142	149	152.8	1
S50	1.65	7.05	789	477	4.4	440	91.35	43	175	110	162	1.1
S51	0.6	6.95	900	553	4.4	800	237	45.1	453	126	210.5	1
S52	0.15	7	520.5	315	5.15	640	156.5	56.05	538.5	101	222.5	1.15
S53	86.1	7.05	784.5	478	3.8	1505	520	46.1	652.5	86.7	172	1.1
S54	69.2	7.55	705	416	4.35	219	36.5	25.5	120.5	89	156.4	1.05
S55	1.25	7.9	815	585	4.1	695	215	9.7	685	255	170.3	1.4
S56	7.05	7.8	1200	749	4	695	125	116.5	470	126	57.3	1.2
Mean	9.029	7.542	894.3	551.7	4.552	727.4	185	92.12	361.8	191.1	150.7	1.012
Min	0.15	6.7	101.8	68	3.55	85	24.5	3.5	25.65	14.7	21.8	0.3
Max	86.1	8.7	2338	1440	6.05	1775	520	229.5	750	913	249	1.8

Electrical Conductivity (EC)

The values of electrical conductivity ranged between 101.8 μmhos/cm to 2338 μmhos/cm. Mean values of EC showed 885.31 μmhos/cm. Higher the concentration of acid, base and salts in water, higher will be the EC. The variability of EC could be explained to the natural concentration of ionised substances present in water (Kataria and Jain, 1995). However the higher values of electrical conductivity (>2000 μmhos/cm), may be due to long residence time and lithology. Ballukraya and Ravi (1999) had proved the variation of the conductivity of the water due to the residential times and the geographical features of the sites.

Total Dissolved Solids (TDS)

It is justified from the analytical report TDS values ranges from 68 mg/L to 1440 mg/L with a mean value of 551.53 mg/L. However groundwater chemistry changes as the water flows through the subsurface and the increase in geological environment and dissolved solids and major ions. Chebotarev (1985), Ramababu and Somashekara Rao (1986) and Joseph (2001) expressed the dissolution of soil particles containing minerals under slightly alkaline condition; favour the TDS concentration in groundwater. However TDS concentration above the permissible limit (1500-ppm) causes gastrointestinal irritation (Shankar and Muttukrishnan, 1994).

Table 40.2: Comparison of Groundwater Quality Data with Drinking Water Standard (BIS, 1998)

Sl.No.	*Parameters*	*BIS (1998)*		*Observed Values*	
		P	*E*	*Minimum*	*Maximum*
1.	Turbidity	5	25	0.15	86.1
2.	pH	6.5	9.2	6.7	8.7
3.	EC	–	–	101.8	2338
4.	TDS	500	1000	68	1440
5	Dissolved oxygen	–	–	3.55	6.05
6.	Total hardness	300	600	85	775
7.	Calcium	75	200	24.5	520
8.	Magnesium	30	100	3.5	229.5
9.	Chloride	250	1000	25.65	750
10.	Alkalinity	200	600	14.7	913
11.	Sulfate	200	400	21.8	249
12.	Fluoride	0.6–1.2	1.5	0.3	1.8

Note: P: Permissible limit; E: Excessive limit.

All parameters are expressed in mg/L except pH, Turbidity (NTU) and conductivity (μmohs/cm).

Dissolved Oxygen (DO)

The present research findings revealed for the values of DO varies from 3.55 mg/L to 6.05 mg/L, and observed mean value is 4.55 mg/L respectively.

In the present study, all the samples showed for DO values were well with in the permissible limits for drinking and domestic purposes. However, the presence of oxygen demanding pollutants (like organic wastes) causes rapid depletion of dissolved oxygen from water. Oxidizable inorganic substances like hydrogen, sulphide, ammonia, nitrites, iron etc., decrease the dissolved oxygen concentration (Sawant *et al.*, 2000). And also due to physical, chemical and biological activities in water the level of DO may vary (Jameel, 2002).

Total Hardness (TH)

Total hardness levels varied from 85 mg/L to 775 mg/L, associated with a mean value of 727.36 mg/L. The BIS (1998) acceptable limit for total hardness is 600 mg/L. In the present study, revealed that 60.7 per cent of total number of water samples crosses the permissible limits of BIS (1998) drinking water standards. Owing to fact that higher amount of hardness in the study area comes mainly from the leaching of igneous rock and carbonate rocks (dolomite, calcite and limestone). Water containing the soluble salts of calcium and magnesium such as chlorides, sulphates and bicarbonates are also governs the quality of water (Ramaswamy and Rajaguru, 1991). The adverse effects of total hardness are formation of kidney stone and the heart diseases (Sastry and Rathee, 1998). Nevertheless, groundwater chemistry is controlled by the composition of its recharge components as well as by geological and hydrological variations (Narayana and Suresh, 1989).

Calcium (Ca^{2+})

Present investigation reports stated for calcium values ranged from 24.5 mg/L to 520 mg/L, with a mean value of 184.99 mg/L. BIS (1998) acceptable limit for calcium is 200 mg/L. However, in the present study 30.2 per cent of water samples crosses the permissible limit. Presence of higher amount of calcium in the study area may be due to groundwater receives the calcium minerals leached from the rocks and other deposits like limestone, dolomites, calcite, gypsum, amphiboles, feldspar, and clay minerals leaching or weathering of igneous rocks. Sewage and domestic waste are also important sources of calcium (Mishra and Saxena, 1989).

Magnesium (Mg^{2+})

Investigation report reveals for the magnesium values ranged from 3.5mg/L to a 229.5 mg/L with a mean value of 92.11 mg/L.

BIS (1998) acceptable limit for magnesium is 100 mg/L and in the present study 48.2 per cent of the water samples crossed the permissible range. Magnesium arises principally from the weathering of rocks contain ferro-magnesium minerals and some carbonate rocks. High concentration of magnesium proves to be diuretic and laxative (Schroeder 1960).

Chloride (Cl^-)

Chloride is also one of the important parameter to know the quality of water. Anthropogenic sources of chlorides include fertilizer, road salt, human and animal waste. Concentration of chlorides is considered to be an indicator of organic pollution of animal origin (Kumara, 2002). Here Chloride values ranged from 25.65 mg/L to 750 mg/L, however the mean value observed in the present study is 361.82 mg/L.

The BIS (1998) acceptable limit for chloride is 1000 mg/L. In the present investigation, the values of chloride for all the sampling sites are with in the permissible range as prescribed by BIS (1998) drinking water standards. However, dissolving of the soil constituents had contributed the chloride into the groundwater and also the soil characteristics play an important role in contributing the chloride content in the groundwater (Shivasankaran, 1997).

Total Alkalinity

In the present study total alkalinity values ranged from a 14.7 mg/L to 913 mg/L and a mean value of 189.31 mg/L. The BIS (1998) acceptable limit for total alkalinity is 600mg/L. In the present study, the data revealed that 10.7 per cent of water samples in the study area crossed the permissible limit. When alkalinity of water exceeds the permissible limits, it is likely to produce incrustation sediment deposits, difficulties in chlorination, certain physiological effects on human systems etc. (Raviprakash and Rao, 1989). The constituents of alkalinity result from dissolution of mineral substances in the soil and atmosphere contributes to alkalinity in groundwater (Mittal and Verma, 1997).

Sulphate (SO_4^-)

It is very interesting to note that, sulphate values ranged from 21.8 mg/L to 249 mg/L with a mean value of 150.2 mg/L. The BIS (1998) permissible range for sulphate is 400 mg/L. In the present investigation, the sulphate values for all the samples are within the prescribed limit of BIS (1998) drinking water standards. Sulphate in groundwater takes place the break down of organic substances in the soil. However, geological, hydrological and geomorphologic characteristics showed a remarkable variation in sulphur content (Alexander, 1961).

Fluoride (F^-)

In the present investigation, fluoride values varied from 0.3 mg/L to 1.8 mg/L with a mean value of 1.01 mg/L. The BIS (1998) acceptable limit for fluoride is 1.5 mg/L. In present study 7.14 per cent of total number of water samples in present investigations have crossed the permissible range as prescribed by BIS (1998) drinking water standards.

Degree of weathering and leachable fluoride in terrain is of great significance for the fluoride present in groundwater than the mere presence of fluoride bearing minerals in rocks (Kumar *et al.*, 2000).

Intake of excess fluoride causes dental, skeletal and nön-skeletal fluorosis. The non-skeletal fluorosis can be observed such as gastrointestinal complaints, intermittent diarrhoea and flatulence in expectant and lactating mothers hard working young adults and children. Therefore, fluorosis has been considered as one of the incurable diseases. Hence, prevention is the only solution for the disease (Hem, 1985).

Conclusion

Currently carried research investigation should give more precise answer on influence of Geomorphological condition than anthropogenic activities in the examined groundwater samples of the study area. Local geological settings may supports the increasing concentration of physico-chemical characteristics in groundwater. The factors like slow circulation, longer period of contact between aquifer and water, dissolving of minerals at the time of weathering, residential time, drainage pattern and surface water link. Porosity of the soil and rock also alters the characteristics of the groundwater.

The high level contents of the parameters observed may be minimized if the groundwater is recharged with the available water in the rainy season. This not only dilutes the constituents of the groundwater but also raises the groundwater level that depletes due to large-scale exploitation.

Groundwater is extremely important to the future economy and growth of rural India. If the resource is to remain available as high quality water for future generation it is important to protect from possible contamination. Hence it is recommended that suitable water quality management is essential to avoid any further contamination.

References

Alexander. 1961. *Introduction to Soil Microbiology*. Wiley, New York, London, p. 171–172.

APHA. 1995. *Standard Methods for the Examination of Water and Wastewater*; 18th edition, AWWA, WPCF, New York, pp. 1120.

Ballukraya, P.N. and R. Ravi. 1999. Characterization of Groundwater the unconfined aquifers of Chennai city. *Indian Journal of Geological Society of India*, 54: 1–11.

BIS. 1998. *Specifications for Drinking Water*, New Delhi, p. 171–178.

Chebotarev. 1985. Metamorphism of natural waters in the crust of weathering. *Geochem. Cosmochim. Acta*, 8: 22–28.

Fareeddudin, A.S. Janardhan and B. Basavalingu. 1988. Sedimentology, minerology and geochemistry of the Kalasapura conglomerate. *Geol. Soc. India. Mem.*, 9: 65– 82.

Gazetteer of India, Karnataka State, Chikmagalure district. 1981. p. 8–15 and 630–633.

Hem, J.D. 1985. *Study and Interpretation of the Chemical Characteristics of Natural Water,* 3rd ed. U.S. Geological Survey, Water Supply Paper, 225: 263.

Jameel, A. 2002. Physico-chemical studies in Vyakondan channel water of river Cauvery. *Pollution Research,* 17(2): 111–114.

Joseph, W. 2001. Groundwater chemistry in the valley De Yabucoa alluvial aquifer, South-eastern Pnerto Rico. *AWRA 3rd International Symposium on Tropical Hydrology,* San Juan U.S.A.

Kataria, H.C. and O.P. Jain. 1995. Physico-chemical analysis of river Ajhar. *Indian Journal of Environmental Protection,* 5: 569–571.

Kumar, P. Kumari and L.K.R. Singh. 2000. *Environmental Biology.* P.G. Dept. of Zoology, S.K. University, Dumka, Jharkand, p. 180–182.

Kumar. 2002. *Ecology of Polluted Waters.* 1: 144–180.

Maiti, T.C. 1982. *Science Reporter,* 19: 360–361.

Miller, D.R. 1984. Chemicals in the environment. In: *Effects of Pollutant at the Ecosystem Level.* John Wiley and Sons, Chinchester, pp. 7.

Mishra, S.R. and Saxena. 1989. Industrial effluent pollution at Birla Nagar, Gwalior. *Pollution Research,* 8(2): 77–86.

Mittal, S.K and N. Verma. 1997. Critical analysis of groundwater quality parameters. *Indian Journal of Environmental Protection,* 17: 426–429.

Mohapatra, T.K. and K.M. Purohit. 2000. Qualitative aspects of surface and groundwater for drinking purpose in Paradeep area. *Ecology of Polluted Waters,* 1: 144.

Naqvi, S.M. 1983. Geochemistry of gneisses from Hassan district and adjoining areas, Karnataka, India. In: *Precambrian of South India,* (Eds.) Naqvi, S.M. and J.J.W. Rogers. Geological Society of India Mem., 4: 401–416.

Narayana, A.C. and G.C. Suresh. 1989. Chemical quality of groundwater of Mangalore city, Karnataka. *Indian Journal of Environmental Health,* 31: 228–236.

Ramaswamy, V. and P. Rajaguru. 1991. *Indian Journal of Environmental Health,* 33(2): 187–191.

Rambabu, C. and K. Somashekara Rao. 1986. Studies on the quality of bore well water by Nuzuid. *Indian Journal of Environmental Protection,* 16(7).

Raviprakash, S. and K.G. Rao. 1989. The chemistry of groundwater, Parvada area with regard to their suitability for domestic and irrigation purposes. *Indian Journal of Geochem.,* 4: 39–54.

Sastry, K.V. and P. Rathee. 1988. Physico-chemical and microbiological characteristics of water of village Kanneli, (Distt. Rohtak) Haryana. *Proc. Academic, Environmental Biology,* 7(1): 103–108.

Sawant, C.P., Saxena, G.C. and Shrivastava, V.S. 2000. Trace Metals in and around the Industrial belt. *Ecology Environment and Conservation,* 6(6): 135–137.

Schroeder, H.A. 1960. Relation between hardness of water and death rates from certain chronic and degeneration diseases in the US. *J. Chron. Dis.,* 12: 568–573.

Shankar and Muttukrishan. 1994. *In situ* bioremediation of contaminated groundwater. *Proceedings of National Seminar on EPCR–04,* UBDT Engineering College, Davangere, pp. 32.

Shivashankaran, M.A. 1997. Hydrogeochemical assessment and current status of pollutants in groundwater of Pondicherry region, South India. *Ph.D. Thesis*, Anna University, Chennai, p. 80–87.

Swaminath, J. and M. Ramkrishnan. 1981. In: *Early Precambrian Supracrustals of South India*, (Eds.) Swaminath, J. and M. Ramakrishnan. Geol. Surv. Ind. Mem., 3: 23–38.

Chapter 41

Effect of Sugar and Distillery Wastes Application on Different Crops: A Review

V. Davamani and A. Christopher Lourduraj*

Department of Environmental Sciences, Tamil Nadu Agricultural University, Coimbatore – 641 003, India

ABSTRACT

The pressmud and molasses are industrial byproducts of sugarcane industry in India. Spentwash is the effluent obtained from the distillery industry. In this article, the response of rice, wheat, maize, sorghum, sugarcane, cotton, oilseeds, legumes, fruit and vegetable crops to pressmud and spentwash are discussed. In general, application of pressmud compost and treated and diluted spentwash significantly enhanced crop yields.

Introduction

In India there are about 450 sugar mills and the number is assuming an increasing trend owing to the growing demand at domestic and export front. The sugar production in this century is 15 million tonnes annually, for which about 150 million tonnes of sugarcane is crushed. The pressmud and molasses are industrial byproducts of sugarcane industry in India, and the production is to the tune of 7.0 Mt and 7.5 Mt released annually (Ramaswami, 1999). In the beginning of twentieth century, the pressmud and molasses were used as soil amendments. Dhar and Mukherjee (1938) were the pioneers in using pressmud and molasses for the reclamation of alkali soils. The distillery spentwash is non toxic. After treatment, with a proper dilution and systematic application, it would not cause any harm to soil and in some cases, the nutrients which are exhausted by the crop would be brought back to soil (Devarajan *et al.*, 1994).

* Corresponding Author: E-Mail: christoens2000@yahoo.com.

Rice

Virendra Kumar and Mishra (1991) reported that the use of pressmud cake increased the uptake of N, P and K significantly, compared to control in rice-maize crop sequence. They also indicated that sulphitation pressmud cake appeared to enhance yield more than the carbonation pressmud cake. Arora *et al.* (1992) stated that the distillery effluent can be converted into biogas and the residue can be utilized as a fertilizer if it is detoxified. The detoxified distillery effluent was used in rice plant culture. The treated rice plant showed better growth and development as compared with control plants growth in full nutrient solution. Devarajan and Oblisami (1995) reported that the distillery spentwash application in rice increased the availability of N, P, K, Ca, Mg, micronutrients and organic matter contents in soil. Chinnusamy *et al.* (1998) reported that the highest grain yield was recorded with 75 times diluted distillery effluent treatment which was on par with 100 times diluted plot. Among the organic amendments, application of pressmud at 12.5 tonnes ha^{-1} recorded the highest grain yield. Rajukkannu and Shanmugam (1998) observed that the spentwash when diluted suitably (1 : 10 to 1 : 50) there was reduction in its BOD and EC, hence it can be made amenable for irrigation to rice crop. They also visualized increase in nutrient content of soils and the yield of rice crop without any adverse effect on soil physical, chemical and biological properties.

Annadurai *et al.* (2001) concluded that in the areas where distillery effluent is available, the effluent should be diluted 50 times with irrigation water along with application of FYM to increase the rice yield. The maximum grain yield was recorded in rice variety ADT-42 due to 75 times diluted distillery spentwash treatment, which was on par with 100 times diluted spentwash applied plot (Chinnusamy *et al.*, 2001).

Wheat

Application of pressmud compost at the rate of 10 tonnes ha^{-1} increased the maize and wheat yield by 129.4 and 62.2 per cent, respectively (Datta and Gupta, 1983). Mishra *et al.* (1982) observed that phospho-compost application increased the phosphorus use efficiency of wheat (20.48 per cent) and green gram (12.90%), as compared to single superphosphate. Further, the results of Mishra *et al.* (1984); Singh and Yadav (1986) and Bangar *et al.* (1989) have shown that the rock phosphate enriched compost is as effective as single superphosphate with regard to crop yields and P uptake. Significant increase in the yield of green gram (*Vigna radiata* L.), wheat (Bangar *et al.*, 1989), cluster beans and red gram (Mishra and Bangar, 1986) were reported due to the application of phospho compost. It was also reported that the phospho compost increased the quality of grains by increasing the protein and Ca contents (Singh and Yadav, 1986). Pathak *et al.* (1998) reported that soil amendment with diluted post methanation distillery effluent increase the yield of wheat and rice grown in sequence.

Maize

Asija *et al.* (1984) reported that composts enriched with fertilizer (urea, single superphosphate and rock phosphate) had significant effect on shoot and root yield of maize. The application of enriched composts also markedly increased nitrogen and phosphorus uptake. The total uptake of N and Zn by corn crop was appreciably increased when pressmud was added along with zinc (Gupta *et al.*, 1986). From the results of greenhouse experiment with maize crop, Patil and Shinde (1995) indicated that application of FYM and 3 : 1 spent slurry-pressmud compost increased plant height as well as dry matter yield. They also reported that the uptake of macro and micro-nutrients enhanced significantly by spent slurry-pressmud compost application. Patil and Shinde (1995) reported that the dry matter yield of maize was significantly increased by application of FYM, spent slurry and spentwash composts.

In the treatment supplied with spent wash liquid, only dry matter yield was significantly decreased. Khruslova and Kolomiets (1974) stated that spentwash application in grasses, maize fodder, beet increased the yield by 45–100 per cent. Shinde and Trivedi (1981) studied the effect of distillery wastewater irrigation on agronomical characteristics in case of *Zea mays*. Inhibition of germination was observed in 50 per cent diluted spentwash and also there was reduction in height of seedling in 50 per cent and10 per cent diluted spentwash. Reduction in the dry matter was found significant at 50 per cent dilution and in other dilutions the dry matter reduction was recorded more or less the same. Increased yield of maize was also seen with the use of distillery spentwash and hence, maize was recommended as the tolerant crop for growing in distillery spentwash. Singh and Raj Bahadur (1998) reported that twelve pre sowing irrigations with distillery effluent had no adverse effect on the germination of maize but improved the growth and Yield. Mallika (2001) showed that the application of spentwash at a rate of 150 kilo liter ha^{-1} produced higher grain Yield in maize than other levels.

Sorghum

Dongale and Savant (1979) observed that there was an increase in organic carbon and available N, P, K in soil with land application of distillery spentwash in sorghum. Zalawadia and Raman (1994) revealed that the irrigation of diluted distillery waste water supplied with 75 per cent of the recommended NPK fertilizer gave a similar yield as the treatment irrigation with well water and supplied with the recommended NPK fertilizer rate.

Sugarcane

There was a significant increase in nitrogen uptake and dry matter production of sugarcane by incorporation of pressmud (normal or enriched with *Pleurotus* or *Trichoderma viride*) into soil (Anon, 1991). Yaduvanshi and Yadav (1995) reported that addition of sulphitation pressmud significantly increased the cane yield and the quality of cane juice due to the increased residual soil fertility. Continuous application of pressmud and nitrogenous fertilizer increased significantly the cane and sugar yield of sugarcane (Tiwari *et al.*, 1998). Venkateswara Rao *et al.* (2000) reported significantly higher per cent juice sucrose of 19.95 and higher cane yield or 138.5 tonnes ha^{-1} with application of bio-compost @ 10 tonnes combined with 20 kg ha^{-1} each of *Azospirillum* and Phosphobacteria.

Optimum dose of spentwash application in sugarcane was found to be 419 m^3 ha^{-1} with N, P, K supplementation (Guimarati *et al.*, 1968). Distillery spentwash when applied at the rate of 135 m^3 ha^{-1} in sugarcane was found to increase the cane yield (Cooper, 1975). Agarwal and Dua (1976) recommended the application of 20 times diluted spentwash to sugarcane to get an yield increase of 12.5 tonnes ha^{-1}. The application of diluted spentwash having a BOD value of 1000 mg l^{-1} in sugarcane increased the yield on an average by 5.1 mt ac^{-1}.

In Australia, spentwash at the rate of 35–50 m^3 ha^{-1} was recommended as optimum dose for higher cane yield (Gloria, 1977; Usher and Wellington, 1979). In Australia and in Brazil, optimum dose for land application of distillery spentwash in sugarcane was found to be 35–50 m^3 ha^{-1} (Orlando *et al.*, 1993). Maximum level of juice purity was recorded by spentwash application to cane in Brazil (Monterrio *et al.*, 1981).

In Philippines, spentwash application at the rate of 80–240 mt ha^{-1} in addition to chemical fertilizers increased the cane yield by 10–12 per cent and the sugar yield by 13–16 per cent than normal irrigation (Gonzales and Tiano, 1982). In Cuba, spentwash application at the rate of 90–150 mt ha^{-1} increased the potassium content in soil, with increased cane yield and sugar recovery (Vierira, 1982). Increased juice quality, brix, pol per cent and yield were recorded in 50 times diluted spentwash in sugarcane CO-140 (Jagdale and Savant, 1979). 50 to 75 times diluted spentwash application was found beneficial to sugarcane crop as reported by Jagdale (1976) and Pawar (1984). It was found that the optimum dose for land application of distillery spentwash to get higher yield in sugarcane was 20 m^3 ha^{-1}, which supplied 186 kg of K ha^{-1}. Devarajan *et al.* (1993) reported that sugarcane CO-8021 gave a normal yield with 50 times diluted spentwash and also found that the sugarcane yield was not affected with one time application. Devarajan *et al.* (1994) observed that the sugarcane crop gave normal yield with 50 times dilution, with an increase in soil pH, EC, macro and micronutrients, and there was no suppression in the soil microbial population and its enzyme activities. Devarajan and Oblisami (1995) recorded decreased juice quality in sugarcane and increased soil pH and salinity by spentwash irrigation.

Cotton

Sundaramurthy (1998) reported that the raw distillery effluent from sugar industry was found to lower the incidence of two major sap feeding insects in cotton: the plant louse and white fly. The incidence of bollworm was also significantly decreased on cotton crops treated with an insecticide in combination with spentwash; the crop yield of seed cotton was significantly enhanced by this treatment, from 851 to 1270–1360 kg ha^{-1}.

Oilseeds

Trivedi *et al.* (1995) observed that phosphatic fertilizers (22 kg ha^{-1}) applied in conjuction with pressmud (5.0 t ha^{-1}) significantly increased pod and haulm yield of groundnut as well as total P uptake and proved superior over FYM. Lesser nodulation and pod setting were observed in the case of groundnut receiving raw distillery effluent (Juwarkar *et al.*, 1990). Field studies were conducted by Devarajan *et al.* (1998) to study the irrigational effect of distillery spentwash at different dilutions *viz.*, water, 50, 40, 30, 20 and 10 times on soil fertility status, yield and quality of oilseed crops. Gingelly (TMV-5), groundnut (TMV-7), soyabean (CO-1) and sunflower (CO-2) performed well under spentwash irrigation. Higher seed yield of the oilseed crops was obtained with spentwash irrigation up to 30 times dilution than water. The oil content was not affected by spentwash irrigations.

A field experiment with groundnut crop was conducted to evaluate the manurial potential of three distillery effluents: Raw spentwash (RSW), Biomethanated spentwash (BSW) and lagoon sludge (LS). It was found that all the three distillery effluent increased total chlorophyll content, crop growth rate, total dry matter and final seed yield. Among the three distillery effluent, BSW produced highest seed yield. However, the distillery effluents did not influence protein and oil contents (Ramana *et al.*, 2002). Murugaragavan (2002) revealed that the seed hardening with spentwash (10 per cent and 20 per cent) significantly improved the germination and growth parameters of crops like ragi (CO-11), groundnut (VRI-2), gingelly (CO-1), sorghum (APK-1) and greengram (Pusa bold) and also remarkably increased the N, P and K contents in soil.

Legumes

Tiwari *et al.* (1988) reported that addition of compost enriched with rock phosphate and microbial cultures increased the nodulation and the Yield of green gram crop. Legumes were found to be more sensitive to application of spentwash (Sahai and Neelam, 1987). Mukherjee and Sahai (1988) reported reduction in the seed germination and seedling growth of *Phaseolous radiatus* and *Cajanus cajan* due to spentwash application. Distillery effluent applied beyond 10 per cent v/v inhibited root growth in *Pisum sativam* (Rani and Shrivastava, 1990). The high BOD load and excess soluble salts in the effluent were responsible for their toxic effect on legumes. The protein content in peas registered a decreasing trend with increasing concentration of spentwash (Rani and Shrivastava, 1991). Karande and Ghanvat (1994) reported that the distillery effluent had significant deleterious effect on seed germination and early seedling growth in pigeon pea. Ghosh *et al.* (1999) studied the effects of 0–100 per cent distillery on germination of peas, *Cicer arietinum* and *Phaseolus mungo.* Percentage of germination increased with upto 75 per cent effluent in *Cicer arietinum* and peas and upto 50 per cent effluent in *P. mungo.* Plumule and radicle growth generally increased upto 50 or 75 per cent effluent concentration and then decreased. Root : shoot ratio decreased with increasing effluent concentration.

Fruit Crops

Jeyabaskaran *et al.* (2001) reported that the performance of different organic manures in banana ratoon in terms of yield was in the order of Poultry manure = Rice husk ash > Pressmud > FYM > control. It was found that nearly 20 per cent of NPK could be saved by adding either 15 kg poultry manure or 15 kg rice husk ash per plant to produce 18.5 per cent more yield than at 100 per cent NPK+ no organic manure in ratoon of poovan and this would lead to an additional profit of either Rs. 9500/ acre or Rs. 13700/acre, respectively. Application of pressmud or poultry manure or rice husk ash @ 1.5 kg plant^{-1} can save 40 per cent NPK to produce statistically same yield as 100 per cent NPK + no organic manure in ratoon of Poovan. Devarajan and Oblisami (1994) reported that 50 times diluted distillery spentwash application in banana improved the available nutrient in soil. The cultivation of banana (poovan) using distillery spentwash increased the soil pH, EC, available N, P, K, Ca and Mg contents (Rajannan *et al.*, 1998).

Vegetable Crops

In bhendi the germination percentage was increased by 15 per cent with tap water and 25 per cent with spentwash (Hari Om *et al.*, 1994). Rathinasamy and Lakshmi Narashimhan (1995) reported that the use of 50 to 75 times diluted spentwash increased the crop quality and yield of bhendi PKM^{-1}. Haritha Josephine *et al.* (2002) reported that onion responded well to application of sulphitation pressmud and sulphur. The highest growth and yield was noticed when super phosphate @ 60 kg S ha^{-1} was applied along with sulphitation pressmud (5 t ha^{-1}) compared to other treatments in the absence of sulphitation pressmud. Ramana *et al.* (2002) reported that the effect of the distillery effluent was crop-specific and due care should be taken before using the distillery effluent for pre-sowing irrigation purposes in vegetable crops. The results are 5 percent concentration of distillery effluent was critical for seed germination in tomato and bitter gourd, and 25 per cent incase of chilli, cucumber and onion.

The organic waste management for the supply of plant nutrients is becoming more essential both for meeting the nutrient requirement and sustaining the soil health through modifications of physical, chemical and biological properties of soils. The use of organics as part of integrated nutrient management practices is on the increase, in view of the escalating cost of fertilizers and from the view point of stabilizing the crop yields and sustaining soil productivity.

References

Agarwal, M.L. and Dua, S.P., 1976. In: *Seminar on Treatment and Disposal of Effluent from Sugar and Distillery Industries*, held at Bangalore, pp. 26–29.

Annadurai, K. *et al.*, 2001. In: *National Seminar on Use of Poor Quality Water and Sugar Industrial Effluents in Agriculture*, held at ADAC and RI, Tiruchirapalli, February 5, p. 88.

Anonymous, 1991. *Newsletter–Sugarcane Breeding Institute*, 19(2): 1–3.

Arora, M. *et al.*, 1992. *Res. Cons. And Recycling*, 64: 347–353.

Asija, A.K. *et al.*, 1984. *J. Indian Soc. Soil Sci.*, 32: 323–329.

Bangar, K.C. *et al.*, 1989. *Biol. Fert. Soils*, 8: 339–342.

Chinnusamy, C. *et al.*, 1998. In: *National Seminar on Application Treated Effluent for Irrigation*, held at REC, Tiruchirappalli, March 23, p. 6.

Chinnusamy, C. *et al.*, 2001. In: *National Seminar on Use of Poor Quality Water and Sugar Industrial Effluents in Agriculture*, held at ADAC and RI, Tiruchirappalli, February, 5, p. 84.

Cooper, B.R., 1975. *Annual Report of Research*, Caroni Research Station, p. 10.

Datta, M. and Gupta, R.K., 1983. *J. Indian Soc. Soil Sci.*, 31: 511–516.

Devarajan, L. and Oblisami, G., 1994. *South Indian Horti.*, 42: 324–326.

Devarajan, L. and Oblisami, G., 1995. *Madras Agric. J.*, 82: 397–399.

Devarajan, L. *et al.*, 1993. *SISTA Sugar. J.*, 20(3): 23–25.

Devarajan, L. *et al.*, 1994. *Kissan World*, 6: 48–49.

Devarajan, L. *et al.*, 1998). In: *National Seminar on Application of Treated Effluent for Irrigation*, held at REC, Tiruchirappalli, March, 23, p. 19.

Dhar, N.R. and Mukherjee, S.K., 1938. *Prog. Sug. Tech. Agroc.*, India, 5: 15–24.

Dongale, J.H. and Savant, N.K., 1979. *J. Maharashtra Agric. Univ.*, 3: 138–139.

Ghosh, A.K. *et al.*, 1999. *Neo Botanica*, 7(1): 27–32.

Gloria, N.A., 1977. Codstil International Publication, Dedibni, Sao Paulo, Brazil, p. 76.

Gonzales, M.Y. and Tiano, A.P., 1982. In: *Proc. 29th Ann. Conv.*, Philippines Sugar Tech. Assoc., pp. 467–490.

Guimarati, B.R. *et al.*, 1968. *O. Solo*, p. 102.

Gupta, A.P. *et al.*, 1986. *J. Indian Soc. Soil Sci.*, 34: 810–814.

Haritha Josephine, V. *et al.*, 2002. In: *National Seminar on Emerging Trends in Horticulture*, held at department of horticulture, Annamalai University, Annamalai Nagar, February, 14–15, p. 130.

Hari Om *et al.*, 1994. *J. Environ. Biol.*, 15: 171–175.

Jagdale, H.N., 1976. *M.Sc. Thesis*, M.P.A.U., Rahuri.

Jagdale, H.N. and Savant, N.K., 1979. *Indian Sugar*, 29: 433–440.

Jayabaskaran, K.J. *et al.*, 2001. *South Indian Horti.*, 49: 105–108.

Juwarkar, A. *et al.*, 1990. *Environ. Monit. and Asses.*, 15(2): 200–210.

Karande, S.M. and Ghanvat, N.A., 1994. *Proc. Acad. Environ. Biol.*, 3(2): 165–169.

Khruslova, T.N. and Kolomiets, D.M., 1974. *Ferment. Spirit Prom.*, 4: 40.

Mallika, 2001. *M.Sc. (Env. Sci.)*, Tamil Nadu Agricultural University, Coimbatore, India.

Mishra, M.M. and Bangar, K.C., 1986. *Biol. Agric. Hort.*, 3: 331–340.

Mishra, M.M. *et al.*, 1982. *Indian J. Agric. Sci.*, 52(10): 674–678.

Mishra, M.M. *et al.*, 1984. *Trop. Agric.*, 61: 174–176.

Monterrio, H. *et al.*, 1981. *Brazil Aciucareiro.*, 97(4): 226–231.

Mukherjee, U. and Sahai, R., 1988. *Acta Botanica Indica*, 14: 21–25.

Murugaragavan, R., 2002. *M.Sc. (Env. Sci.)*, Tamil Nadu Agricultural University, Coimbatore, India.

Orlando, F.J. *et al.*, 1993. *Sugarcane*, 2: 4–8.

Pathak, H. *et al.*, 1998. *J. Indian Soc. Soil Sci.*, 46: 155–157.

Patil, G.D. and Shinde, B.N., 1995. *J. Indian Soc. Soil Sci.*, 43(4): 700–702.

Patil, G.D. and Shinde, B.N., 1995. *J. Living World*, 1(2): 120–125.

Pawar, R.B., 1984. *M.Sc. Thesis*, M.P.A.U., Rahuri.

Rajannan, G., *et al.*, 1998. In: *National Seminar on Application of Treated Effluent for Irrigation*, held at REC, Tiruchirapalli, March, 23, p. 56.

Rajukkannu, K. and Shanmugam, K., 1998. In: *National Seminar on Applications of Treated Effluent for Irrigation*, held at REC, Tiruchirapalli, March, 23, p. 19.

Ramana, S. *et al.*, 2002. *Biores. Technol.*, 31(2): 117–121.

Ramana, S. *et al.*, 2002. *Biores. Technol.*, 82(3): 273–275.

Ramaswami, P.P., 1999. *J. Indian Soc. Soil Sci.*, 47: 661–665.

Rani, R. and Shrivastava, M. M., 1990. *International Eco. Environ. Sci.*, 16: 126–132.

Rani, R. and Shrivastava, M.M., 1991. In: *Advances in Himalayan Ecology, Recent Researchers in Ecology, Environment and Pollution*. Today and Tomorrow's Printers and Publishers, New Delhi, India, 61: 31–139.

Rathinasamy, A. and Lakshmi Narashimhan, C.R., 1995. *Madras Agric. J.*, 85: 403–405.

Sahai, R. and Neelam, 1987. *Indian J. Ecol.*, 14: 21–25.

Shinde, D.B. and Trivedi, R.K., 1981. *Distillery Effluent Problems and Solutions*, pp. 20–25.

Singh, R.D. and Yadav, D.V., 1986. *Agric. Wastes*, 180: 73–79.

Singh, Y. and Raj Bahadur, 1998. *Indian J. Agric. Sci.*, 68(2): 70–74.

Sundaramurthy, V.T., 1998. *International Sugar J.*, 100(1199): 566–567.

Tiwari, R.J. *et al.*, 1998. *J. Indian Soc. Soil. Sci.*, 46(2): 243–245.

Tiwari, V.N. *et al.*, 1988. *J. Indian Soc. Soil Sci.*, 36(7): 280–283.

Trivedi, B.S. *et al.*, 1995. *J. Indian Soc. Soil Sci.*, 43(4): 627–629.

Usher, J.F. and Wellington, I.P., 1979. *Sugarcane Fertil. Proc., Absct.*, 1: 143.

Venkateswara Rao, V. *et al.*, 2000. *SISSTA Sugar J.*, 25: 53–55.

Vierira, D.B., 1982. *Saccharum Apc. Sao Paulo*, 5: 21–26.

Virendra Kumar and Mishra, B., 1991. *J. Indian Soc. Soil Sci.*, 39: 109–113.

Yaduvashi, N.P.S. and Yadav, D.V., 1995. *J. Indian Soc. Soil Sci.*, 44(1): 158–160.

Zalawadia, N.M. and Raman, S., 1994. *J. Indian Soc. Soil Sci.*, 42(4): 575–579.

Chapter 42

Toxicological Effects of the Effluent of a Chlor-alkali Industry on a Cyanobacterium Under Controlled Conditions and its Ecological Significance

Priyadarshini Hotta and Ashok K. Panigrahi

Environmental Science and Biotechnology Research Division, Department of Botany, Berhampur University, Berhampur – 760 007, Orissa, India

ABSTRACT

Westiellopsis prolifica, Janet was exposed to graded series of concentrations of the amended effluent and the lethal concentration values were determined. Experiments were conducted at MAC, LC_{10} and LC_{50}. Significant stimulation of growth was not very much marked at lower concentration of the toxicant when compared to the control value. Inhibition of growth was recorded at higher concentrations of the toxicant, as all growth parameters showed significantly low values. The exposed alga, at higher concentrations could not recover, in recovery studies, which indicated that the toxicant causes permanent damage to the exposed system. Insignificant stimulation at low concentrations and significant inhibition at higher concentrations indicate the behaviour of the amended toxicant. With the increase in exposure period and concentration of the toxicant, both OD and dry weight decreased showing a negative correlation. Insignificant increase in chlorophyll content was recorded at lower effluent treatment, when compared to the control value. Significant decrease in chlorophyll content was marked at higher effluent concentration, when compared to the control value. Significant decrease in phaeophytin content was recorded in effluent treatment, when compared to the control value. Carotenoid content decreased in all the exposed concentrations, when compared to the control value. During the experimental period, bleaching of the effluent exposed algal filaments was marked at higher concentrations. It was clearly visible that colour of the algal culture showed variation at higher concentrations of the effluent and at higher exposure periods. Interestingly, during recovery period, when the effluent exposed alga was transferred to effluent free medium, tiny coloured bead like structures appeared after 30 days of recovery. Ultimately, the entire bleached algae in the recovery culture turned coloured, indicating that the exposed alga was not dead but by some mechanism avoided the stress period.

Keywords: *Toxicity, Effluent, Chlor-alkali industry, Chlorophyll, Phaeophytin, Carotene, Photosynthetic efficiency.*

Introduction

Pollution of surrounding biota through the discharges of effluents and solids wastes from chlor-alkali industries has been amply demonstrated (Shaw *et al.*, 1986 and Hall *et al.*, 1987). Bouveng (1968) studied the problem of discharge of mercury along with the effluent from Caustic soda industry and its control. Shaw *et al.* (1985 and 1986) reported the residual mercury accumulation in different biotic systems available in and around a chlor-alkali industry. Shaw *et al.* (1991) reported the changes in aquatic primary productivity of the estuary contaminated with the effluent of the chlor-alkali industry.

The chlor-alkali industry M/S Jayashree Chemicals Pvt. Ltd., is situated at Ganjam, on the Bank of Rushikulya estuary about 1.5 km. away from the sea, Bay of Bengal, on the East and 30 km. North of Berhampur city on the south-eastern side of India at 84°53'E longitude and 19°16'N latitude. In the process of manufacture of chemicals the Chlor-alkali factory discharges the effluent containing mercury and chlorine, into the estuary and deposits solid waste collected from the effluent canal and primary treatment tank on the adjacent land areas. Mercury is thus; discharged into the environment through effluent and solid waste routes, contaminating the adjacent aquatic and terrestrial ecosystems, respectively. Algal bioassays are ecologically significant since algae are the dominant primary producers in most freshwater environments. Furthermore, algal assays have proven to be very sensitive indicators of contaminant stress (Miller *et al.*, 1978). In particular, algae are now days extensively used as bioassay materials. It has been reported that mercurial compounds were toxic to algae, mercuric ion prolonged the lag phase or growth and at low concentrations of mercurial compounds stimulatory effects were reported (De Filippis and Pallaghy, 1976a; Rath, 1984; Shaw, 1987; Sahu and Panigrahi, 2002 and Rath, 1991). Industrial discharges occur both as effluents and solid wastes and cause hazardous effect on living organisms (Agarwal and Kumar, 1978). Although the toxic effect of the effluent have been studied on different algae (Tewari *et al.*, 1990 and Rath, 1991). Much work has yet to be done on the effects of the effluent, solid wastes and its leached chemicals on crop field inhabiting blue-green algae. Several studies demonstrated the fact that mercurial compounds have the ability to accumulate in living organisms (Jernelov, 1974). It is well established that the effluent from the chlor-alkali industry contains mercury as the potential pollutant, which makes an entry into the paddy fields, where these tiny, nitrogen fixing BGA inhabits enriching the fertility of the paddy fields. The present piece of work was designed to study the effect of the effluent of a chlor-alkali industry on the nitrogen fixing BGA, *Westeillopsis prolifica,* Janet grows abundantly and thereby, is exposed to toxic stress of mercury.

Materials and Methods

Collection and Storage of the Effluent

Effluent from the discharge channel of the chlor-alkali industry was collected in 5 liter capacity stoppered glass bottles from three different points during the discharge of cell house washings and brought to the laboratory. All the effluent samples were mixed in a bigger glass container and stored for experimental use. In the present piece of study the physico-chemical properties of the effluent was characterized by mixing the effluents of three consecutive mercury cell washings collected at different points of the discharge channel and analyzing in the laboratory. The different characteristics of the effluent are as follows:

Table 42.1: Physico-chemical Properties of the Effluent

Sl.No.	Parameter	Data Mean of 5 Estimations with Standard Deviation
1.	Temperature (°C)	26.4±1.8
2.	pH	9.2±0.2
3.	Alkalinity (as $CaCO_3$ in mg l^{-1})	238.6±16.1
4.	Hardness (as $CaCO_3$ in mg l^{-1})	432.4±12.4
5.	Chlorinity (in mg l^{-1})	1087.2±22.6
6.	Dissolved oxygen (in mg l^{-1})	2.9±0.4
7.	BOD (in mg l^{-1})	29.3±2.8
8.	COD (in mg l^{-1})	301.2±5.2
9.	Suspended solids (in mg l^{-1})	141.8±14.2
10.	Total nitrogen (in mg l^{-1})	5.2±0.3
11.	Total phosphorus (in mg l^{-1})	0.24±0.04
12.	Total mercury (in mg l^{-1})	3.14±0.11

Experimental Procedure

Each time the effluent before use was thoroughly hand-shaken and the suspended particles were allowed to settle. The decanted clear supernatant was used for the experiments. Graded concentrations of the effluent were prepared using the culture medium as diluents and were expressed in terms of per cent (v/v). Three concentrations of the effluent *viz.* 0.2 per cent, 0.4 per cent and 0.8 per cent were selected from the algal bioassay and were named as treatment 'PH-2 (MAC value)', 'PH-3 (LC_{10})' and 'PH-4 (LC_{50})' respectively in all the experiments. Treatment 'PH-2' was the maximum allowable concentration (MAC), treatment 'PH-3' was LC_{50} and treatment 'PH-4' was LC_{90} deduced from toxicity test. Estimation of different parameters was made after the 3rd, 6th, 9th, 12th and 15th days of exposure. After 15 days of exposure, three sets of flasks from each treatment and control, were centrifuged, the algal residues were thoroughly washed in double distilled water, centrifuged again and were re-suspended in freshly prepared sterilized culture medium for recovery studies. Allen and Arnon's (1955) nitrogen free medium with trace elements (Fogg, 1949) as modified by Patnaik (1964) was found most suitable for the growth of the alga and used as the basic culture solution throughout all the experiments. Growth was measured by optical density studies and dry weight measurements in a single pan electrical balance. The amount of total chlorophyll and total phaeophytin was estimated and calculated by using the formula given by Vernon (1960). The amount of carotenoid was calculated by using the formula given by Davies (1976). All the obtained values were statistically analysed.

Results

During the experimental period, bleaching of the effluent exposed algal filaments was marked at higher concentrations, when compared to control algal filaments. At lower concentration of the toxicant, stimulatory growth was noticed, when compared to the control set. It was clearly visible that colour of the mass algal culture showed variation in colour at different concentrations of the toxicant. At higher effluent concentrations, the bleached filaments remained as such without showing any change. Interestingly, during recovery period, when the effluent exposed alga was transferred to toxicant free medium, tiny coloured bead like structures appeared after 15 days of recovery. The growth measured in terms of optical density at 550 nm of the alga exposed to different concentrations of effluent and control set at different days of exposure and recovery was presented in Table 42.2. In optical density study the "PH-2" concentration value was less than the control value on all days of exposure and recovery, and the two values showed a constant significant decrease with the increase in days of exposure up to 15th day (Table 42.2). The highest per cent of increase in O.D. value over the control value was seen at "PH-3" concentration on 6th day of exposure (47.8 per cent) and highest per cent decrease was seen on 3rd day of exposure in "PH-3" concentration. In concentration PH-3, with the increase in exposure period, the per cent decrease decreased showing improvement in the medium. A maximum of 25 per cent decrease was recorded in case of conc. PH-2 at 15th day of exposure. Where as, in case of "PH-3" the optical density value decreased from 0.20 to 0.15 on 15th day and than the value increased with the increase in recovery period and a maximum of 0.18 value was reached on 15th day of recovery. Little change in value was marked in recovery period. A maximum decrease by 21.43 per cent was recorded on 15th day of exposure in case of concentration-PH-4 when compared to the control value. The correlation coefficient analysis indicated the existence of positive significant correlation between days of exposure and optical density values in control ($r = 0.969$, $p \leq 0.01$) and Conc. PH-2 ($r = 0.986$, $p \leq 0.001$), however, negative non-significant correlation was observed in Conc. PH-3 ($r = -0.466$; p = NS). Recovery studies showed highest recovery in "PH-2" concentration at 15th day of recovery, when compared to the 15th day exposure value of the control set. Significant recovery was marked in case concentration 'PH-3' when exposed to a longer duration of time. Interesting recovery was noted in concentration PH-4.

Table 42.2: Changes in Optical Density Values in Control and Effluent Exposed Blue-green Algae at Different Exposure and Recovery Periods

Conc. of the Effluent (A)	*Exposure in Days*						*Recovery in Days*
	0	*3*	*6*	*9*	*12*	*15*	*15*
Control PH-1	0.04	0.105	0.115	0.125	0.14	0.20	0.35
PH-2 (0.2%)	0.04	0.06	0.075	0.085	0.13	0.18	0.21
PH-3 (0.6%)	0.04	0.05	0.06	0.07	0.12	0.15	0.18
PH-4 (0.8%)	0.04	0.025	0.035	0.055	0.11	0.15	0.175

Data are mean of samples.

The data presented in Table 42.3 depicted the change in dry weight of the alga *Westiellopsis prolifica*, at different days of exposure and recovery, when exposed to different concentrations of effluent. The dry weight of the control alga increased linearly with the increase in the exposure period, showing a perfect sigmoid curve *i.e.* showing normal growth under controlled conditions. At concentration "PH-2" (*i.e.* 0.2 per cent of toxicant) the dry matter value was greater than the control value up to 9th day of exposure and than decreased up to 15th day of exposure and recovery. Higher rate of growth was marked in case of concentration "PH-2" than "PH-3". Highest increase in percentage of growth (31.2 per cent) was marked on 6th day of exposure over the control value. The "PH-3" concentration showed a significant decreasing trend in growth when compared to control, and "PH-

2" in which a maximum decrease of 66.77 per cent was marked on 15th day of exposure. No increasing trend was marked up to 6th day of exposure. With the increase in exposure period, the per cent decrease increased showing a positive correlation. In case of PH-4 concentration, highly significant decrease (80 per cent) in dry weight was noted. Complete bleaching of the algal mass inside the test solution (PH-4), was observed from 6th day of exposure onwards. Gradually tiny blue-green particles started making their appearance in the white turbid mass as observed in the naked eye during recovery period. These particles grew in size with time. It was probably due to the appearance of photosynthetic pigments, which disappeared due to the effluent stress on the alga. Slowly the entire white mass got converted into a blue-green mass with the increase in exposure period 60 days of recovery. Hence, at "PH-4" [0.8 per cent concentration (LC_{90})] a drastic decrease in dry weight value was noticed on 9th day of exposure and afterwards it showed continuous decrease up to 15th day of exposure, however, the values were far less than control, and "PH-2". The dry wt of control, and "PH-2" concentration increased significantly with the increase in days of exposure, which were significantly with the increase in days of exposure. Where as in case of 'PH-2' an initial increase up to 19.8 mg/100 ml culture was marked on 9th day of exposure. In case of conc. PH-3 the value of dry weight significantly decreased to 10.8 mg/100 ml culture and when the alga was transferred to toxicant free medium, partial recovery was marked. During recovery period, significant recovery was not marked in PH-2 concentration when compared to control. But in case of PH-3 and 4, significant recovery was noted when compared to control and PH-1. This might be due to decrease in stimulatory power of the toxicant at lower concentration. Higher rate of recovery at higher concentration might be due to decrease in the inhibitory power of the toxicant. The dry weight values decreased in conc. PH-3 and 4 significantly at all exposure days at $p \leq 0.01$ levels except on 3rd day with the increase in concentration of the toxicant. The per cent change value of the dry weight over its control also increased significantly with the increase in days of exposure at concentration "PH-2" except at concentration "PH-3 and 4". At the initial exposure period up to 6th day all the inoculated BGA in PH-3 and PH-4 turned white turbid mass indicating death of the cells but after 9th day of exposure slowly tiny bead like structures appeared and on 12th day clear pigments appeared and it seemed that the alga started growing with the increase in exposure period. The exposed alga could recover interestingly during recovery period. However, it could not reach to its respective control value. The two-way analysis of variance ratio test indicated that there was significant difference between rows and columns.

Table 42.3: Changes in Dry Weight (mg/100 ml culture) of Control and Effluent Exposed Blue-green Alga, at Different Exposure and Recovery Periods, and at Different Concentrations of the Effluent

Conc. of the Effluent (A)	*Exposure in Days*						*Recovery in Days*
	0	*3*	*6*	*9*	*12*	*15*	*15*
Control PH-1	4.2	9.5	12.5	18.8	25.1	32.5	56.6
PH-2 (0.2%)	4.2	11.6	16.4	19.8	22.9	31.2	39.2
PH-3 (0.6%)	4.2	5.1	5.8	8.5	9.1	10.8	28.2
PH-4 (0.8%)	4.2	4.3	4.3	4.6	5.2	6.5	14.9

Data are mean of samples.

The total chlorophyll content increased from 0.122 to 1.112 mg/100 ml culture within 15 days of exposure in the control set and the value increased to 1.904 mg/100 ml culture on 15th day of recovery. The total chlorophyll content increased significantly ($p \leq 0.01$) in the exposed set, PH-2 after 6th day of exposure. A maximum of 16.4 per cent increase was recorded on 9th day and on 15th day 18.2 per cent increase over the control value was marked (Table 42.4) in conc. PH-2 set. Where as, in concentration PH-3, set the total chlorophyll content increased from 0.122 to 0.178 mg/100 ml culture on 3rd day of exposure and the value then decreased with the increase in exposure period and a lowest value of 0.914 mg/100 ml culture was recorded on 15th day of exposure in conc. PH-3. When the exposed alga was transferred to toxicant free medium, partial significant recovery was marked. The total chlorophyll content increased to 1.122 mg/100 ml culture within 15th days of recovery. Conc. PH-2 set showed an increase in total chlorophyll content at higher exposure periods. In case of concentration PH-3, the total chlorophyll content though increased on 3rd day, when compared to 0 day value but 58.6 per cent decrease was recorded, when compared to the control value. The correlation coefficient analysis between days of exposure verses total chlorophyll indicated the existence of significant positive correlation in control ($r = 0.965, p \leq 0.01$) and Conc. PH-2 ($r = 0.986, p \leq 0.001$). A negative correlation ($r = -0.711$, p = NS) was marked in Conc. PH-3. A significant negative correlation ($r = -0.981, p \leq 0.01$) was marked in Conc. PH-4. The two-way analysis of variance ratio test indicated that there was significant difference between rows and columns.

Table 42.4: Changes in Total Chlorophyll Content (mg/100 ml Culture) of Control and Exposed BGA at Different Exposure and Recovery Periods, and at Different Concentrations of the Effluent

Conc. of the Effluent (A)	*Exposure in Days*						*Recovery in Days*
	0	*3*	*6*	*9*	*12*	*15*	*15*
Control, PH-1	0.122	0.430	0.473	0.689	0.851	1.112	1.904
PH-2 (0.2%)	0.122	0.328	0.457	0.802	0.914	1.314	1.980
PH-3 (0.6%)	0.122	0.178	0.382	0.412	0.585	0.914	1.122
PH-4 (0.8%)	0.122	0.104	0.280	0.318	0.435	0.562	0.973

Data are mean of samples.

Table 42.5 indicates the changes in total phaeophytin content in control and amended effluent exposed blue-green alga at different exposure and recovery periods. The total phaeophytin content increased from 0.314 to 2.316 mg/100 ml culture in a period of 15 days. The value further increased to 2.850 mg/100 ml culture within 15 days of recovery. The phaeophytin content interestingly declined from control value at all exposure and recovery periods in conc. PH-2. No doubt, the phaeophytin content increased from 0.314 to 1.90 mg/100 culture, showing a positive correlation ($p \leq 0.001$), with the increase in exposure period, like the control set but the values were interestingly less than the respective control value for each exposure and recovery period, except on 6th day where an increase over the control value was noted. In concentration 'PH-3', at 6th day of exposure, the phaeophytin content increased from 0.314 to 2.345 mg/100 ml culture and than the value significantly declined with the increase in exposure period. The value declined from 2.345 to 1.714 on 15th day of exposure. When the exposed alga of concentration 'PH-3' was transferred to toxicant free nutrient medium, partial recovery was marked. The value increased to 1.936 mg/100 ml culture, on 15th day of recovery, indicating no significant damage and destruction of the pigment. In case of concentration PH-4, the phaeophytin content increased slowly from 0.314 to 1.516 mg/100 ml culture within 15 days of exposure. However, the pigment amount was far less than the control and other amended effluent concentrations. The correlation coefficient analysis between phaeophytin content and days of exposure indicated the existence of positive and significant ($p \leq 0.01$) correlation in control ($r = 0.990$) and Conc.

PH-2 (r = 0.988). But in case of Conc. PH-3 and PH-4 a significant negative correlation was marked (r = –0.964, p ≤ 0.01). The two-way analysis of variance ratio test indicated that there was significant difference between rows and columns.

Table 42.5: Changes in Total Phaeophytin Content (mg/100 ml culture) of Control and Exposed Blue-green Alga, at different Exposure and Recovery Periods and at Different Concentrations of the Effluent

Conc. of the Effluent (A)	*Exposure in Days*						*Recovery in Days*
	0	*3*	*6*	*9*	*12*	*15*	*15*
Control, PH-1	0.314	0.803	1.802	1.984	2.103	2.316	2.850
PH-2 (0.2%)	0.314	0.639	2.227	1.722	1.815	1.990	2.214
PH-3 (0.6%)	0.314	0.591	2.345	1.459	1.512	1.714	1.936
PH-4 (0.8%)	0.314	0.591	1.316	1.366	1.411	1.516	1.810

Data are mean of samples.

Changes in carotenoid content of control and effluent exposed blue-green alga was demonstrated in Table 42.6, at different exposure and recovery periods. The carotenoid content of the control alga increased from 0.0009 to 0.0072 mg/100 ml culture within a period of 15 days and the value further increased to 0.0156 mg/100 ml culture at 15 days of recovery. The carotenoid content significantly decreased at concentration PH-2 at all exposure and recovery periods except on 3rd day, where an insignificant increase was noted. The per cent change in carotenoid content, where, dichotomous behaviour of the toxicant was not clearly evinced. At concentration PH-2, the carotenoid content increased at 3rd day of exposure, when compared to the control value. A maximum of 10 per cent increase over the control value was marked on 3rd day of exposure and 43.06 per cent decrease was recorded on 15th day of exposure. In contrast, at concentration PH-3, the per cent change significantly and linearly decreased showing a maximum of 54.17 per cent decrease on 15th day of exposure. At concentration PH-4, the per cent change significantly and linearly decreased showing a maximum of 56.95 per cent decrease on 15th day of exposure. However, higher per cent decrease was marked at earlier periods of exposure. Higher decrease at lower exposure period was due to instant shock to the BGA by the stress and lower depletion value at higher exposure period indicate slow acclimatization by the BGA to the stress. A partial recovery was recorded in concentration 'PH-2' and no recovery was marked in concentration "PH-3 and PH-4" when the exposed alga was transferred to toxicant free, nutrient medium. The correlation coefficient analysis between carotenoid content of control and exposed alga versus days of exposure indicated the existence of a positive correlation in the control set (r = 0.996, p ≤ 0.001) and the values were highly significant. In Conc. PH-2 positive correlation (r = 0.981, p ≤ 0.01) was marked. A positive (r = –0.914, p ≤ 0.05) correlation was marked in Conc. PH-3 and a non-significant correlation was marked at PH-4. The two-way analysis of variance ratio test indicated that there was significant difference between rows and columns.

Discussion

Shaw (1987) reported that the alga *Westiellopsis* could tolerate only 40 per cent of the supernatant effluent concentration of a chlor-alkali industry, the mercury concentration of which was 1.1 mg/50 ml culture. Shaw (1987) did some basic line of work on the liquid waste of the industry, which lacked a detailed work. His work was mostly on toxicity testing and probable detoxification waste by available environmental chemicals. Further, Shaw (1987) amended the nutrient media with the chemicals present in the medium itself, and tried to find out the possibility of detoxification. Sahu (1987) also tried to use the extra-cellular products of the blue-green algae as agents to mask the toxic effects of the effluent/ waste of the Chlor-alkali industry. In all the experiments tried in our laboratory, we got sufficient indications regarding, the possible use of the BGA as an agent, for detoxifying the contaminated environments polluted by mercury. The supernatant was more toxic (Shaw, 1987) than the solution of inorganic form of mercury ($HgCl_2$) (Rath, 1984). Another important thing to note was that, the supernatant was much more toxic than the effluent as a whole. The effluent allowed a considerable growth of the alga even at 50 per cent concentration, whereas the supernatant did not allow the alga to grow at and beyond 40 per cent concentration. Further, stimulation in the growth at 10 and 20 per cent concentrations was much more in the effluent than the supernatant.

Table 42.6: Changes in Carotenoid Content (mg/100 ml Culture) of Control and Exposed Blue-green Alga, at Different Exposure and Recovery Periods, and at Different Concentrations of the Effluent

Conc. of the Effluent (A)	*Exposure in Days*						*Recovery in Days*
	0	*3*	*6*	*9*	*12*	*15*	*15*
Control, PH-1	0.0009	0.0020	0.0032	0.0039	0.0044	0.072	0.0156
PH-2 (0.2%)	0.0009	0.0022	0.0031	0.0032	0.0035	0.0041	0.0069
PH-3 (0.6%)	0.0009	0.0012	0.0015	0.0021	0.0024	0.0033	0.0068
PH-4 (0.8%)	0.0009	0.0012	0.0010	0.0013	0.0016	0.0031	0.0064

Data are mean of samples.

In the present piece of investigation, during the experimental period, bleaching of the amended effluent exposed algal filaments was marked at higher concentrations, when compared to control algal filaments. At lower concentration of the toxicant, stimulatory growth was noticed, when compared to the control set. It was clearly visible that colour of the mass algal culture showed variation in colour at different concentrations of the toxicant. At higher effluent concentrations, the bleached filaments remained as such with out showing any change. Interestingly, during recovery period, when the effluent exposed alga was transferred to toxicant free medium, tiny coloured bead like structures appeared after 15 days of recovery. At the initial exposure period up to 6th day all the inoculated BGA in PH-3 and PH-4 turned white turbid mass indicating death of the cells but after 9th day of exposure slowly tiny bead like structures appeared and on 12th day clear pigments appeared and it seemed that the alga started growing with the increase in exposure period. The exposed alga could recover interestingly during recovery period. However, it could not reach to its respective control value.

The total chlorophyll content increased with the increase in exposure period in the control set and the value further increased on 15th day of recovery. The total chlorophyll content increased significantly ($p \leq 0.01$) in the exposed set, PH-2 after 6th day of exposure. The value increased on 15th day of exposure. A maximum of 16.4 per cent increase was recorded on 9th day and on 15th day 18.2 per cent increase over the control value was marked in conc. PH-2 set. The chlorophyll pigment data showed a similar trend with the dry weight data. At lower concentrations increase in the value when compared to control indicated stimulation. With the increase in exposure period and toxicant concentration, the parameters decreased which indicated inhibition. The same toxicant behaved differently at different concentrations. We can conclude here that the toxicant showed dual or dichotomous behaviour during exposure. Significant recovery at lower concentrations and exposure periods is interesting. Prolonged exposure at higher concentrations of the toxicant induced death of the organisms might be due to higher accumulation of the toxicant and masking effect of the chemical(s), No recovery or partial recovery was recorded at higher concentrations.

De Filippis and Pallaghy (1976b,c) investigated tolerance of *Chlorella* to zinc. They noticed a gradual development of tolerance, which became more pronounced after 40 series of cell divisions, when the division rate was quite similar to that of the control. Reduction in number of zinc binding sites and inhibition of a temperature sensitive component of zinc uptake was evident after the development of tolerance. However, mercury showed lethal effects on blue-green algae. In the present investigation, the exposed alga was transferred to toxicant free nutrient medium and was allowed to grow for the same period of exposure, where partial recovery only was marked. After 10 minimum series of cell multiplication further increase in tolerance was not marked. This indicated that algal tolerance to mercury-contained effluent was probably physiologic rather than genetic. Hence, the toxicity of mercury to BGA may be related to availability of binding sites and membrane permeability of cells (Sahu, 1987). The findings of De Filippis and Pallaghy (1976a,b) and the suggestions of Rai *et al.* (1981 a,b) are, therefore, suggestive of physiological development of an exclusion mechanism. Whitton (1970) however, points to the possibility of a genetic adaptation as the concentration of zinc employed by De Filippis and Pallaghy (1976c) was quite high. Some physiological and morphological cases of increased tolerance may also be of importance. Sahu (1987) suggested involvement of some internal detoxifying mechanism. In the present investigation, it was observed that the alga *Westiellopsis prolifiica*, Janet can detoxify the contaminated aquatic environment. It has been observed that *Westiellopsis* can reduce the toxicity of the effluent by way of volatilization of mercury from the effluent. Higher rate of detoxification was marked at lower concentration of mercury in the effluent. This result indicated the need of initial dilution of the effluent to a particular level, where volatilization and absorption become faster. This system can be well utilized in conditions of environment, where mercury passes through the soil strata by leaching and percolation. Sahu (1987) and Shaw (1987) probably never tried to test the rate of detoxification by way of volatilization but concentrated more on the dual behaviour of the toxicant at different concentrations.

Walsh and Alexander (1980) reported that a liquid industrial waste might affect the algal growth in any of the three ways–Stimulation, inhibition or stimulation at lower concentration and subsequent inhibition at higher concentration. The present result agrees with the findings of Walsh and Alexander (1980) but the effluent's behavior is totally dose dependent. Rath (1984), Shaw (1987), Sahu (1987) and the present investigation agrees with the third line possibility. Prasad and Prasad (1982) observed stimulation of the algal growth at low concentration of Cadmium, Lead and Nickel. Neither Cd, Pb and Ni have been reported to be essential micro-nutrients for algae (O'Kelley, 1974) nor the pure solution of these heavy metals are expected to act as growth regulator. Further, inorganic forms are active only when these are present in free ionic state (Spencer and Nichols, 1983). Gadd and Griffith (1978) and Sahu (1987) reported that biological activity of heavy metal ions is markedly affected by the presence or absence of other ions. Several authors (Sahu *et al.*, 1986; Shaw *et al.*, 1988,89 and Rath *et al.*, 1986) have found that heavy metals cause prolongation of lag phase more or less in proportion to doses, followed by normal growth (Bartlett *et al.*, 1974 and Zingmark and Miller, 1975). The common explanation for stimulation at lower concentration is probably that the medium is modified during the first part of the experiment either by exudation from living cells or by leaching from dead cells to render the heavy metal less toxic by some sort of chelation (Nielsen and Andersen, 1970 and Shaw, 1987). The prolongation of lag period in exposed cells can be explained. The explanation is probably in presence of toxicants; the medium is not favourable for growth due to alteration of osmotic balance and availability of high concentration of ions in the medium. This explanation was supported by the findings of Sahu (1987), Rath (1991), Mohapatra (1992) and Sahu and Panigrahi (2002) of our laboratory. Similar findings were observed in the present study. At lower concentration of the effluent, interestingly no such lag was observed. However, at higher concentrations of the effluent prolongation of the lag phase leading to death of the alga was marked. Even when the alga was transferred to the toxicant free nutrient medium for recovery, it could not recover at all. This indicated the total death of the exposed alga. The growth inhibition by the effluent has been reported (Sahu, 1987) to be due to some alterations in the permeability of cell membranes, inhibition of enzyme systems, inhibition of photosynthesis, nucleic acid synthesis and numerous other metabolic processes.

Kashyap and Gupta (1981), Rath (1984), Rath *et al.* (1986), Sahu (1987), Shaw (1987), Sahu *et al.* (1988) and Shaw *et al.* (1988,1989 a, b) observed stimulation by different mercurial compounds, however, the mechanism of growth stimulation cannot easily be explained. Such stimulation might be due to the presence of some growth regulating compounds and/or trace elements (Hufford, 1971). The effluent of the chlor-alkali industry may only contain an insignificant or negligible amount of trace elements or growth regulating compounds in it. Hence we do not agree at this stage with the reports of Hufford (1971) and Dustan *et al.* (1975). Some workers suggested the uptake and interference of the toxicant with the cellular biochemicals which might be producing some growth regulating chemicals or chemicals which can induce stimulation in growth as the possible mechanism for the growth stimulation (Sahu, 1987). Here, we have observed an increase in the final yield following effluent treatment. Since, an increase in the final yield in the effluent treated alga was observed in the present study, the possibility of stimulation either by the absorbed metal or by some other mechanisms looks more appropriate than the metabolization of the effluent with mercury. The only speculation left, to account for the enhancement was the likely presence of some growth regulator(s), which might have influenced climax of the test alga. This type of speculation is not acceptable at this stage and is not valid for this type of effluent treatment, where most of the fractional constituents at higher concentrations, are independent poisons/toxicants and in combination might show antagonistic or synergistic effects. The peculiar behavior of the algal organisms under stress to avoid the stress is an interesting feature in toxicological studies. Due to exudation, the medium might be changing or the exuding chemicals might be reacting with mercury and the other chemicals of the effluent forming a hard cyst, which must be providing an adhering surface for the heavy metal. The cyst might be restricting the heavy metal's entry into the cell, due to the formation of a barrier. It has been reported that with the increase in exposure period, the mercury concentration increased in exposed algae (Rath, 1984; Sahu, 1987 and Shaw, 1987). The same authors opined that with the increase in residual mercury concentration the growth decreased significantly. In the present study, at higher exposure period, depletion in growth rate was observed. Growth rate studies by optical density method showed inconsistent data in exposed cultures. However, consistency was observed in the dry weight measurement studies.

Dry weight has been considered by good number of workers as a parameter of growth. The change observed in optical density study exactly does not reflect the real changes induced by the pollutant, but an approximation can be made out of this data. Since growth is a summation of all cellular metabolisms, any inhibition of growth reflects toxic effects on a number of metabolic processes. Also, the use of growth rates allows one to observe, if the bioassay organism has the capability of recovery from the toxic effect, over extended periods of time. There is the disadvantage that as the algal cells increase in number, the concentration of the toxicant per cell decreases from the original value. Walsh and Alexander (1980) demonstrated that algal species that were sensitive to pesticides in monoculture were less sensitive in the presence of resistant species, perhaps because, the resistant species grew quickly and absorbed the pesticide, thus, reducing its concentration in growth medium. Industrial waste caused a reversal in species numerical dominance in mixed algal cultures.

Photosynthetic pigments of the plant systems play a vital role in trapping solar energy. The pigments are known to participate in generation of energy for CO_2 fixation (Kashyap and Gupta, 1981). The chlorophylls have long been recognized as the primary light acceptors in plants and they are invariably present in every organism, which carries out photosynthesis with absorption of CO_2 and evolution of molecular oxygen. Shaw (1987) reported that with the waste treatment, the chlorophyll content increased to a great extent in lower concentrations and virtually no decrease in the level in higher concentration. Zingmark and Miller (1975) and De Filippis and Pallaghy (1976b) reported that heavy metals inhibit photosynthesis. De Filippis and Pallaghy (1976b), Rai *et al.* (1981b) and Rath *et al.* (1986a,b) reported that heavy metals reduce chlorophyll content. De *et al.* (1985) suggested that at 20 ppm the Hg(II) lowered the chlorophyll by decreasing the synthesis of chlorophyll, as well as possibly by increasing the synthesis of chlorophyll as well as possibly by increasing chlorophyllase activity in *Pistia.* De Filippis and Pallaghy (1976b) reported that all the heavy metal solutions inhibited the rate of chlorophyll synthesis in the cultures, will, PMA and $ZnCl_2$ causing the greatest and least inhibition respectively. O'Kelly (1974) reported that copper inhibits growth as well as photosynthesis of alga. Photosynthesis was also severely affected by zinc. Many studies on lead point to its weak toxic effect on photosynthesis, respiration and cell division of various algae. Shaw (1987) opined that when mercury has been reported to be toxic, an increase in the level of chlorophyll in the alga exposed to the effluent highly concentrated with mercury could hardly be explained. At lower concentration of the effluent, stimulatory growth was noticed, when compared to the control set. Carotenoids play a vital role as a protector of photosynthetic tissues against photosensitized oxidation. The decrease in carotenoid content in algal cells exposed to heavy metal stress lead to a decrease in protection from the stress to the photosynthetic tissue. When the nutrients in the medium are exhausted or a toxicant is introduced into the medium, this ratio rises due to decrease in chlorophyll content. The wide spread occurrence, as well as certain chemical properties of chlorophyll pigments *in vivo* suggest that these pigments play an active role in photosynthesis functioning as a photo enzyme and the mercurial compounds are toxic for the biosynthesis of chlorophyll pigments. Chlorophylls are known to be converted to phaeophytins as a consequence of exposure to weak acids, by replacement of Mg^{2+} with two atoms of hydrogen and thereby changing the spectral properties. The phaeophytin content of an algal system is very important, since any unfavorable change in the environment inhabited by the alga is reflected through a change in phaeophytin level. Degradation to phaeophytins might be the first step towards break down of chlorophylls. This is clearly evident from the fact that increased levels of phaeophytins are found in toxicant exposed algae (Sahu and Panigrahi, 2002). At higher concentration of toxicant, phaeophytins are further broken down, showing a decrease in phaeophytin level (Sahu, 1987). The importance of BGA in crop field ecosystem, as primary producer and nitrogen fixer maintaining the nitrogen economy of crop fields is already known. Sahu and Panigrahi (2003) reported that this alga can detoxify a mercury contaminated environment by way of volatilization of mercury from the environment. This can be a possible mechanism by which the alga could tolerate a higher concentration of mercury. The possible consequences of contamination of crop fields with waste chemicals of industry have also been discussed emphasizing their effects on the friendly BGA. The findings of the present investigation indicated the high toxic nature of the effluent, which is reflected from the extent of damage that it caused to different growth parameters of the test organism, besides inhibiting the growth. In spite of the high toxicity of the effluent, the alga *Westiellopsis prolifica,* Janet was however, able to withstand or tolerate relatively high concentration of the mercury contained effluent. Thus, the resistant nature of the BGA may be used for decontaminating the mercury contained crop fields, none the less at the same time increasing the fertility of the fields by nitrogen fixation.

Acknowledgements

Authors wish to thank the Head, Department of Botany for laboratory facilities and DOD (OSTC) for financial support in the form of a project.

References

Agarwal, M. and H.D. Kumar, 1978. Physico-chemical and phycological assessment of two mercury polluted effluents. *Ind. J. Environ. Health*, 20: 141–155.

Allen, M.B. and D.I. Arnon, 1955. Studies on nitrogen fixing blue-green algae. Growth and nitrogen fixation by *Anabaena cylindrica*, Lemm. *Pl. Physiol., Lancaster*, 30: 366–372.

Bartlett, L. and Rabe, F.W., 1974. Effects of copper, zinc and cadmium on *Selenastrum capricornutum. Water Res.*, 8: 179.

Bouveng, H.O., 1968. The chlorine industry and mercury problem. *Modern. Kemi.*, 3: 45.

Davies, B.H., 1976. In: *Chemistry and Biochemistry of Plant Pigments*, Vol. 2, (Ed.) T.W. Goodwin. Academic Press Inc., London, p. 38.

De, A.K., A.K. Sen, P. Modak and S. Jana, 1985. Studies of toxic effects of Hg (II) on *Pistia stratiotes. Water, Air, Soil Pollut.*, 24: 351–360.

De Filippis, L.F. and C.K. Pallaghy, 1976a. The effects of sub-lethal concentrations of mercury and zinc on *Chlorella*. 1. Growth characteristics and uptake of metals. *Z. Pflanzen-physiol. Bd.*,78: 197–207.

De Filippis, L.F. and C.K. Pallaghy, 1976b. The effects of sub-lethal concentrations of mercury and zinc on *Chlorella*. II. Photosynthesis and pigment compositions. *Z. Pflanzen-physiol. Bd.*, 78: 314–322.

De Filippis, L.F. and C.K. Pallaghy, 1976c. The effects of sub-lethal concentrations of mercury and zinc on *Chlorella*. III. Development and possible mechanism of resistance to metals. *Z. Pflanzen-physiol. Bd.*, 78: 323.

Fogg, A.E., 1949. Growth and heterocyst production in *Anabaena cylindrica,* Lemm. in relation to carbon and nitrogen metabolism. *Ann. Bot. N.S.*, 13: 241–259.

Gadd, G.M. and A.J. Griffith, 1978. Microorganism and metal toxicity. *Microbial Ecol.*, 4: 303– 317.

Hall, A., A.C. Duarte, M.T.M. Caldeira and M.F.B. Lucas, 1987. *The Science of the Total Environment*, 64: 75–87.

Hufford, G.L., 1971. The biological response of oil in the marine environment, a review. Background report, Washington, D.C., U.S. Coast guard Oceanographic Unit, Office of Research and Development, U.S. Coast guard Headquarters.

Jernelov, A., 1974. Factors in the transformation of mercury to methylmercury. In: *Environmental Mercury Contamination*, (Eds.) R. Hortung and B.D. Dinman. Ann. Arbor Science Publisher Inc., p. 167.

Kashyap, A.K. and S.L. Gupta, 1981. Effects of dichlone, macromolecular synthesis and photosynthetic pigments in blue-green algae. *Acta Bota. Indica*, 9: 265–271.

Miller, W.E., J.C. Greene and T. Shiroyama, 1978. The *Selenastrum capricornutum* Printz algal assay a bottle test, EPA-600/9-78-018. USEPA, Corvallis, OR.

Mohapatra, A., 1992. Eco-physiology, resistance and ecological implications of a mercurial compound on a blue-green alga. *Ph.D. Thesis*, Berhampur University.

Nielsen, E.S. and W. Wium-Anderson, 1970. Copper ions as poison in the sea and in fresh water. *Mar. Boil.*, 6: 93–97.

O'Kelley, J.C., 1974. Organic nutrients. In: *Algal Physiology and Biochemistry*, (Ed.) W.D.P. Stewart. Blackwell Scientific Publications, Oxford, pp. 610–635.

Pattnaik, H., 1964. Studies on nitrogen fixation by *Westiellopsis prolifica*, Janet. *Ph.D. Thesis*, University of London.

Prasad, P.V.D. and P.S.D. Prasad, 1982. Effect of cadmium, lead and nickel on three fresh water green algae. *Water, Air and Soil Pollut.*, 17: 263–268.

Rai, L.C., J.P. Gaur and H.D. Kumar, 1981a. Phycology and heavy metal pollution. *Biol. Rev.*, 56: 99–151.

Rai, L.C., J.P. Gaur and H.D. Kumar, 1981b. Protective effects of certain environmental factors on the toxicity of zinc, mercury and methyl mercury to *Chlorella vulgaris*. *Env. Res.*, 25: 250–259.

Rath, P., 1984. Toxicological effects of pesticides on a blue-green algae, *Westiellopsis prolifica, Janet. Ph.D. Thesis*, Berhampur University, India.

Rath, P., A.K. Panigrahi and B.N. Misra, 1983a. Residual mercury level in a blue-green alga, *Westiellopsis prolifica*, Janet. *Curr. Sci.*, 52: 611–612.

Rath, P., A.K. Panigrahi and B.N. Misra, 1983b. Effect of inorganic mercury on the growth of *Westiellopsis prolifica*, Janet. *J. Environ. Biol.*, 4: 103–109.

Rath, P., A.K. Panigrahi and B.N. Misra, 1985. Effect of pesticides on the photosynthetic efficiency of a BGA, *Westiellopsis prolifica*, Janet. *Microbios Letters*, 29: 25–29.

Rath, P., A.K. Panigrahi and B.N. Misra, 1986a. Effect of 2-methoxy ethylmercury chloride Emisan-6 on the nitrogen metabolism. *Microbios Letters*, 31: 15–20.

Rath, P., A.K. Panigrahi and B.N. Misra, 1986b. Effect of both inorganic and organic mercury on ATPase activity of *Westiellopsis prolifica*, Janet. *Environ. Pollut.*, 42: 143–149.

Rath, S.C., 1991. Toxicological effects of a mercury contained toxicant on *Anabaena cylindrica*, L. and its ecological implications. *Ph. D. Thesis*, Berhampur University, India.

Sahu, A., 1987. Toxicological effects of a pesticide on a blue-green alga. *Ph.D. Thesis*, Berhampur University, Orissa.

Sahu, A., B.P. Shaw, A.K. Panigrahi and B.N. Misra, 1986. Effect of phenyl mercuric acetate on ATPase activity of a blue-green alga, *Westiellopsis prolifica*, Janet. *Microb. Lett.*, 33: 45–50.

Sahu, A., B.P. Shaw, A.K. Panigrahi and B.N. Misra, 1988. Effects of phenyl mercuric acetate on oxygen evolution in a nitrogen fixing blue-green alga, *Westiellopsis prolifica*, Janet. *Microb. Lett.*, 37: 119–123.

Sahu, A. and A.K. Panigrahi, 2002. Eco-toxicological effects of a chlor-alkali industry effluent on a cyanobacterium and its possible detoxification. In: *Algological Research in India*, pp. 351–430, (Ed.) N. Anand. Bishen Singh Mahendra Pal Singh, Dehra Dun, India.

Sahu, A. and A.K. Panigrahi, 2003. Possible detoxification of mercury contained effluent of a chlor-alkali industry by a cyanobacterium. *J. Curr. Sciences*, 3(2): 315–322

Shaw, B.P., 1987. Eco-physiological studies of industrial effluent of a chlor-alkali factory on biosystems. *Ph.D. Thesis*, Berhampur University, Orissa.

Shaw, B.P., A. Sahu and A.K. Panigrahi, 1988. Effects of the effluent from a chlor-alkali factory on a blue-green alga: Changes in the oxygen evolution rate. *Microb. Lett.*, 37: 89–96.

Shaw, B.P., A. Sahu, A.K. Panigrahi, 1989a. Effect of the effluent from a chlor-alkali factory on a BGA. *Bull. Environ. Conta., Toxicol.*, 43: 618–626.

Shaw, B.P., A. Sahu and A.K. Panigrahi, 1989b. Effect of effluent from a chlor-alkali industry on nitrogen fixation ability of *Westiellopsis prolifica*, Janet. *Microb. Lett.*, 42: 91–96.

Spencer, D.F. and L.H. Nichols, 1983. Free nickel ion inhibits growth of two species of green algae. *Environ. Pollut.*, 31(A): 97–104.

Tewari, A., S. Thampan and H.V. Joshi, 1990. Effect of chlor-alkali industry effluent on the growth and biochemical composition of two marine macroalgae. *Mar. Pollut. Bull.*, 21(1): 33–38.

Vernon, 1960. Spectrophotometric estimation of chlorophylls and pheophytins in plant extracts. *Analyt. Chem.*, 32: 114–150.

Walsh, A.E. and S.V. Alexander, 1980. A maine algal bioassay method: Result with pesticide and industrial wastes. *Water, Air and Soil Pollution*, 13: 45–55.

Zingmark, R.C. and T.G. Miller, 1975. The effects of mercury on the photosynthesis and growth of estuarine and oceanic phytoplankton. In: *Physiological Ecology of Estuarine Organisms*, (Ed.) F.J. Vernberg. Belle W. Baruch Library of Marine Science, University of South Carolina Press, Columbia, 3: 45–57.

Chapter 43

Histopathological Alterations Induced by Aquatic Pollutants in *Glossogobius giuris* from Avalapalli Dam

G.V. Venkataraman[1], P.N. Sandhya Rani[2], M.B. Nadoni[2] and P.S. Murthy[2]

[1]*Department of Studies in Environmental Science, University of Mysore, Manasagangotri – 570 006, India*

[2]*Department of Zoology, Bangalore University, Bangalore – 560 056, Karnataka, India*

ABSTRACT

The present study is aimed to assess the histological damage caused to the fish *Glossogobius giuris* by various aquatic pollutants present in polluted waters of Avalapalli dam. Light microscopic studies exhibited severe histopathological changes in the gills, thyroid, kidney and blood cells due to the impact of the pollutants. The significance of the results was discussed in relation to physiological stress leading to the development of anaerobic conditions at the tissue level in pollutant stressed fish.

Keywords: *Histopathology, Pollutants, Gill, Kidney, Thyroid, Haematopoietic, G. giuris.*

Introduction

Pollution of the aquatic environment generally causes changes in the physiological and structural aspects of the inhabitant organisms, particularly the fishes. Histopathological effects of pollutants vary with the body parts, nature of the pollutant, medium and duration of exposure (Vijayamadhawan and Iwai, 1975 and Venkataramana *et al.*, 2001). Water quality characteristics influence histopathological manifestations of toxic effects (Galat *et al.*, 1985).

Avalapalli dam is a heavily polluted area situated between Bangalore and Hosur in Karnataka. With tremendous growth of urbanization and human settlement around the dam and the beautification programme, the water spread is reducing at a faster rate. Besides the sewage from the surrounding slums and colonies and untreated Industrial effluents are continuously dumped into the dam. This has led to the hypereutrophication of the dam adding to the nutrients (nitrates and phosphates) as well as toxic elements, which lead to frequent fish kills in the dam. The liquid and gaseous wastes from the above industrial complexes discharged without standard treatment have caused heavy pollution of water and air in the surrounding area.

The dam water is toxic today. Low water levels, high temperature of water and air, increased turbidity and decreased sulphates, absence of dissolved oxygen, presence of nitrates, presence of poisonous gases and occurrence of heavy algal blooms in the summer add to the insanitary conditions. As the freshwater fishes in our country constitute an important part of animal protein in rural as well as in urban areas; their direct or indirect pollution and the consequent toxicity are of great importance to the environmental biologists. The present report describes our observations on the histomorphological changes in vital tissues such as gills, thyroid, kidney and blood cells of the freshwater teleost *Glossogobius giuris*.

Materials and Methods

Study Area

Avalapalli dam in between Bangalore and Hosur is located at 13°20′58″ to 13°21′39″ S latitude and 77°06′22″ to 77°06′38″ W longitude, about 23 miles North East of Bangalore. The dam covers an area of 200 acres and has a catchments area of about 2.5 km^2.

The freshwater gobiid fish *G. giuris* (body length 8–10 cm, body wt. 30–40 gm) from other water bodies (Shanky tank) as control fish and polluted water body (Avalapalli dam) fish were collected by using cost and gill net (mesh: 10 mm). Both (control) and polluted water (experimental) live fishes brought to the laboratory and maintained in separate glass aquaria (60″ × 30″ × 20″) with respective of their tank water. The tissues of gills of both control and experimental fishes removed immediately and fixed in Boun's fluid, thyroid in buffer neutral formalin dehydrated in ethanol and embedded in paraffin (58–60°C) and mid-longitudinal sections of 5-6 μ thickness were cut and stained Ehrlich haematoxylin stain for histopathological studies. Imprints of head kidney were used for haemopoiesis. A thin slice of excised organ of both was collected and placed on a methanol coated clean microslides for preparing smears of head kidney by adapting the technique as described by Ashley and Smith (1963). The smears are air dried and stained with Giemsa for cytomorphological studies. The blood smears were also made immediately by severing the caudal peduncle on clean microslides. The slides were air dried and blood smears were fixed in methanol and stained with Giemsa for cytomorphological studies.

Results and Discussion

Histology of Gill

Control

Healthy and control gill is characterized by the presence of primary lamellae alongwith the secondary lamellae confirming to the general architectural design of the tissue. It also shows mucous cells lying scattered on both the sides of epithelium. The microridges are visible on the peripheral side of the epithelial lining. The pillar cells are scattered and basement membrane traverses throughout the secondary gill lamella.

Effect of Pollutants

Gills, mainly concerned with respiration and osmoregulation showed certain interesting degenerative changes like the fusion of secondary lamellae, bulging of tips of gill filaments and atrophy of the lamellae which results in shifting of energy metabolism. In some cases a complete degeneration of gill filaments and respiratory epithelium was observed. Hyperplasia of interlamellar cells and necrosis of secondary lamellae was also observed in few cases. These changes reduce the respiratory area thereby reducing the respiratory and osmoregulatory potential. It also indicates a decrease in energy metabolism due to degeneration of respiratory epithelium and the damage of the gill tissue may finally result in tissue hypoxia.

Gardner and Yevich (1970) and Pandey *et al.* (1997) observed impairment of respiration and external functions in the gills of an estuarine teleost fish (*Fundulus heteroclitus* and *Liza parsia*) due to such similar degenerative changes in the respiratory epithelium. These pathological changes in respiratory gill might have resulted in such shift from aerobic to anaerobic pathway in tissues of fish under pollutant stress.

Histology of Thyroid

Control

Histologically the thyroid gland is composed of a large number of follicles each of which is in the form of hallow ball consisting of a single layer of epithelial cells enclosing fluid filled space. These follicles vary in shape and size and are bound together by connective tissue. The epithelium surrounding the follicle may be thick or thin and the height of the cells depends upon its secretory activity. The less active follicles generally show a thin epithelium. The lumen of each follicle is full of colloid which may be basophilic or acidophilic depending upon the secretory activity of the follicle.

Effect of Pollutants

The thyroid follicles exhibited marked cytological changes. These changes are accompanied by cellular hypertrophy and follicular hyper plasia. The epithelial cells were found to be cuboidal with dense colloid, thus many vacuoles are frequently recognized in the colloid adjacent to the epithelium (arrow). Most of the follicles were devoid of colloid. In some follicles, the follicular lumen was totally obliterated with total loss of colloid, few follicles with little amount of colloid were also noticed.

The follicular lumen was obliterated with total loss of colloid, similar observations were made by Van Overbreeke and McBride (1971) in Sockey salmon *Oncorhynchus nerka,* Hyperactivity of thyroid follicles of few teleost were also reported by Baker Cohen (1961), and Rangeker and Latey (1977). In the light of present findings on the *G. giuris* cause damage on the structure of follicle, amount of colloid, and function of the cell.

Histology of Kidney (Haematopoietic tissue)

Control

The normal head kidney smears is characterized by the presence of stem cells called haeomoblast and interstitial tissue. Haemoblast is the pluripotant nature, which gives rise to unipotent precursor cells. These are in clumps and scattered in the imprints of head kidney. The developmental stages of erythroblast, lymphoblast, granuloblast and monoblast are spherical, oval or round with distinct cell membrane.

Effect of Pollutants

The histopathological alterations due to the pollutants in kidney smears are the changes in the cellular nature of haemoblast, granuloblast, acidophilic erythroblast, lymphoblast cells, interstitial tissue and desquamation of the epithelium. In majority of the fish severe vacuolar degeneration of haemoblast cells, monoblast and local necrosis was observed. Chronic inflammation of the interstitial tissue was also noticed.

The degeneration of the cells suggests an impairment in the developing cell lines and infiltration process leading to imbalance in osmotic regulation of the body fluids. Dheer *et al.* (1986) have suggested that the penultimate developmental stages of both erythrocytes and leucocytes in the haematopoietic tissue, is decreased significantly during the stress conditions. These pathological changes may be due to the preferential accumulation of pollutants that causes degeneration leading to lymphocytic infiltration as a measure of resistance to the toxicants and tissue susceptibility (Eaton, 1974; Mount and Stephan, 1967; Kumari and Kumar, 1997).

Histology of Blood Cells

Control

The blood cells of *G. giuris* are broadly classified into erythrocytes, thrombocytes and leucocytes. The mature erythrocytes are oval, oblong or elliptic with distinct rounded or oval nuclei containing clumped chromatin material. Thrombocytes varied from round to horseshoe and spiked form. They are differentiated by their scant, pale grey to colourless cytoplasm. The leucocyte population composed of lymphocytes, neutrophils, monocytes and macrophages. The lymphocytes are amoeboid or spherical in shape with pseudopoid formation, nucleus contains coarsely clumped chromatin material which stained deeply with Giemsa. Neutrophils are spherical, oval and fairly large with distinct cell membrane and the nucleus appears to be spherical or dumb-bell in shape. Manocytes and monophases are the rare type of cells with larger size and eccentric irregular nucleus.

Effects of Pollutants

Erythrocytes were extensively degranulated and cellular hypertrophy was evident, some cells showed cytoplasmic vacuolization with dark, condensed nucleus (arrow). The shrinkage of erythrocyte cell membrane was also noticed. The cell membrane was indistinct and the degranulation of the thrombocytes was evident. The lymphocytes showed indistinct cell membrane and the cytoplasm was chromophobic. Neutrophil cells showed hypertrophy, the nuclei exhibited marked cytological changes. In most of these cell are granules were compact, deeply stained and confined to the periphery of the cell membrane as a result of accumulation of chromatin on one side of the nucleus. The cell membrane became shrinkened or in serrated in nature, monocytes and macrophages were hypertrophied, boundaries of cell were indistinct and these cells are large and pycnotic nuclei with diffused chromatin material.

Erythrocytes are oxygen carrying devices, the quantitative decrease in their levels might have led to the dearrangement of the oxidative metabolism with concomitant decrease in the tissues of respiratory potential. The decrease in erythrocytes leading to anaemia as a result of inhibition of erythropoiesis, haemosynthesis and increase in the rate of erythrocyte destruction in the haemopoietic organs. Similar reports have been made by Goel and Kalpana (1995) and Sampath *et al.* (1998). Natarajan (1991) reported the reduction of haemoglobin (Hb) content, erythrocyte count and haematocrit values, resulting in hypochromic anaemia, due to deficiency of iron and decreased utilization for Hb synthesis. The increase of neutrophil may be due to tissue damage and lymphocytes could lead to loss of immune mechanisms of fish. (Kurde and Singh 1995 and Mc Leay and Gordon 1977). The present study indicates that produce deformities in the cytomorphology of blood cells.

From the present study it is inferred that histopathological changes in fishes would serve useful purpose in evaluating the toxic effects of various pollutants present in large amounts in a heavily polluted Avalapalli Dam.

Acknowledgements

The authors are grateful to the Chairman, Department of Environmental Science, University of Mysore, Manasagangotri and Department of Zoology, Bangalore University, Bangalore for providing laboratory facilities.

References

Ashley, L.M. and Smith, C.E., 1963. Advantages of tissue imprints over tissue sections of blood cell formation Progve. *Fish. Cult.*, 25: 93–96.

Baker-cohen, K.F., 1961. *The Role of the Thyroid in the Development of Platyfish*. Zoological Society, New York, 46: 181–222.

Dheer, J.M.S., T.R. Dheer and C.L. Mahajan, 1986. Haematological and haemopoietic response to sodium chloride stress in a fresh water air breathing fish *Channa punctatus* (Bloch). *J. Fish Biol.*, 28: 119–128.

Eaton, J.G., 1974. Chronic cadmium toxicity to the blue gill *Lepomis macrochirus* rafinesque. *Trans. Am. Fish. Soc.*, 103: 729–735.

Galat, D.L., G. Post, T.J. Keefe and G.R. Boucks, 1985. Histopathological changes in the gill, kidney and liver of Lohonta cut throat trout, *Salmo clarki Henshawi*, living in lakes of different salinity alkalinity. *J. Fish. Biol.*, 27: 533–552.

Gardner, G.R. and P.P. Yevich, 1970. Histological and haematological responses of an estuarine teleost to cadmium. *J. Fish. Res. Bd. Can.*, 27: 2185–2196.

Goel, K.A and G Kalpana. 1995. Haematological characteristics of *Heteropneustes fossilis* under the stress of zinc. *Indian J. Fish.*, 36: 256–259.

Kumari, A.S and Kumar, R.S.N., 1997. Histopathological alterations induced by aquatic pollutants in *Channa punctatus* from Hussainsagar lake. *J. Environ. Biol.*, 18(1): 11–16.

Kurde, S. and R. Singh, 1995. Effects of two samples of textile effluents and dyes on total erythrocyte count and related parameters of Oster rats. In: *Proc. Acad. Environ, Biol.*, 4: 177–181.

McLeay, D.J. and M.R. Gordon, 1977. Leucocrit a simple haematological technique for measuring acute stress in Salmonid fish, including stressful concentrations of pulp mill effluent. *J. Fish. Res. Bd. Can.*, 34(1): 2164–2175.

Mount, D.L. and C.E. Stephan, 1967. A method for detecting cadmium poisoning in Fish. *J. Wild L. Mgt.*, 31: 168–172.

Natarajan, G.M., 1991. Changes in the biomodal gas exchange and some blood parameters in the air-breathing fish, *Channa striatus* (Bloch) following lethal exposure to metasystox (Dimeton). *Curr. Sci.*, 50: 40–41.

Pandey, A.K., K.C. George and P.M. Mohamed, 1997. Histopathological alteration in the gill and kidney of an estuarine mullet in *Liza parsia* (Ham) caused by sub lethal exposure to lead. *Indian. J. Fish.*, 44(2): 171–180.

Rangeker, P.S. and A.M. Latey, 1977. Influence of the thyroid gland on gonadal activity in the teleost, *Tilapia mossambica. J. Anim. Morpho. Physiol.*, 23(2): 344–352.

Sampath, K., R. James and K.M. Akbar Ali, 1998. Effects of copper and zinc on blood parameters and prediction of their recovery in *Oreochromis mossambicus. Indian. J. Fish.*, 45(2): 129–139.

Van Overbeeke, A.P. and J.R. McBride, 1971. Histological effects of 11-ketotestosterone, 17-methyl testosterone, Estradiol, Estradiol cypionate and cortisol on the inter renal tissue, thyroid gland and pituitary gland of gonodectomised Sokey Salmon (*Oncorhynchus nerka*). *J. Fish Research Board of Canada*, 28(4): 21–28.

Venkataramana, G.V., D.U. Anandhi, and P.S. Murthy, 2001. Effect of malathion on the haematology of gobild fish, *Glossogobius giuris* (Ham). *Indian. J. Trends in Life Sciences*, 16(1): 19–26.

Vijayamadhawan, K.T., and T. Iwai, 1975. Histochemical observations on the permeation of heavy metals into taste buds of gold fish. *Bull. Jap. Soc. Fish*, 41: 631–639.

Chapter 44

The Assessment of the Soil Pollution Parameters of the Various Soil Samples of Sanganer Town of Pink City, Rajasthan

*Dinesh Kumar, H.S. Shivran, M. Prasad and R.V. Singh**

Department of Chemistry, University of Rajasthan, Jaipur – 302004

ABSTRACT

Southern part of the Jaipur city is an industrially developing area in Rajasthan. It is mainly a rural area where agriculture is main substance. A large number of industries are operated in this area. Mostly water resources are engaged as in industrial water sources. Thus, the industrial effluents and domestic sewage are mainly being used for the irrigation purposes. The recent investigating survey was carried out in the industrially polluted zone of Jaipur city to evaluate the quantitatively as well as qualitatively of soil pollution. For that, different physico-chemical parameters of four contaminated soil profiles soils have been analyzed during the July 2004. Analyzed results showed that the industrial effluents being discharged in to the Amanishah nallah, which is being used for the agricultural purposes, pollute the soils of the study area.

Keywords: Assessment, Soil pollution, South part, Agriculture, Irrigation and Physico-chemical.

Introduction

Organic manures in addition to supplying essential nutrients to the current crops, very often leave substantial residual effects on the succeeding crops in the cropping system and this residual

* E-mail: singh-rv@uniraj.ernet.in; kudiwal@datainfosys.net, Fax: +91-141-2700451.

effects lasts for several seasons by Bhattacharajee *et al.* (2003). At present, there has been a shift in the research priority from crops to cropping system considering the effects of N, P and K. Soil is extremely complex medium and its variability causes many problems in analyses and in the interpretation of the analytical results. Plants are the producers in the biosphere and all the animals, including man are absolutely dependent on them for food in one way or the other. Since soil is the basis of the rooted plants, soil pollution can adversely affects the quality and quantity of the crops grown over it (Pande *et al.*, 1999, Dasgupta *et al.*, 2001).

Materials and Methods

The soil samples were collected from, Sanganer at the different depths by post hole auger carefully from four different points. Samples were collected from the depth of 0", 12", 24", 36", 48" and 60" in the polythene bags for analytical studies. The soil samples were dried and then crushed till we taken powdered soil for analyses. Intended for the measuring all the physico-chemical parameters of the soil (except N, P, K and organic carbon). 10 gm of well-powdered soils were taken in a conical flask; at this time 100 ml of double distilled was added for the making 1 : 10 suspension. The N, P, K, and organic carbon of the soil were estimation through the procedure' suggested by the methods, Manual ICAR (1999).

Table 44.1: Physico-chemical Characteristics of the Various Soil Samples Collected from the Sanganer Town of Pink City

Parameter	*S-I*						*S-II*					
	Depth						*Depth*					
	0"	*12"*	*24"*	*36"*	*48"*	*60"*	*0"*	*12"*	*24"*	*36"*	*48"*	*60"*
pH	7.5	7.4	7.3	7.3	7.2	7.1	7.8	7.5	7.6	7.7	7.7	7.7
Total Hardness (mg/kg)	38	30	28	24	24	23	450	350	240	190	135	130
Calcium Hardness (mg/kg)	12	10	14	16	13	10	190	110	95	160	145	130
Magnesium Hardness (mg/kg)	26	20	14	8	11	13	60	40.	45	50	70	80
Calcium (mg/kg)	4.8	4.0	5.6	6.4	5.2	4.0	156	124	78	56	34	32
Magnesium (mg/kg)	6.2	4.8	5.6	1.9	2.6	3.1	14.4	9.6	10.8	12.0	12.0	12.0
Total Alkalinity (mg/kg)	42	59	48	40	25	19	40	40	35	30	20	10
Fluoride (mg/kg)	0.15	0.21	0.20	0.18	0.18	0.18	0.15	0.19	0.69	0.83	1.00	1.15
Nitrage (mg/kg)	360	347	339	326	331	299	335	330	295	282	244	244
Sulphate (mg/kg)	48	42	38	36	35	35	125	115	90	62	60	55
Chloride (mg/kg)	68	31	24	23	23	23	140	130	125	125	120	120
Electrical Conductivity (μScm^{-1})	370	329	279	260	215	190	1020	960	890	800	715	700
Organic Carbon (%)	0.19	0.17	0.18	0.I8	0.15	0.14	0.18	0.19	0.14	0.12	0.10	0.10
Nitrogen (%) N	0.39	0.36	0.32	0.29	0.30	0.28	0.29	0.28	0.26	0.25	0.22	0.20
Phosphorus (kg/ha) P	19	18	24	23	20	18	26	26	26	19	25	18
Potassium (kg/ha) K	270	190	240	200	190	170	290	285	287	260	240	230

Results and Discussion

Analytical results of different samples were taken from the four farms of Sanganer are given in Tables 44.1 and 44.2. It was found that the pH values were ranging from 7.1 to 8.0. The minimum value was obtained at 60″ depth from the soil sampling station (S-1) and the maximum value was observed at 36″ and 60″ depths in the soil sample station (S-II). The minimum concentration of total hardness was 18 mg/kg at 60″ depth from the sample station (S-III) and the maximum amount of total hardness was 250 mg/kg at 0″ depth from the sample station (S-II) thorough the study. During the entire study, the calcium hardness values were reported in the range of 10 mg/kg to 190 mg/kg. The estimated magnesium hardness contents were ranging from 6 mg/kg to 80 mg/kg.

Table 44.2: Physico-chemical Characteristics of the Various Soil Samples, which were Collected from the Sanganer Town of Pink City

Parameter	*S-III*						*S-IV*					
	Depth						*Depth*					
	0″	*12″*	*24″*	*36″*	*48″*	*60″*	*0″*	*12″*	*24″*	*36″*	*48″*	*60″*
pH	7.8	7.3	7.4	7.4	7.4	7.4	7.4	7.4	7.3	7.3	7.4	7.4
Total Hardness (mg/kg)	756	342	220	100	85	80	100	92	80	75	72	70
Calcium Hardness (mg/kg)	589	279	180	70	55	50	56	50	46	48	31	32
Magnesium Hardness (mg/kg)	167	63	40	30	30	30	44	42	34	27	41	38
Calcium (mg/kg)	235	171	72	28	22	20	40	36	32	30	29	28
Magnesium (mg/kg)	40.8	15.1	9.6	7.2	7.2	7.2	10.6	10.0	8.1	6.5	9.9	9.1
Total Alkalinity (mg/kg)	28	49	31	23	27	19	18	18	14	15	12	12
Fluoride (mg/kg)	0.06	0.09	0.13	0.19	0.20	0.22	0.10	0.08	0.09	6.10	0.09	0.10
Nitrate (mg/kg)	130	120	126	134	139	128	300	228	220	235	240	225
Sulphate (mg/kg)	110	110	100	75	70	70	125	105	110	95	55	45
Chloride (mg/kg)	76	71	69	58	48	37	58	50	39	32	24	23
Electrical Conductivity (μScm^{-1})	1680	720	690	630	550	490	800	840	620	680	570	410
Organic Carbon (%)	0.10	0.09	0.08	0.08	6.07	0.06	0.18	0.12	0.16	0.12	0.11	0.10
Nitrogen (%) N	0.24	0.22	0.20	0.19	0.18	0.18	0.32	0.30	0.26	0.24	0.22	0.21
Phosphorus (kg/ha) P	26	22	21	20	20	18	0.20	22	17	21	18	16
Potassium (kg/ha) K	210	175	160	130	125	120	225	200	230	290	185	170

During the entire investigations, the total alkalinity values were reported in the range of 10 mg/kg to 59 mg/kg. The minimum and maximum values were observed at 24″–60″ depths from the sample station (S-II) and at 12″ depth from sample station (S-I), respectively. The estimated sulphate amounts were ranging from 35 mg/kg to 125 mg/kg. The minimum value was observed at 48″–60″ depths from sample station (S-I) and the maximum value was noticed at 0″ depth from the sample station (S-IV). During the entire study; the chloride contents were recorded in the range of 13 mg/kg to 105 mg/kg. The lowest amount and highest amount was observed at 60″ depth from samples station (S-IV) and at 48″ depth from sample station (S-II), respectively. The content of chloride decrease with the increase in the depth of all the soil profiles by Yin *et al.* (1996). The reported fluoride concentrations were found in the range of 0.06 mg/kg to 1.15 mg/kg. The least value was noticed at 0″ depth from the (S-III) sample station and the highest value was observed at 60″ depth from the sample station (S-II). The minimum content of nitrate was 30 mg/kg at 12″ depth from the sample station (S-II) and the maximum concentration of nitrate was 155 mg/kg at 60″ depth from sample station (S-IV).

During the entire investigations, the electrical conductivity amounts were estimated in the range of 190 μScm^{-1} to 1680 μScm^{-1}. The minimum value was noticed at 60" depth from the sample station (S-I) and the minimum amount was observed at 0" depth from the sample station (S-III). The reported calcium contents were estimated in the range of 4.0 mg/kg to 76 mg/kg. At 60" and 0" the minimum and maximum contents of the calcium were noticed from the sample stations S-I and S-III, respectively. The amounts of the magnesium estimated are in the range of 1.5 mg/kg to 19.2 mg/kg. The minimum and maximum values were found from the sample stations (S-I, S-III) at 60" and 0" depths. The observed available organic carbon contents were ranging from 0.06 per cent to 0.19 per cent. The minimum value was noticed at 60" depth from the sample station (S-III) and the maximum amount was observed at 0" depth from the (S-I) sample station. The results shows that the available organic carbon values decreased with increase the depth in the soil profile. The estimated available nitrogen content was ranging from 0.28 per cent to 0.45 per cent. The minimum value was noticed at 60" depth from the sample station (S-I) and the maximum amounts were observed from the sample station (S-III) at 0" depth. The minimum value of available phosphorous was 16 kg/ha at 60" depth from the sample station (S-IV) and the maximum content of available phosphorous was 26 kg/ha from the sample station (S-II) at 0" and 12" depths.

During the whole study, the potassium values were reported in the range of 120 kg/ha to 290 kg/ha. The minimum content was recorded at 60" depth and the maximum value was noticed 0" depth from the sample stations (S-Ill and S-II) respectively.

Conclusion

The results indicated that most of the parameters decreased with the increased in the depth with some exception in nitrate value which may be due to the leached out. From the observation it concluded that the available organic carbon percentage very low in all the soil samples which is due to the use of inorganic manures in much extent. It has been suggested to the farmers the use of organic manures (excreta of animals) in the soil at the time of cultivation. The high phosphorous content in the mostly soil samples may be attributed to continuous addition of fertilizers. Available nitrogen's content is decreased with the increase in the depth probably due to the presence of nitrifying bacteria, which make the surface soil enriched with available nitrogen.

Acknowledgement

One of the author (D.K.) thanks, the University of Rajasthan, Jaipur for providing the University Fellowship.

References

Bhattacharjee, K.G., Choudhary, S.K. and Sharma, P.C., 2003. Effect on pH and EC of soil with respect to extent of degradation of petroleum hydrocarbons in soil under natural environment, *Res. J. Chem. Env.*, 7(3): 28–32.

Pande, K.S. and Sharma, S.D., 1999. Distribution of organic matter and toxic metals in the sediments of Ramganga River at Moradabad. *Poll. Res.*, 18(1): 43.

Dasgupta, M. and Purohit, K.M., 2001. Studies on soil profiles for investigation of soil pollution in the vicinity of mini cement plants in Mandiakudar, Orissa. *J. Env. Poll.*, 8(3): 293–298.

Singh, D., Chhankar, P.K. and Pandey, R.N., 1999. *Analyses: A Methods Manual.* ICAR, pp. 6–21.

Yin, Y., Allen, H.E., Li, Y., Haey C.P. and Sauders, P.F., 1996. Heavy metals in the environment adsorption of mercury (II) by soil: effect of pH, chloride and organic matter. *J. Environ. Qual.*, 25: 837–846.

Chapter 45

Accumulation of Heavy Metal Concentrations in Indian and Foreign Cigarettes

P. Martin Deva Prasath[1], J. Samu Solomon[1] and M. Palanisamy[2]*

[1]Department of Chemistry, TBML College, Porayar – 609 307, Nagapattinam District, Tamil Nadu, South India

[2]Department of Chemistry, Poompuhar College, Melaiyur – 609 107

ABSTRACT

Accumulation or heavy metals in both Indian and foreign made cigarettes were carried out in the analytical laboratory using atomic absorption spectrophotometer (AAS) of Perkin Elmer Model 373. The results obtained revealed that arsenic, lead and nickel were high and exceeded the WHO limits. Iron levels were detected only in consulate and in all Indian cigarettes. Similarly nickel levels were detected only in Indian cigarettes. In general, the heavy metal concentrations in cigarettes was found in the order of As > Fe > Cu > Cr > Ni > Pb > Cd. The results also revealed that Indian made cigarettes contain higher concentrations of heavy metals than those of foreign made.

Keywords: Cigarettes, Heavy metals, Smoking and Cancer.

Introduction

In India tobacco is a vital cash crop and it is also sold in local markets. The leaf on the plant is an imperative living organ containing 85 to 90 per cent water. When dried up, it can be stored.

* Corresponding Author: E-mail: martinprasath@rediffmail.com.

According to Snell (1974) the probable connection between tobacco smoking and health has been recommended and studied by many scientists during the past seventy years. Tobacco is the chief cause of lung cancer. He further stated that it destroys more people through many diseases including heart disease, stroke, emphysema and chronic lung diseases. Babies born to women who smoke during pregnancy as well as those infants exposed to Environmental Tobacco Smoke (ETS) have a considerably greater risk of dying of Sudden Infant Death Syndrome (SIDS) (Awake 1998). Tobacco is detrimental to male reproductive function; sexual intercourse was found to be lower among smokers (Cendnon *et al.*, 1970). Tobacco is habit forming because of the upshot it has on the nerves: first, it subdues pressure, then stimulates the body; when people are under pressure they desire more than something to soothe their tattered nerves and that tranquility theory seem to find in lighting up a cigarette (Tony, 1995). According to Jill (1997) of the Centre for Disease Control, each day in America, 6,000 teenagers light up their first cigarette, 3,000 teens enter the ranks of regular smokers, meaning they have smoked at least one cigarette a day for a month, and 1,000 adults die ahead of time as a direct result of a decision made in adolescence to take up smoking. He further said that 400,000 Americans die each year from smoking related illnesses. Worldwide, about one million smokers smoke five trillion cigarettes, making tobacco the most serious pollutant of modern times far out weighing other hazard, passive smoking–the inhaling of cigarettes smoked by the smoker leads to cancer, glue ear in children, cut death, pneumonia and bronchitis. In addition, children, whose parents smoke end up with lower intelligence quotient trian their mates from non-smoking homes (Tony, 1995). The smokers of pipes and cigars were found to acquire much less health risk. However, the incidence of cancer and heart diseases among them was found to be greater than among non-smokers (Encyclopedia Americana, 1980). Around the world, three million people a year–six every minute–die from smoking according to the hook mortality from smoking in developed countries 1950–2000. If modern smoking patterns persists, then by the time the young smokers of today reach middle or old age, there will be about 10 million deaths a year from tobacco–one death every three second (Awake, 1995).

Tobacco is known to contain more than 4,000 chemicals, including 50 substances known to cause cancer. Endotoxins are a group of poisonous substances produced by bacteria and naturally occur in the air and elsewhere. The level of the toxic substances in the air is 120 higher in a smoky room than in a smoke free room. This can be one reason why smokers so often suffer from respiratory ailments. The fact that passive smoking entails exposure to extremely high concentrations of endotoxins is an entirely new breakthrough Lennart (2004).

In addition to an increased risk of cancer and heart disease, smokers suffer more frequently from colds, gastric ulcers, chronic bronchitis and high blood pressure than non-smoker (Awake, 1999). Children who are exposed to cigarette smoking have an increased chance of developing asthma (Olukayode, 1988). As a form of environmental pollution, smoke coming from a burning cigarette is dangerous due to chemical reaction (Amodu, 1995).

There had been emphasis on the danger of smoking cigarettes in various media. The main point on various media has been that "smoking is dangerous to health and smokers are liable to die young". Therefore, the objective of this work is to find the concentration of some selected toxic metals in various brands of Indian and foreign cigarettes and to compare the levels of toxic metals with the WHO limits.

Materials and Methods

Fresh packets of six brands of Indian cigarettes and three brands of foreign cigarettes were bought from commercial stores. The cigarette brands are Benson and Hedges, Consolate, Rothmans, Wills Filter, Gold Flake Filter, Scissors Filter, Gold Flake Plain, Scissors Plain and Charminar Plain. In this study, the method used by S.D. Nwajei and G.E. Nwajei (2001) was adopted for the estimation of heavy metals in cigarettes. The cigarettes were stored in dry plastic containers. The cigarette was crushed to smooth level with mortar and pestler.

Five grams of each sample was weighed and placed in a digestive flask. Equal ratio 1 : 1 (20 ml HNO_3 and 20 ml $HClO_4$) of acid mixture was added. This was placed on a hot plate for few minutes until the nitro oxide gas was evolved and the digestive solution was transferred into the fume cupboard where the digest was made to stay over night. On cooling, it was filtered and the filtrate was made up to 100 ml in a volumetric flask with deionized water. The solutions were kept in a refrigerator before the heavy metal analysis. The samples were analyzed with atomic absorption spectrophotometer: (AAS). Perkin Elmer 373.

Results

The results obtained on the heavy metal concentrations in nine samples of cigarettes are shown in Table 45.1. The units of the values shown below are in µg/g dry weight.

Table 45.1: Heavy Metal Concentration (µg/g) in 9 Samples of Cigarettes

Sl.No.	*Brands of Cigarettes (Samples)*	*Heavy Metals (µg/g)*						
		Cu	*Fe*	*Ni*	*Pb*	*As*	*Cd*	*Cr*
1.	Benson and Hedges	6.40	ND	ND	1.20	149.40	0.76	1.60
2.	Consolate	18.16	130	ND	2.46	316.20	0.64	2.96
3.	Rothmans	12.24	ND	ND	ND	291.04	0.41	1.40
4.	Wills Filter	18.96	112.45	0.65	2.04	152.14	1.56	1.95
5.	Gold Flake Filter	20.13	10.97	7.25	0.56	485.12	1.36	2.87
6	Scissors Filter	24.52	12.82	4.26	1.23	712.46	1.87	2.45
7.	Gold Flake Plain	2145	12.87	0.18	0.75	512.13	0.95	2.69
8	Scissors Plain	16.25	17.59	1.23	0.45	685.12	0.75	1.54
9.	Charminar Plain	18.43	12.45	0.96	0.65	485.12	0.91	2.09

ND: Not detected.

Discussion

Copper (Cu) levels were detected in all the nine samples analyzed. The results obtained are arranged in the descending order as follows: 24.52 µg/g (Scissors Filter) > 21.45 µg/g (Gold Flake Plain) > 20.13 µg/g (Gold Flake Filter) > 18.96 µg/g (Wills Filter) >18.43 µg/g (Charminar Plain) > 18.16 µg/g (Consolate) > 16.25 µg/g (Scissors Plain) > 12.24 µg/g (Rothmans) > 6.40 µg/g (Benson and Hedges).

Iron (Fe) levels were detected in Consolate (130 µg/g) and were below detectable level in Rothmans and Benson and Hedges. In all Indian cigarettes, considerable amount of iron concentrations were present. The highest level in Indian cigarette was obtained in Gold Flake Plain (124.87 µg/g) and the lowest level was obtained in Gold Flake Filter (10.97µg/g).

Nickel (Ni) concentration was not detected in all the three samples of foreign cigarettes. The highest level of Nickel was detected in Gold Flake Filter (7.25 µg/g) and the lowest level was found in

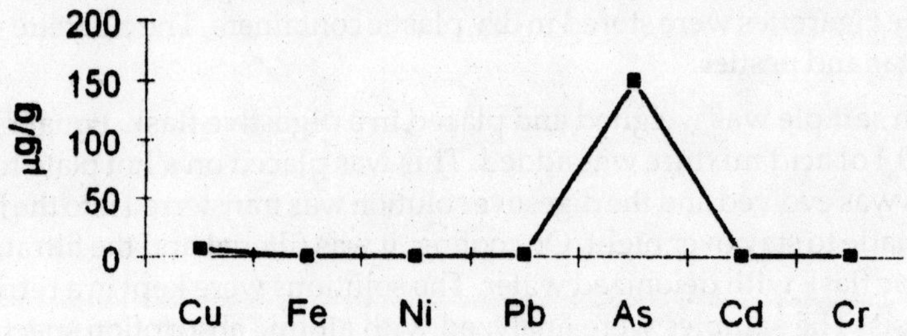

Figure 45.1: Heavy Metal Concentrations (µg/g) in Benson and Hedges

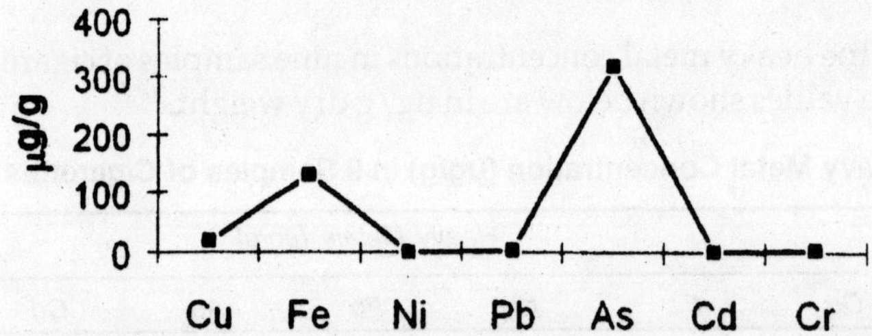

Figure 45.2: Heavy Metal Concentrations (µg/g) in Consolate

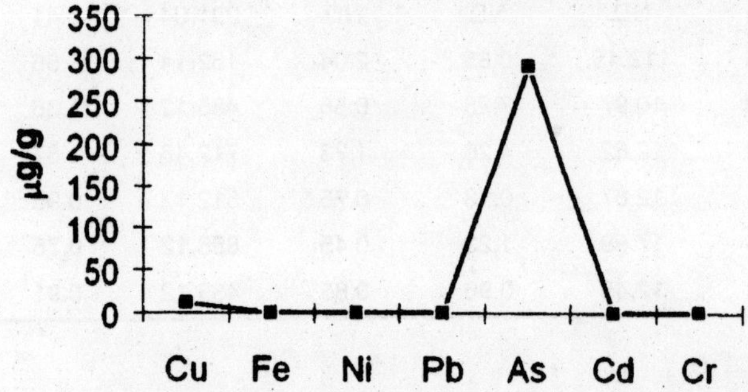

Figure 45.3: Heavy Metal Concentrations (µg/g) in Rothams

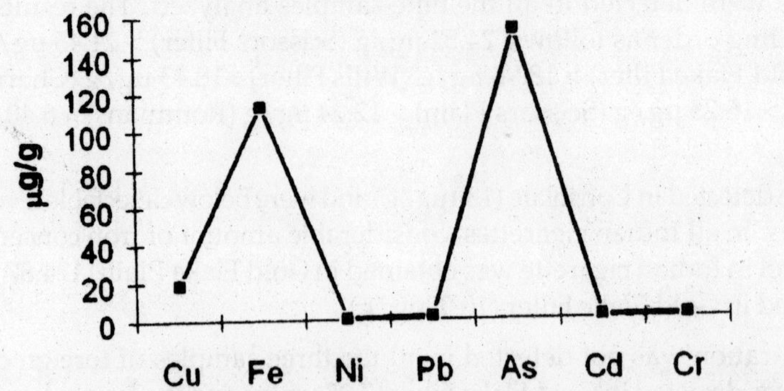

Figure 45.4: Heavy Metal Concentrations (µg/g) in Wills Filter

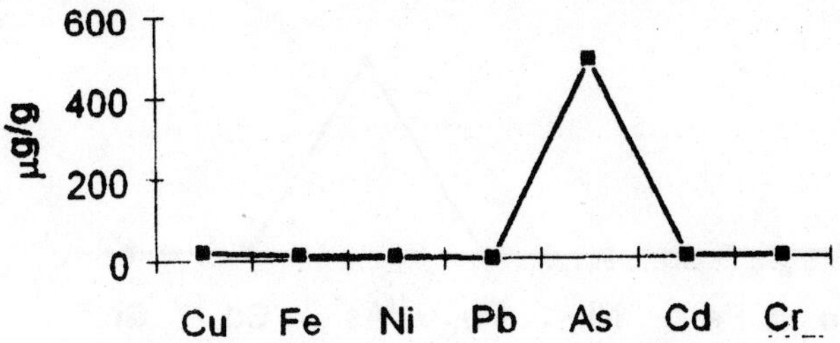

Figure 45.5: Heavy Metal Concentrations (µg/g) in Gold Flake Filter

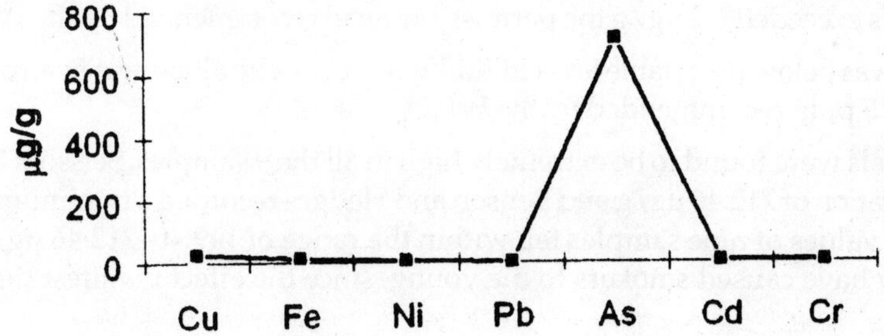

Figure 45.6: Heavy Metal Concentrations (µg/g) in Scissors Filter

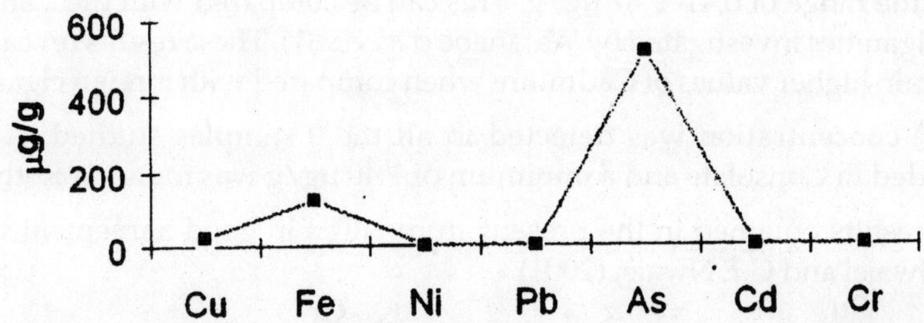

Figure 45.7: Heavy Metal Concentrations (µg/g) in Gold Flake Plain

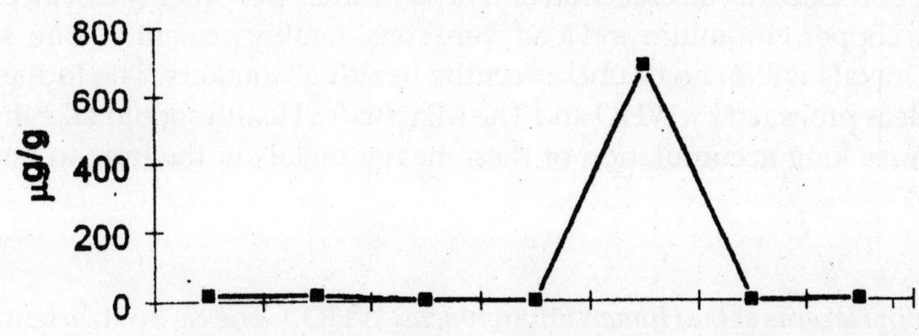

Figure 45.8: Heavy Metal Concentrations (µg/g) in Scissors Plain

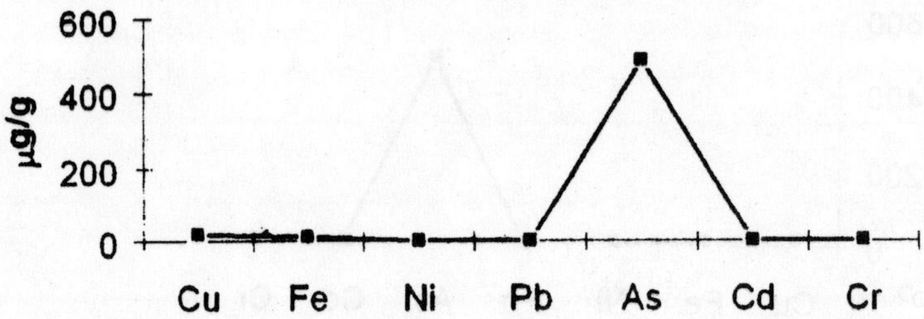

Figure 45.9: Heavy Metal Concentrations (µg/g) in Charminar Plain

Gold Flake Plain (0.18 µg/g) in Indian cigarettes. Except Gold Flake Plain, all the other five brands of Indian cigarette levels exceeded 0.2 µg/g the permissible level recommended by the WHO.

Lead (Pb) level was below detectable level in Rothmans cigarette alone and in all other 8 samples it had exceeded 0.0015 ppm recommended by the WHO.

Arsenic (As) levels were found to be extremely high in all the 9 samples. Scissors Filter recorded a maximum concentration of 712.46 µg/g and Benson and Hedges recorded a minimum concentration of 149.40 µg/g). The values of nine samples fell within the range of 149.40–712.46 µg/g. These high levels of arsenic may have caused smokers to die young, since the effect manifest through gradual accumulation.

Cadmium (Cd) levels were obtained in all the 9 samples analyzed. Scissors Filter has the highest level (1.87 µg/g), while Rothmans has the lowest value (0.41 µg/g). Therefore, the entire results obtained fell within the range of 0.41–1.87 µg/g. This can be compared with the Cadmium content (0.25–2.70 µg/g) in cigarettes investigated by Watanabe *et al.* (1981). These results revealed that Indian made cigarettes contain higher values of Cadmium when compared with foreign cigarettes.

Chromium (Cr) concentration was detected in all the 9 samples studied. A maximum of 2.96 µg/g was recorded in Consolate and a minimum of 1.40 µg/g was found in Rothmans.

In general, the results obtained in the present study align in good agreement with the study conducted by S.D. Nwajei and G.E Nwajei (2001).

Conclusion

It is healthier to stop smoking for our personal health and for the health of other people around us. Most people smoke as a result of their peer groups, without knowing the implications. The results obtained in this study revealed that arsenic, lead and nickel values were highly elevated and exceeded the WHO limits. Iron, copper, chromium and lead were considerably present in some samples. These high levels of heavy metals will in no doubt, affect the health of smokers. The Indian Government should support the ideas projected by WHO and The Ministry of Health should take immediate steps to ban smoking because long accumulation of these heavy metals in the human body may cause premature death.

References

Amodu, F., 1995. *Health Hazards of the Human Environment*. WHO, Geneva, Switzerland.

Awake, 1998. Are they spreading. *Watchtower and Tract Society*, p. 14–15.

Awake, 1995. Killing millions to make millions. *Watchtower and Tract Society*, p. 3–13.

Awake, 1999. Is your life style killing you? *Watchtower and Tract Society*, p. 4–7.

Cendnon, H. and Valley, M.J., 1970. Les effect de la'age Sur Lactive Sexuelle Masculine, Jundence de quelques facterus dou Le tabac, Presse Med., 78: 1765.

Encyclopedia Americana, 1980. *Smoking and Health*. Grolier International Inc., 25: 70–74.

Jill, S., 1997. Big tobacco takes a hit. *Daily Times*, Nigeria, p. 10–23.

Lennart Larson, 2004. Poison in the Air. *The Hindu*, September 5, 2004, India, p. 6

Nwajei, S.D. and Nwajei, G.E., 2001. Heavy metal contaminations in cigarettes, medicated stuff and tobacco leaf. *Journal of Ecotoxicology and Environmental Monitoring*, 11: 65–69.

Olukayode, O., 1988. Taming the world wide tobacco, epidemic. *Guardian*, June 28, Nigeria, p. 26.

Revelle, C. and Revelle, P., 1974. *Source Book on the Environment: The Scientific Perspective*. Boston, Houghton Miffin.

Snell, E., 1974. *Encyclopedia*, 2nd edition, Vol. 20. McGraw Hill, Philippines, p. 503–525.

Tony, I., 1995. Tobacco the edible poison. *Daily Times*, Nigeria, p. 15.

Watanebe, T., Koisumi, A., Fjimoto, H., Ishimori, A. and Ikeda, M., 1982. Tohoku. *J. Expt. Med.*, 138: 443–444.

Chapter 46

Influence of Nitrogen and Spacings on Growth and Yield of the Medicinal Plant: Kasturibenda (*Abelmoschus moschata*)

M.M. Naidu and G. Narasimha Murthy

Acharya N.G. Ranga Agricultural University, Regional Agricultural Research Station, Chintapale – 531 111, Visakhapatnam Dist., A.P.

ABSTRACT

A filed experiment was conducted at Regional Agricultural Research Station, Chintapalle to study the influence of nitrogen and spacings on growth and yield of Kasturibenda (*Abelmoschus moschata*) during kharif seasons of 2001 and 2002. Factorial randomized block design was adopted with five levels of N (0, 20, 40, 60 kg/ha and 10 t FYM/ha) and four spacings (45 cm × 15 cm, 45 cm × 30 cm, 60 cm × 15 cm, 60 cm × 30 cm) with basal dose of 40 kg P_2O_5 and 50 kg K_2O per ha.

Among twenty treatmental combinations S_4N_3 combination (60 cm × 30 cm and N at 60 kg/ha) recorded significantly higher yield of 2166 kg/ha and is at par with S_3N_3 combination (1986 kg/ha) (60 cm × 15 cm and N at 60 kg/ha). The higher yield in S_4N_3 is due to higher number of capsules/plant. It can be concluded that a nitrogen level of 60 kg N/ha coupled with a spacing of 60 cm × 30 cm is optimum to realize higher productivity in Kasturibenda.

Introduction

Kasturibenda or Musk (*Abelmoschus moschatus medic*) is an aromatic plant, which is distributed all over India. It is universally known as "Ambrette" and oil extracted from its seed is called ambrette oil. The crop is native to India and comes up well throughout the tropical regions. Presently India has started exporting ambrette seeds to France, Germany, Japan and several other countries. The ambrette seed oil is used in preparing high-grade perfumes and cosmetics. It is also used in manufacture of medicines, chewing tobacco, agarbatties and tea products. The information regarding production technology of Ambretta is meagre (Anonymous, 2003). Keeping this in view a study was conducted at Regional Agricultural Research station, Chintapalli, Visakahapatnam Dist. to find out the spacing and nitrogen requirement of the crop.

Materials and Methods

Field experiment was carried out during kharif seasons of 2001 and 2002. The treatments were 5 levels of nitrogen (0, 20, 40, 60 kg/ha and 10 tonnes FYM/ha and 4 spacings (45 cm × 15 cm, 45 cm × 30 cm, 60 cm × 15 cm, 60 cm × 30 cm) comprising a total no. of twenty treatmental combinations. These treatments were laid out in Randamized block design with three replications. Sowing was takenup in the first week of August with basal, application of 40 kg P_2O_5, and 50 kg K_2O per ha along with 1/3 dose of nitrogen as per the treatment. The remaining 2 split doses of nitrogen were applied at 40 days and 80 days after sowing.

Results and Discussion

Data on plant height, number of branches/plant, number of leaves/plant are presented in Table 46.1. The interaction effect of plant height, number of leaves and number of branches per plant are non-significant. However, individual effect of nitrogen (Table 46.2) revealed that nitrogen application from 20 kg/ha to 60 kg/ha increased plant height significantly (82.0 cm) when compared to other treatments. There was no influence of FYM on plant height as N_4 treatment (FYM 10 t/ha) recorded lowest plant height. Srinivasan *et al.* (2003) also reported increase in plant height with application of Nitrogen.

The interaction effect with regards to number of branches/plant is non-significant. However, individual effect of Nitrogen (Table 46.1a) was significant and recorded higher number of branches/plant (4.7–5.7 branches/plant). Similarly among different spacings (Table 46.1b) 60 cm × 30 cm, 60 cm × 15 cm, 45 cm × 30 cm recorded significantly higher number of branches (5 to 6 branches/plant).

Table 46.1: Interaction Table for Plant Height, No. of Branches and No. of Leaves/Plant

Sl.No.	*Treatment Combination*	*Plant Height (cm)*	*No. of Branches/ Plant*	*No. of Leaves/ Plant*
1.	S_1N_0	70.0	3.5	26.5
2.	S_1N_1	75.3	6.0	31.2
3.	S_1N_2	83.3	5.6	38.0
4.	S_1N_3	79.6	4.7	33.5
5.	S_1N_4	54.3	3.7	29.6
6.	S_2N_0	65.6	4.3	27.5
7.	S_2N_1	85.6	5.5	38.4
8.	S_2N_2	84.0	6.0	45.2
9.	S_2N_3	83.3	5.4	38.6
10.	S_2N_4	69.3	4.8	33.4
11.	S_3N_0	61.6	3.6	30.0
12.	S_3N_1	94.0	5.8	37.7

Contd...

Table 46.1–Contd...

Sl.No.	Treatment Combination	Plant Height (cm)	No. of Branches/ Plant	No. of Leaves/ Plant
13.	S_3N_2	70.3	5.5	42.3
14.	S_3N_3	80.0	5.7	50.7
15.	S_3N_4	67.6	4.5	35.0
16.	S_4N_0	55.0	5.2	37.6
17.	S_4N_1	74.0	6.9	53.5
18.	S_4N_2	69.6	5.6	42.0
19.	S_4N_3	83.6	7.2	50.0
20.	S_4N_4	67.6	6.0	34.4
CD at 5% for spacings x 'N' levels	N.S	N.S	N.S	–
CV%	22.3	30.3	35.4	–

Table 46.1a: Individual Effect of 'N' Levels on Plant Height, No. of Branches and No. of Leaves/Plants

Sl.No.	Characters	Nitrogen Levels					SEM	CD at 5%	CV%
		N_0	N_1	N_2	N_3	N_4			
1.	Plant height	63.0	82.2	76.8	81.6	64.7	6.7	13.8	22.3
2.	No. of branches/plant	4.2	6.0	5.6	5.7	4.7	0.6	1.4	30.3
3.	No. of leaves/plant	30.4	32.0	41.8	43.2	33.0	5.4	NS	35.4

Table 46.1b: Individual Effect of Spacings on Plant Height, No. of Branches and No. of Leaves/Plant

Sl.No.	Characters	Spacings				SEM	CD at 5%	CV%
		S_1	S_2	S_3	S_4			
1.	Plant height	72.5	77.5	74.7	69.9	6.0	NS	22.3
2.	No. of branches/plant	4.7	5.2	5.0	6.2	0.5	1.2	30.3
3.	No. of leaves/plant	31.7	36.6	39.0	43.5	4.9	NS	35.4

With regards to number of leaves/plant, there was no significant difference in treatmental combinations and individual effect of 'N' levels and spacings were also non-significant.

During kharif 2001 (Table 46.2) among twenty treatment combinations, S_4N_3 combination (60 cm × 30 cm and N at 60 Kg/ha) recorded significantly higher yield of 2166 kg/ha followed by S_3N_3 (60 cm × 15 cm and N at 60 kg/ha) which recorded 1986 k/ha. Similarly during 2002 (Table 46.2) also same treatment S_4N_3 combination recorded significantly higher yield of 1800 kg/ha. The higher yield in S_4N_3 combination is attributed to higher number of branches and capsules per plant. Among 'N' levels individual effect of Nitrogen (N at 60 kg/ha) is significant in both the years recording 1738 kg/ha (Table 46.2a) and 1426 kg/ha (Table 46.2b) during 2001 and 2002 respectively.

Table 46.2: Interaction Table for Number of Capsules/Plant and Yield/ha

Sl.No.	Treatmental Combination	2001–02		2002–03	
		No. of Capsules/Plant	Yield (kg/ha)	No. of Capsules/Plant	Yield (kg/ha)
1.	S_1N_0	31.5	986.6	22.3	781.6
2.	S_1N_1	21.8	1533.3	22.0	1150.0
3.	S_1N_2	35.7	1633.3	29.6	930.0
4.	S_1N_3	34.6	1400.0	22.3	1191.6
5.	S_1N_4	33.2	866.6	25.6	1273.3
6.	S_2N_0	29.8	1233.3	27.0	973.3
7.	S_2N_1	33.3	1100.0	34.6	855.0
8.	S_2N_2	21.2	1950.0	28.6	1633.0
9.	S_2N_3	30.2	1400.0	36.6	1275.0
10.	S_2N_4	21.7	1906.6	37.3	1548.0
11.	S_3N_0	35.8	700.0	30.0	1175.0
12.	S_3N_1	30.7	1866.6	22.3	953.0
13.	S_3N_2	30.6	1050.0	29.0	1100.0
14.	S_3N_3	28.8	1986.6	22.3	1440.0
15.	S_3N_4	30.0	1106.6	25.0	1316.0
16.	S_4N_0	23.4	983.3	22.0	971.0
17.	S_4N_1	31.0	1100.0	34.6	1186.0
18.	S_4N_2	29.1	1400.0	31.0	1280.0
19.	S_4N_3	37.3	2166.6	42.3	1800.0
20.	S_4N_4	27.4	1100.0	37.3	1011.0
21.	CD at 5% for spacings	1.1	89.3	9.0	116.8
22.	CD at 5% for N levels	1.3	99.8	8.1	104.5
23.	CD S × N at 5%	2.6	199.7	4.0	51.7
24.	CV%	5.2	8.7	4.5	9.1

Table 46.2a: Individual Effect of 'N' Levels on No. of Capsules/Plant and Yield (kg/ha) (2001–02)

Sl.No.	Characters	Nitrogen Levels					SEM	CD	CV%
		N_0	N_1	N_2	N_3	N_4			
1.	No. of capsules/plant	30	29.2	29	32.7	28	0.6	1.3	5.2
2.	Yield kg/ha	975.8	1399.9	1508	1738	1244	49	99.8	8.7

Individual Effect of Spacings on No. of Capsules/Plant and Yield kg/ha

Sl.No.	Characters	Spacings				SEM	CD	CV1
		S_1	S_2	S_3	S_4			
1.	No. of capsules/plant	31.4	27.2	31	29.6	0.5	1.1	5.2
2.	Yield kg/ha	1284	1517	1441	1350	43.9	89.3	8.7

Table 46.2b: Individual Effect of 'N' Levels on No. of Capsules/Plant and Yield kg/ha (2002–03)

Sl.No.	Characters	Nitrogen Levels					SEM	CD	CV%
		N_0	N_1	N_2	N_3	N_4			
1.	No. of capsules/plant	25.3	28.3	29.5	30.8	31.3	3.9	8.1	45
2.	Yield kg/ha	975	1036	1235.7	1426.6	1287	51	104.5	90.7

Sl.No.	Characters	Spacings				SEM	CD	CV1
		S_1	S_2	S_3	S_4			
1.	No. of capsules/plant	24.21	32.8	25.7	33.4	4.4	9	45
2.	Yield kg/ha	1065	1256.8	1196.8	1249.6	57.2	116.8	90.7

Conclusion

Study conducted on the spacings and N levels in Kasturibenda revealed that among twenty treatment combinations S_4N_3 (60 cm × 30 cm and N at 60 kg/ha) recorded significantly highest yield of 2166 kg/ha and 1800 kg/ha during 2001 and 2002 respectively. Therefore, farmers are advised to apply Nitrogen @ 60 kg/ha by adopting a spacing of 60 cm × 30 cm to realise optimum yields in Kasturibhenda.

References

Anonymous, 2003. *Proceedings of International Conference on Medicinal and Aromatic Crops*, Hyderabad.

Srinivasan, R. and Vadivel, 2003. Package of practices of Kasturibenda. In: *Proceeding of International Conference on Medicinal and Aromatic Crops*, Hyderabad.

Chapter 47

Studies on the Management of Sunflower Necrosis Disease

P. Dhevagi, S.K. Manoranjitham, M. Ramaiah and P. Vindhiyavarman

Department of Oilseeds, Tamil Nadu Agricultural University, Coimbatore – 641 003

ABSTRACT

A field survey has been carried out to find out the occurrence of major diseases of sunflower. The survey results indicated that in and around Coimbatore district to assess the disease severity. The necrosis incidence varied from 0.0 to 36.6 per cent at 45–60 days. Necrosis disease incidence was high in July/August sown crop irrespective of inbred/hybrid. Necrosis incidence was more and coinciding with the incidence of thrips. Seed treatment with Imidacloprid @ 5g/kg of seed followed by spraying of confidor 0.01 per cent at 15 days interval (15, 30, 45 and 60th days) reduced necrosis virus incidence. There was a positive correlation between weather parameters and necrosis virus incidence.

***Keywords**: Sunflower, Necrosis, Management, Chemical control.*

Introduction

Sunflower (*Helianthus annuus L.*) necrosis disease came into prominence in 1997, at Bagepally in Kolar District near Bangalore (Singh *et al.*, 1997). The disease incidence has been observed as very eratic which varies from place to place, genotype to genotype and season to season. The disease intensity varied from 5 to 70 per cent. Now the disease is prevalent in almost all sunflower growing areas of India. Sunflower Necrosis disease virus was suspected as peanut bud necrosis (PBNV) (Jain *et al.*, 2000), tobacco streak virus (TSV) (Brunt *et al.*, 1996, Reddy *et al.*, 2002), Tobacco streak virus (Brunt *et al.*, 1996; Prasada Rao *et al.*, 2000; Bhat *et al.*, 2002), cucmo, umbra virus (Brunt *et al.*, 1996),

Poty and tospo virus (Anon, 1997; Jain *et al.*, 2000). Maximum incidence of the disease (upto 80 per cent) occurred in Kharif crops and less intensity during rabi crops (Ramiah *et al.*, 2001).

Symptoms include necrotic spots which coalesce giving a scorched appearance of leaves. The necrosis then spreads further along the petiole and the stem causing necrotic streaks. Late the necrosis also spreads to the floral organs leading to the collapse of the plant. Early infected plants manage to produce heads that contain chaffy, ill filled and poor quality seeds. SND mainfested various types of external symptoms on sunflower plants *viz.*, Yellow mosaic, Local lesions, necrosis of tissues, mottling, chlorosis, leaf enations, narrowing and mixed infection (Anon, 2001 and Manoranjitham and Muralidharan, 2003).

Earlier studies revealed that SND was not seed borne in nature. However the presence of virus was noticed in the seed during the seed development stage but not in the fully developed/mature seeds. The following experiment was planned to find out whether any relation is there between thrips population, date of sowing with sunflower necrosis incidence and how far the vector control helps to reduce the disease incidence.

Materials and Methods

Influence of Date of Sowing on Necrosis Incidence

Three cultivars of sunflower *i.e.* Morden, PAC 1091 and TCSH–1 were used for the experiment. The cultivars were sown at every month *i.e.* June 2004 to February 2005. All agronomic practices except plant protection measures were followed. The population of thrips, other insects and weather parameters were observed regularly on 10 randomly selected plants at monthly interval throughout the crop growth period. The disease incidence was expressed as per cent infection by counting the total number of plants and the number of plants infected.

Management of Necrosis

Two sets of management trial were conducted with eight treatments (Table 47.1) and three replications in split plot design. Per cent incidence of the disease and thrips population were recorded at fortnightly interval up to sixty days.

Table 47.1: Treatment Details of the Management of Necrosis Disease

Treatment	*Treatment Details (Plot Size 3 × 4.2)*
Main Plots	*With borders*
	Without borders
Sub Plots	
T_1	Control (no spray *j* no seed treatment)
T_2	Seed Treatment with Imidacloprid @ 5 g/kg of seed
T_3	Foliar Spray with Confider 200 SL (0.01 per cent) at 15, 30 and 45 DAS
T_4	Foliar Spray with Confider 200 SL (0.01 per cent) at 15 and 45 DAS
T_5	T_2 + Foliar Spray (0.01 per cent) at 15 DAS interval
T_6	$T_2 + T_3$
T_7	$T_2 + T_4$
T_8	T_2 + Rouging of diseased plant

On Farm Trial

Based on the result obtained from the above study an on farm trial was conducted near Navavoor privu at Sulthanipuram, Coimbatore to assess how far the border crop and vector control helps to reduce the disease incidence. The trial was laid out with five treatments in a replicated trial with Randomized Block Design and the variety used for the trial was CO_4.

Results and Discussion

A field survey has been carried out to find out the occurrence of major diseases of sunflower. The survey results indicated that in and around Coimbatore district to assess the disease severity. The necrosis incidence varied from 0.0 to 36.6 per cent at 45–60 days. The disease incidence in relation to date of sowing gives better understanding of the disease spread. Since the disease was noticed very recently the information about the influence of weather factors, vector population will be very much useful.

Influence of Date of Sowing on Necrosis Incidence

Only one popularly released hybrid of Coimbatore centre TCSH–1 and two susceptible cultivar (PAC1091) and morden was taken up for sowing in a replicated trial at monthly sowing in plot size of 3 × 4.2 m. Per cent incidence of the disease and thrips population was recorded at fortnightly interval upto sixty days. The entries showed no necrosis virus disease incidence had low thrips population (Table 47.2).

Table 47.2: Testing the Popular Released Hybrids Against Necrosis Virus Under Different Sowing Dates (2004–05)

Entries	*July*	*Aug.*	*Sept.*	*Oct.*	*Nov.*	*Dec.*	*Jan.*	*Feb.*
Morden	6.9	2.5	0	0	5.1	4.4	4.65	5
PAC 1091	1.2	9.3	0	1.96	1.2	1.2	0	0
TCSH 1	4.4	0	0	0	5	–	5	2.4
Thrips popn/plant	13–16	3–7	3–4	5–8	12–15	3–5	3–12	4–7
Mean/max temp	31.2	31.5	31.6	29.9	28.3	29.5	31.6	31.8

Mean of four replications.

Apparently the incidence of necrosis disease was more in morden as compared to TCSH–1 and PAC 1091 as the former had 6.9 per cent incidence as against 5 per cent in the TCSH–1. Most of the plants showed yellow mosaic type symptoms during young stage and plants remained stunted and did not reach to reproductive stage leading to yield loss. Necrosis disease incidence was high in July/August sown crop irrespective of inbred/hybrid. Necrosis incidence was more and coinciding with the incidence of thrips. Dry weather with moderate temperature was conducive to thrips incidence (Singh, 2005).

Management of Necrosis

Earlier studies revealed that SND is not of seed borne nature. However, the presence of virus was noticed in the seeds during the seed development stage but not in the fully developed/mature seeds. To find out is there any relation between thrips population, with SND incidence, a management trial was laid out and the results were presented in Tables 47.3 to 47.5. The weather parameters prevailed during the crop growth was recorded.

Table 47.3: Management of Sunflower Necrosis Virus Disease at 15 Days Interval with Border Crop Sorghum

Treatments	*Days 1 (15)*		*Days 2 (30)*		*Days 3 (45)*		*Days 4 (60)*	
	Thrips Popn	*Disease Incidence*	*Thrips Popn*	*Disease Incidence*	*Thrips Popn*	*Disease Incidence*	*Thrips Popn*	*Disease Incidence*
T_1	0	1.43	0	1.43	3	4.83	2	4.83
T_2	1	0.33	2	0.33	2	0.00	3	0.00
T_3	8	1.80	1	1.80	4	2.73	2	2.73
T_4	0	0.67	0	0.67	3	1.77	5	1.77
T_5	1	0.00	0	0.33	4	2.40	4	2.40
T_6	0	0.00	1	0.00	5	0.00	3	0.07
T_7	0	0.33	1	0.33	3	0.33	7	0.47
T_8	10	3.13	5	3.50	5	6.63	5	7.17

Value represents mean of three replications.

Table 47.4: Management of Sunflower Necrosis Virus Disease at 15 Days Interval Without Border Crop Sorghum

Treatments	*Days 1*		*Days 2*		*Days 3*		*Days 4*	
	Thrips Popn	*Disease Incidence*	*Thrips Popn*	*Disease Incidence*	*Thrips Popn*	*Disease Incidence*	*Thrips Popn*	*Disease Incidence*
T_1	1	3.07	8	13.13	0	13.13	4	13.13
T_2	0	2.00	5	4.96	1	4.93	5	4.96
T_3	1	0.90	7	2.43	3	2 43	3	2.57
T_4	3	2.47	3	3.70	1	3.70	5	3.73
T_5	2	2.80	5	4.33	2	4.33	3	4.33
T_6	4	1.37	4	1.47	3	1.47	4	1.47
T_7	2	2.37	3	2.84	1	2.84	5	2.84
T_8	5	4.20	12	9.23	0	9.23	7	9.23

Thrips count and nercrosis incidence was taken at every fifteen days interval and the results were given in Tables 47.3 and 47.4. There was no significant difference between the main plots. But in sub plots significant difference was noticed (Table 47.5). In case of plots with border crop sorghum, even though the thrips population was high, the disease incidence was less compared to plots without border crop. In control the disease incidence was 9.6 per cent others had 2.4 to 6.9 per cent since the border crop prevents the entry of thrips into sunflower crop and the remaining thrips were arrested by insecticidal action (Anon, 2002).

Correlation analysis was performed among thrips count, necrosis disease incidence and weather parameters for the standard weeks namely, maximum temperature, minimum temperature, rainfall, sunshine hours and relative humidity (Table 47.5). Simple correlation was worked between the weather parameters, thrips population and disease incidence. The results revealed that significant and positive association between disease incidence and thrips count (0.33**) and between disease incidence and maximum temperature (0.27**). Significant negative association was observed between disease incidence and minimum temperature (–0.25*) (Table 47.6). There was a positive correlation between RH and the thrips population. This was in accordance with the findings of Singh (2005). However rainfall presented a negative correlation with thrips population. This indicates that dry cold weather enhanced the population of the thrips population in turn increased the intensity of necrosis disease (Anon, 2004).

Table 47.5: Results on the Management of Necrosis Disease

Treatment	*Treatment Details*	*Mean Necrosis %*	
		15 DAS	*60 DAS*
Main Plots	With borders	4.5(12.3)	5.3 (13.3)
	Without borders	3.8 (11.2)	4.7 (12.5)
	SEm±	1.0	0.08
	CD (P = 0.05)	NS	NS
Sub Plots			
T_1	Control (no spray/no seed treatment)	6.7 (14.9)	9.6 (18.0)
T_2	Seed treatment with Imidacloprid @ 5 g/kg of seed	4.8 (12.6)	5.2 (13.1)
T_3	Foliar spray with confider 200 SL (0.01%) at 15, 30 and 45 DAS	3.9 (11.4)	4.2 (11.9)
T_4	Foliar spray with confider (0.01%) at 15 and 45 DAS	3.1 (10.1)	3.9 (11.4)
T_5	T_2 + Foliar spray (0.01%) at 15 DAS	3.9 (11.4)	5.4 (13.5)
T_6	$T_2 + T_3$	2.9 (9.8)	2.9 (9.8)
T_7	$T_2 + T_4$	3.0 (9.9)	3.2 (10.3)
T_8	T_2 + Rouging of diseased plant	5.9 (14.1)	6.6 (15.3)
	SEm±	2.3	2.4
	CD (P = 0.05)	NS	6.9

Table 47.6: Simple Correlation Analysis Between SND, Thrips Population and Weather Parameters

X1 Thrips	1.00						
X2 Disease	0.33**	1.00					
X3 Max Temp.	0.19	0.27*	1.00				
X4 Min Temp.	–0.05	–0.25*	–0.87**	1.00			
X5 Rainfall	0.13	–0.07	0.02	0.32**	1.00		
X6 Sunshine hrs	0.07	0.24	0.71**	–0.79**	–0.67**	1.00	
X7 RH Maximum	0.17	0.22	0.93**	–0.80**	0.30*	0.43	
X8 RH Minimum	–0.07	–0.20	–0.52**	0.58**	0.76**	–0.96**	–0.18

Based on the results obtained from the management trial an on farm trial was conducted and the results were given in Table 47.7. Treatment 3 found to be the best treatment compared to control, *i.e.*

Sorghum (barrier crop) sowing on the borders one month before sunflower sowing followed by Seed treatment with Imidacloprid 2g/kg and foliar spray with Imidacloprid 0.01 per cent (15th, 30th DAS) reduced sunflower necrosis virus disease incidence to 0.36 per cent compared to control. In control the disease incidence was 3.22 per cent since the sunflower necrosis virus disease was transmitted by thrips, border crop prevent the entry of thrips into sunflower field to some extent. The remaining thrips population was arrested by the insecticidal action of Imidacloprid (Chander Rao, 2002 and Singh, 2005). Seed treatment with Imidacloprid 70 WS @ 5 g/kg of seed reduced insect vector load during early stage of crop. Prophylactic sprays of Imidacloprid 200SL @ 0.1 ml/l of water between sprays of Phosphamidan (0.03 per cent) and Dimethaoate (0.03 per cent) kept the disease incidence below 5 per cent compared to 60 per cent in the untreated check (Anon, 2000; Kannan *et al.*, 2005).

Table 47.7: Results on the Farm Trial

Sl.No.	*Treatments*	*% Disease Incidence*	*Yield (kg/ha)*	*C : B Ratio*
1.	T_1 Seed treatment of Imidacloprid 2g/kg + Foliar spray with Imidacloprid 0.01 per cent (15th and 30th DAS)	7.80b	710d	1 : 1.45
2.	T_2 Sorghum barrier crop sowing on the borders one month before sunflower sowing	15.30c	555b	1 : 1.25
3.	T_3 $T_1 + T_2$	4.80a	880e	1 : 1.83
4.	T_4 Methyl Demoton 1.5 ml/lit on 15th and 30 DAS (check)	18.30d	600c	1 : 1.35
5.	T_5 Control	28.30e	444a	
	LSD (0.05%)	2.37	25.53	

The thrips and pollen (sunflower and parthenium) individually and in combination released or dusted to sunflower plants grown under insect proof cages could not be successful in transmitting the disease. Through direct transmission could not be established successfully, the serological tests showed positive. That the pollen and thrips collected from the infected plants contained the TSV and negative in the samples collected from uninfected plants (Anon, 2002).

The above study reveals that by reducing thrips population, the necrosis incidence will be reduced. September, October sown crop escapes from necrosis incidence. Seed treatment with Imidacloprid @ 5g/kg of seed followed by spraying of confidor 0.01 per cent at 15 days interval (15, 30, 45 and 60th days) reduced necrosis virus incidence. There was a positive correlation between weather parameters and necrosis virus incidence.

References

Anonymous, 1997. *Annual Progress Report (1996–1997).* Indian Agricultural Research Institute, New Delhi, pp. 24.

Anonymous, 2000. *Annual Progress Report (1999–2000).* Directorate of Oilseeds Research, Hyderabad.

Anonymous, 2001. *Annual Progress Report (2001–2002).* Directorate of Oilseeds Research, Hyderabad.

Anonymous, 2002. *Annual Progress Report (2001–2002).* Indian Agricultural Research Institute, New Delhi, pp. 34.

Anonymous, 2004. *Annual Progress Report (2004–2005).* Directorate of Oilseeds Research, Hyderabad, pp. 55.

Bhat, A.L., Jain, R.K. and Ramaiah, M., 2002. Detection of tobacco streak virus from sunflower and other crops by reverse transcription polymerase chain reaction. *Indian Phytopathology*, 55: 216–218.

Brunt, A.A., Crabtree, K., Dallwitz, M.J., Gibbs, A.J. and Watson, L., 1996. *Viruses of Plants*. CAB International, Wellingfor, UK, pp 484.

Chander Rao, 2002. *Annual Report (2002–2003)*. Indian Agricultural Research Institute, New Delhi, pp. 47.

Jain, R.K., Bhat, A.I., Byadgi, A.S., Nagaraju, Singh, Harvir, Halkeri, A.V., Anahosur, K.H. and Verma, A., 2000. Association of tsopovirus with sunflower necrosis disease. *Current Science*, 79(12): 1730–1705.

Kannan, R., M. Ramaiah, S.K. Manoranjitham, V. Muralidharan and N. Manivannan, 2005. *Transmission and Management of New Necrosis virus Disease of Sunflower* (Accepted for publication in MASU).

Manoranjitham S.K. and V. Muralidharan, 2003. Epidemiology of new necrosis "virus disease of sunflower'. In: *National Seminar on Integrated Plant Disease Management for Sustainable Agriculture* (IPDMSA) held at Annamalai University, Annamalai Nagar, Chidambaram, March 20–21, 2003.

Prasada Rao, R.D.V.J., Reddy, A.S., Chander Rao, S. Varaprasad, K.S., Tirumala Devi K., Nagaraju, Muniappa, V., Reddy, D.V.R., 2000. Tobacco streak virus as a causal agent of sunflower necrosis disease in India. *Journal of Oilseeds Research*, 17: 400–401.

Ramiah, M., A.J. Bhat, Jain, R.K., R.P. Pant, Y.S. Ahlawat, K. Prabhakar and A. Verma, 2001. Partial characterization of on isometric virus causing sunflower necrosis disease. *Indian Phytopathol.*, 54: 246–250.

Reddy, S., Prasada Rao, R.D.V.J., Tirumala Devi, K. Satyanarayana, T., Suramanium, K. and Reddy, D.V.R., 2002. Occurrence of tobacco streak virus on peanut (*Arachis hypogea* L) in India. *Plant Disease*, 86: 173–178.

Singh, H., 2005. Thrips incidence and necrotic disease in sunflower *Ielianthus annuus* L. *J. Oilseeds Res.*, 22: 90–92.

Singh, S.J., Nagaraju, Krishna Reddy, K.M., Muniappa, V. and Virupashappa, K., 1997. Symposium, Economically important disease of Crop plants, 18–20 December, Bangalore.

Chapter 48

Distribution and Ecology of Zoobenthos in the Anchar Lake of Kashmir (India)

M. Jeelani[1], H. Kaur[1]* and S.G. Sarwar[2]

[1]Department of Zoology, Punjabi University, Patiala – 147 002, Punjab
[2]Hydrobiology Research Laboratory, S.P. College, Srinagar – 190 001, Jammu and Kashmir

ABSTRACT

Due to increasing anthropogenic activities, most of the water bodies in the Kashmir valley have become polluted. The Anchar lake, a shallow basined is situated 12 kms to the northwest of Srinagar within the geographical coordinates of 34°20′ and 34°26′ N latitude and 70°82′ and 74°85′ E longitude at 1584 m a.s.l. receives sewage from the catchments area, agricultural run-off and hospital waste. The present investigation was carried out by analysing some physico-chemical characteristics of water and the effect of increased concentration of these on the distribution of zoobenthos from September 2000 to August 2002.

Keywords: *Distribution, Ecology, Zoobenthos, Anchar lake, Kashmir.*

Introduction

The Anchar lake is monobasined with its catchments from Srinagar city and adjoining villages and a network of channels from the river Sind enters the lake which is the main source of water. During the years the area of the lake has been considerably decreased. The run-off from surrounding agricultural fields and the sewerage from the bordering human settlements are also drained into the

* Corresponding Author.

lake. In addition much of the effluents from SKIMS hospital are also drained into the lake water. All these activities have resulted into the water contamination.

Although the hazardous effect on aquatic environment affecting biota have been reported in India by Ganapati and Alikunhi (1950), Ray and David (1966) and Sinha *et al.*, (1988). But till date no information is available on the effect of sewage and agricultural runoff on zoobenthic fauna in the Anchar lake. Keeping this in view, the present study has been carried out to assess the impact of these discharges on the zoobenthos.

Materials and Methods

Selection of Sites

Three sampling sites were selected in the lake that represents different environmental features. The site-I is located at the point where a channel from the Sind river enters the Anchar lake. It is characterized by flowing water. Site-II is located in the centre of the lake where water is standing while Site-III is located close to the point where hospital (SKIMS) wastes are thrown into lake water. Several sewerage outlets from the catchment area also discharge their contents into water at this site.

Collection of Samples

The sampling was carried out for a period of two years extending from September 2000 to August 2002. The physico-chemical analysis of water samples were carried out according to standard methods of Golterman and Clymo (1969) and APHA (1989). For biological study *i.e.* the collection of the zoobenthos, sampling was carried out after every three months by Ekman's grab. The mud samples collected was passed through a series of sieves with addition of water on the spot. The animals were picked up with forceps and pipettes and also by hand picking and preserved in 70 per cent alcohol. The density of organisms was calculated per square meter of bottom area and the identification was done with the help of the keys given by Ward and Whipple (1959), Mellanby (1963), Pennak (1978), Tonapi (1980) and William and Feltmate (1992).

Results and Discussion

The physico-chemical features of water are summarized in Table 48.1. The water temperature in general ranged from 3.0–28.0°C showing usual seasonal trend with maximum values in summer and minimum values in winter. The pH values fluctuated from 6.9–8.6 indicating that water is well buffered. Dissolved oxygen ranged from 0.4–9.5 mg/L with minimum at site-III and maximum at Site-I. The conductivity values of Anchar lake indicate high ionic concentration. The range fall between 205 μS/cm to 580 μ S/cm. As per Oslen (1950), the water of the lake can be categorised as of B-mesotrophic type. Calcium concentration varied from 9 to 43 m/L and that of magnesium from 4.3 to 28.5 mg/L indicating the lake water to be nutriently rich with cation progression as Ca > Mg. Total alkalinity ranged from 59–315 mg/L in the investigated lake. Philipose (1960) suggested that a water body with alkalinity values more than 100 mg/L is nutriently rich. On this basis, the lake water too seems to be rich in its nutritive contents. The chloride content of water ranged from 9.0–9.5 mg/L with maximum at site-III showing that water here is polluted. Krecker (1939) also observed increase in chloride contents in water receiving wastes via sewage outfalls. The content of nitrate-nitrogen ranged from 245 to 611 μg/L and that of phosphorus from 200–505 μg/L showing that lake is nutriently rich. This is evident by the profuse growth of macrophytes at site-III and site-II. Various physico-chemical characteristics reveal that lake is rich in Ca, Mg, Cl, NO_3-N and PO_4-P showing it to be nutriently rich and polluted.

Table 48.1: Range in Physico-chemical Characteristics in Anchar Lake

Parameter	*Unit*	*Range*
Temperature	°C	3.0–28
pH	–	6.9–8.6
Conductivity	µS/cm	205–580
Dissolved oxygen	mg/L	0.4–9.5
Calcium	mg/L	9.0–43.0
Magnesium	mg/L	4.3–28.5
Total Alkalinity	mg/L	59–315
Chloride	mg/L	9.0–95.0
Nitrate nitrogen	µg/L	245–611
Total phosphorus	µg/L	200–505

Qualitative and quantitative analysis of zoobenthic invertebrate fauna revealed three groups *viz.* Annelida (5.5 per cent) represented by *Placobdella* sp. (Hirudinea); Crustacea (16.6 per cent) represented by *Gammarus* sp (Amphipoda) and Insecta (77.0 per cent) represented by 4 taxa namely *Stenonema* sp (Ephemeroptera), *Enallagma* sp (Odonata), *Chironomus* sp. (Diptera) and *Micropsectra* sp (Diptera). Insect dominated the overall benthic fauna. A single peak during spring in terms of species diversity and population density of benthos was observed. Dhillon *et al.* (1993a), and Yousuf *et al.* (2002) failed to record any seasonal trend in the population density of zoobenthic organisms whereas Jindal and Singh (2005) recorded a single peak during monsoon in freshwater ecosystems of Punjab.

Per cent Contribution of Zoobenthic Taxa in the Anchar Lake

Taxa	*Site-I*	*Site-II*	*Site-III*
Enallagma sp.	37.5	–	–
Gammarus sp.	25	12.5	–
Stenonema sp.	37.5	–	–
Chironomus sp.	62.5	62.5	100
Micropsectra sp.	12.5	12.5	–
Placobdella sp.	12.5	12.5	–

At site-I *Gammarus* sp. (25 per cent), *Enallagma* sp. (37.5) and *Stenonema* sp. (nymphs only, 37.5 per cent) were found while annelids were absent. It was observed that the population density and species diversity was lower than site-II. This site is located where the river Sind feeds the Anchar lake and water here is fast flowing. The lake bed at this site is sandy and the macrophytic vegetation is scarce. The silt accumulated here seems to be detrimental to the aquatic organisms as also reported by Kaul *et al.* (1978) and Pennak (1978). The nutrients available at this site are lowest of all the sites. Kumar (1996b) also observed the paucity of nutrients to be responsible for low density and diversity of zoobenthic fauna.

The zoobenthic fauna recorded at site-II included one annelid represented by *Placobdella* sp. (12.5 per cent), one crustacean *Gammarus* sp. (12.5 per cent) and larvae of two dipteran insects both belonging

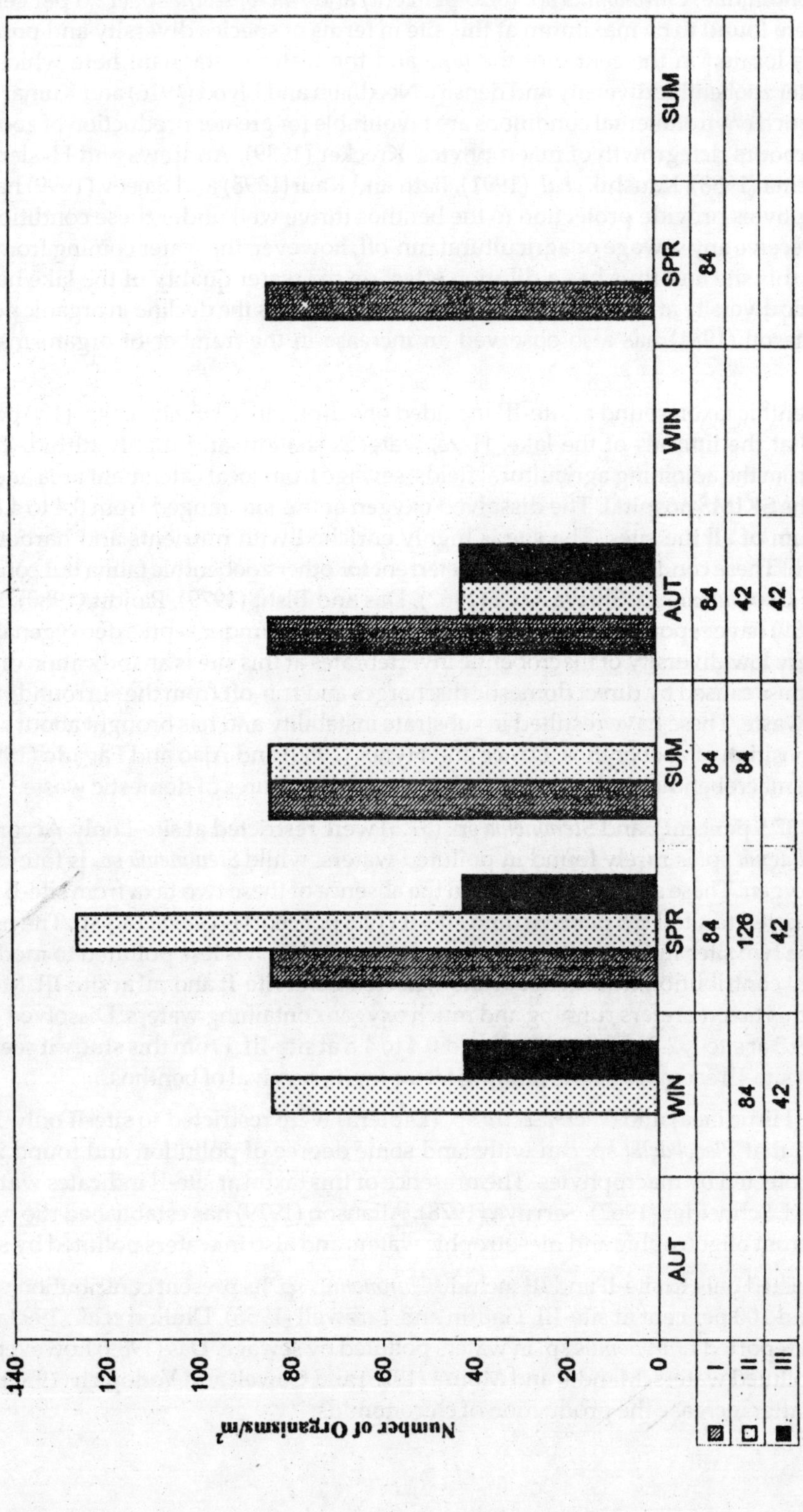

Figure 48.1: Seasonal Variation in the Population Density (Number of organisms/m²) of the Zoobenthos at the Investigated Sites of the Anchar Lake

to the family Chironomidae, *Chironomus* sp. (62.5 per cent) and *Micropsectra* sp. (12.5 per cent). The zoobenthic taxa were found to be maximum at this site in terms of species diversity and population density. This site is located in the centre of the lake and the water is stagnant here which seems favourable for greater zoobenthic diversity and density. Needham and Llyod (1916) and Kumar (1996b) also observed that such environmental conditions are favourable for greater production of zoobenthic fauna. This site harbours rich growth of macrophytes. Krecker (1939), Andrews and Hasler (1943), Tonapi (1980), Sharma (1988), Kaushik *et al.* (1991), Bath and Kaur(1998) and Sajeev (1999) have also opined that macrophytes provide protection to the benthos thrive well under these conditions. The site itself does not receive any sewage or agricultural run-off, however, the water coming from site-III and site-I mixes at this site and thus has a dilution effect on the water quality of the lake here. The increased zoobenthic diversity at this site as compared to site-III shows the decline in organic pollution load at this site. Mason (1981) has also observed an increase in the number of organisms at less polluted sites.

The only zoobenthic taxon round at site-III included one dipteran, *Chironomus* sp. (100 per cent). This site is located at the littorals of the lake. Here, water is shallow and highly turbid. This site receives effluents from the adjoining agricultural fields, sewage from local catchment area and above all effluents from the SKIMS hospital. The dissolved oxygen at this site ranged from 0.4 to 4.8 mg/l, which was minimum of all the sites. The site is highly enriched with nutrients and harbours rich macrophytic growth. These conditions seem to be deterrent for other zoobenthic fauna but conducible for survivability of *Chironomus* sp. Shrivastava (1962), Das and Bisht (1979), Pandit (1980), Sharma (1986), and Das (1989) have reported that *Chironomus* sp. can survive under septic, deoxygenated and polluted waters. Very low diversity of macrobentic invertebrates at this site is an indication of impact of anthropogenic stress caused by direct domestic discharges and run-off from the surrounding areas including hospital waste. These have resulted in substrate instability and has brought about a drastic change in the composition of zoobenthos. Dance and Hynes (1980) and Ajao and Fagado (1990) also observed decline in macrobenthos in waters receiving complex mixtures of domestic waste.

Enallagma sp. (37.5 per cent), and *Stenonema* sp. (37.5) were restricted at site-I only. According to Pennak (1978) *Enallagma* sp. is rarely found in polluted waters, while *Stenonema sp.* is found where there is plenty of oxygen. These suggestions explain the absence of these two taxa from site-II and III. The only taxon restricted to site-I (25 per cent) and site-II (12.5 per cent) is *Gammarus* sp. The presence of this taxon at these two sites indicates its preference for water which is less polluted to moderately polluted. Its per cent contribution was higher at site-I, decreased at site-II and nil at site-III. Mellanby (1963) opined that this taxon prefers running and much oxygen containing waters. Dissolved oxygen ranged from 5.8 to 9.5 at site-I, 2.8 to 8.9 at site-II and.0.4 to 4.8 at site-III. From this study it seems that the oxygen range at site-III seems to be the limiting factor for its survival of benthos.

Placobdella sp. (Hirudinea) and *Micropsectra* sp. (Diptera) were restricted to site-II only. Pennak (1978) has reported that *Placobdella* sp. can withstand some degree of pollution and found it at the places which are protected by macrophytes. The presence of this taxon at site-II indicates water to be moderately polluted. Schneider (1962), Serruya (1978), Allanson (1979) has established tlie presence of *Microspectra* sp. from oligotrophic and mesotrophic waters and also in waters polluted by sewage.

The taxon restricted only to site-II and III include *Chironomus* sp. Its present contribution was 62.5 per cent at site-II and 100 per cent at site-III. Gaufin and Tarzwell (1956), Dhillon *et al.* (1993a and b) and Kumar (1996b) reported *Chironomus* sp. in waters polluted by sewage. Das (1989) however, found this both in clean polluted waters. Mandal and Moitra (1975) and Cowell and Vodopich (1981) opined that rich organic matter increase the production of chironomids.

Acknowledgements

Kind acknowledgements are made to the Head, Department of Zoology, Punjabi University, Patiala and Principal, S.P. College for providing laboratory facilities.

References

A.P.H.A., 1989. *Standard Methods for the Examination of Water and Wastewater*, 17th Edition, Washington.

Ajao, E.A. and Fagade, S.O., 1990. A study of sediment and communities in Lagos Lagoon, Nigeria. *Oil and Chemical Pollution*, 19: 123–130.

Allanson, B.R., 1979. Lake Sibaya *Monogr. Biologicae* 36, W. Junk, The Hague, pp. 364.

Andrews, J.D. and Hasler, A.D., 1943. Fluctuations in the animal populations of the littoral zone in lake Nendota. *Trans. Wis. Acad. Sci. Arts. Lett.*, 35: 175–185.

Bath, K.S. and Kaur, H., 1998. Seasonal distribution and population dynamics of aquatic insects in Harike reservoir (Punjab). *J. Ecobiol.*, 10(1): 43–46.

Cowell, B.C. and Vodopich, D.S., 1981. Distribution and abundance of benthic macroinvertebrates in a sub tropical Florida lake. *Hydrobiologia*, 78: 97–105.

Dance, K.W. and Hynes, H.B.N., 1980. Some effects of agricultural land use on stream insect communities. *Environ. Pollut.*, 22(A): 19–28.

Das, S.M., 1989. *Handbook of Limnology and Water Pollution.* South Asian Publishers Pvt. Ltd., New Delhi.

Das, S.M. and Bisht, R.S., 1979. Ecology of some hemiptera and coleoptra of Kumaun lakes. *Indian J. Ecol.*, 6(1): 35–40.

Dhillon, S.S., Kaur, H., Bath, K.S., Mander, G. and Syal, J., 1993a. The impact of effluence on the zooplankton population of Beas river. *J. Ecoloxicol. Environ. Monit.*, 3(2): 177–120.

Ganapati, S.V. and Alikunhi, K.H., 1950. Factory effluents from the Mettur Chemical and Industrial Corporation Ltd. Mettur Dam, Madras, of the river Cauvery. In: *Proc. Nat. Inst. Sci., India*, 16(3): 189.

Gaufin, A.R. and Tarzwell, C.M., 1956. Aquatic invertebrates as indictors of organic pollution in Lytte Creak. *Sewage and Indust. Wastes*, pp. 28–906.

Golterman, H.L. and Clymo, R.S., 1969. *Methods for Chemical Analysis for Freshwater*. IBP Handbook No. 8 Blackwell Scientific Publications, Oxford.

Jindal, R. and Singh, H., 2005. Ecology of benthos of some freshwater ecosystem of Punjab. In: *Proceedings of National Seminar New Trends in Fishery Development in India*, (Ed.) M.S. Johal. Punjab University Chandigarh, pp. 179–184.

Kaul, V., Fotedar, D.N., Pandit, A.K. and Trisal, C.L., 1978. A Comparative study and plankton populations of some typical freshwater bodies of Jammu and Kashmir State. In: *Environmental Physiology and Ecology of Plants*, (Eds.) Sen, D.N. and Bansal, R.P. Bishen Singh and Mahendra Pal Singh, Dehra Dun, India, pp. 249–261.

Kaushik, B., Sharma, S. and Saxena, D.N., 1991. Ecological studies of certain polluted lentic waters of Gwalior region with reference of aquatic insect communities. In: *Current Trends in Limnology–I*, (Ed.) Shastree, N.K., pp. 185–200.

Krecker, F.H., 1939. A comparative study of the animal population of certain submerged aquatic plants. *Ecology*, 20: 553–562.

Kumar, A., 1996b. Impact of organic pollution on macro-zoobenthos of the river Mayurakshi of Bihar. *Poll. Res.*, 15(1): 85–88.

Mandal, B.K. and Moitra, S.K., 1975. Seasonal variations of benthos and bottom soil edaphic factors in a freshwater fish pond at Burdwan, West Bengal. *Trop. Ecol.*, 16: 43–48.

Mason, C.F., 1981. *Ecology of Freshwater Pollution.* Longman, London.

Mellanby, H., 1963. *Animal Life in Freshwater*, 6th ed. Chapman and Hall Ltd., London.

Needham, J.G. and Llyod, J.J., 1916. *The Life in Inland Waters*, New York.

Oslen, S., 1950. *Aquatic Plants and Hydrospheric Factors.* Svensk Bot. Tidskr., 44 pp.

Pandit, A.K., 1980. Biotic factors and food chain structure in some typical wetlands of Kashmir. *Ph.D. Thesis*, Kashmir University, Srinagar.

Pennak, R.W., 1978. *Freshwater Invertebrates of the U.S.*, 2nd edn. John Wiley and Sons Inc., New York.

Philipose, M.T., 1960. Freshwater phytoplankton of inland fishers. In: *Proc. Sym. Algology*, ICAR, New Delhi, pp. 272–291.

Ray, P. and David, A., 1966. Effects of industrial wastes and sewage upon the chemical and biological composition and fisheries of the river Ganga at Kanpur (U.P.). *Environ. Hlth.*, 8(4): 307–339.

Sajeev, K.I.D.K., 1999. Analytical studies on the ecosystem of the Ropar headworks reservoir. *Ph.D. Thesis*, Punjabi University, Patiala.

Schneider, R.F., 1962. Seasonal succession of certain invertebrates in northwestern Florida lake. *Quart. J. Fl. Acad. Sci.*, 25: 127–141.

Serruya, C., 1978. Lake kinneret. *Monogr. Biologicae* 32, W. Junk, The Hague, pp. 501.

Sharma, S., 1988. Effect of physico-chemical factors on benthic fauna of Bhagirathi river, Garhwal Himalayas. *Ind. J. Ecol.*, 13(I): 133–137.

Shrivastava, H.N., 1962. Aquatic fauna as indicator of pollution. *Environ. Hlth.*, 4: 106–113.

Sinha, R.K., Singh, A.K., Sharan, R.K. and Prasad, K., 1988. Impact of sewage on water quality of the river Ganga at Patna. In: *Proc. Nat. Sci. Management of Water and Wastewater Systems*, Patna, pp. 3.1–3.18.

Tonapi, G.T., 1980. *Freshwater Animals of India: An Ecological Approach.* Oxford and IBH Publishing Co., New Delhi.

Ward, H.B. and Whipple, G.C., 1959. *Freshwater biology*, 2nd Edition. John Wiley and Sons, New York, USA.

William, D.D. and Feltmate, B.W., 1992. *Aquatic Insects*, CAB International Walling Ford, Oxon, U.K.

Yousuf, A.R., Gazala, F. and Kousar, J.P., 2002. Ecology and feeding biology of commercially important cyprinid fishes of Anchar lake, Kashmir, with a note on their conservation. In: *Natural Resources of Western Himalaya*, (Ed.) Pandit, A.K. Valley Book House Hazratbal, Srinagar, pp. 243–272.

Chapter 49

Eco-ethology and Conservation of Hanuman Langur, *Semnopithecus entellus*

L.S. Rajpurohit, A.K. Chhangani, R.S. Rajpurohit, N.R. Bhaker, D.S. Rajpurohit and G. Sharma

Department of Zoology, Faculty of Science, J.N.V. University, Jodhpur – 342 005, India
Primate Research Centre (PRC), 396, 3rd C Road, Sardarpura, Jodhpur

ABSTRACT

Hanuman langurs are the most adaptable south Asian Colobine (Primates : Cercopithecide) and live in a wide range of habitats from the Himalayas (ca 4000 m altitude) and peninsular forest, from semi-arid lands (like Jodhpur–western Rajasthan) to the rain forests of northeast. The Hanuman langur around Jodhpur has been studied for last 35 years. The present study highlights their population, demographic parameters and the feeding ecology and threats to this geographical isolated population of about 1800 langurs of Jodhpur. They face problems largely due to habitat destruction and increasing human population in the area. Although they are considered sacred, never been hunted. However, there is need of the restoration of habitat and the awareness among the local people.

Keywords: Sociobiology, Ecology, Population, Conservation, Semnopithecus entellus.

Introduction

The Hanuman langur, (*Semnopithecus entellus* Dufresne, 1797) is the best-studied and most adaptable south Asian Colobine. They live in a wide range of habitats from the Himalayas and peninsular forests to semiarid woodlands, in villages and towns and on cultivated lands (Oppenheimer, 1977; Roonwal and Mohnot, 1977). The species has a highly variable social organization. The two basic types of social groups are bisexual troops and all-male bands. The bisexual troops are matrilineal groups of adult females and offsprings with either on adult male (unimale troops) or more than one adult male (multimale bisexual troops). In Jodhpur the former category is predominant.

In areas around temples and human habitations, the langurs have been provisioned for many centuries. This conditioning of langurs to human as a source of food makes the study sites to undertake a naturalistic study of langur ecology but an excellent location for a series of yearly behavioural observations, since the animals are already habituated.

In historical times the vast plains of northern India were covered with vast tracts of forests are now being treeless plains and to the inherent conflict between survival of large number of people and preservation of forest habitats. Law protects the remaining forests and no live wood may be cut without permission. However, enforcing such legislation is nearly impossible when so many tribal and poor people depend for a living on cutting firewood. And nearly every part of the forest is in the process of making treeless. The highly adaptable and of course, sacred langurs have with stand depredations to the environment more successfully than other species. Increasing secularization, combined with the inescapable fact that langurs do cause damages is undermining the unique protection that monkeys in India enjoy. In addition to crop raiding (Mohnot, 1971a) langurs, wasteful manner of breaking off branches whey they feed, damages ornamental plants and other property (Hrdy, 1977). In addition to ruined roofs, langurs invade kitchens; break food/water containers, destroy/break TV antenna and occasionally cause accidents in city areas.

The present paper provides the details on population ecology and conservation aspects in Hanuman langur in and around Jodhpur.

Materials and Methods

Study Area and Animal

Jodhpur is located in Rajasthan at the eastern edge of the Great Indian Desert. The town was erected on a hilly sandstone plateau of approximately 150 km^2 surrounded by flat semiarid lands. The plateau is inhabited by a geographical isolated population of 1800–1850 langurs which has been studied by various Indian and foreign researchers for more than 35 years.

The climate is dry with maximum temperatures ca 48°C in May/June and maximum around 0°C in December/January. This area receives 90 per cent of its scanty rainfall (average 380 mm) during the monsoon in July–September. The natural open scrub vegetation is dominated by xerophytic plants including *Prosopis juliflora, Acacia senegal, Prosopis cineraria, Caparis deciduas* and *Euphorbia caducifolia.* There are numerous irrigated parks and fields. The langurs feed on about 208 natural and cultivated plant species. For religious reasons local people provision most of the groups with wheat preparations, vegetables, nuts and fruits (Mohnot, 1974; Winkler, 1981; Srivastava, 1989; Chhangani, 2000). Some groups raid crops and orchards (Mohnot, 1971a), but because they are considered to be sacred are never hunted. Apart from feral dogs, there are no natural predators. The animals are easy to observe since they are not shy and spend most of the daytime on the ground (Mohnot, 1974).

The geographically isolated langur population around Jodhpur is comprised of 34–35 unimale bisexual troops and 12–14 all-male bands. Each troop occupies its own home range of about 0.5–1.5 km^2. Females remain for life in their natal troop and males emigrate or expelled usually as juveniles to unisexual all-male bands, whose home ranges can be as large as 20 km^2 (Sommer, 1987; Rajpurohit, 1987; Rajpurohit and Sommer, 1993). All-male bands invade home ranges of bisexual troops in an unpredictable pattern, sometimes resulting in rapid or gradual replacement of the resident male

(Rajpurohit, 1987; Rajpurohit and Mohnot, 1988). New residents may kill infants sired by their predecessors (Mohnot, 1971 b; Sommer, 1987; Agoramoorthy and Mohnot, 1988).

The different behavioural activities were recorded with the help of *ad libitum*, scans and focal animal sampling (Altmann, 1974). The counting of animals was done when groups leave roosting sites in morning or when they return to the preferred site at evening. To minimize the error of counted twice or some animal left uncounted, the large groups were counted two to three times and by at least two observers.

Observations and Results

Observations were made between December 2002 and November 2003 on Hanuman langur population in and around Jodhpur to investigate the different parameters of population dynamics. This geographically isolated pocket population of 1829 langurs organized in about 49 groups (35 unimale bisexual troops and 14 all-male bands). The present study reveals that the troop size varies from 9–154 (average 45.3; total 1635; n = 35). The male female sex ratio at birth is 1 : 1 but due to high mortality of male juveniles in the process of resident male replacement and the expulsion of male juveniles from their natal troops to immigration into male band, the sex ratio changes dramatically and becomes female biased at adult (*i.e.* 1 male : 5.7 females as per present data). The population density increased from 5.7 in 1971 (*i.e.* 870) to 12.2 (*i.e.* 1829 animals present data).

The threats to this langur population and its habitat are the habitat destruction for urbanization and other developmental activities. Part of this hilly habitat is regularly mined for sandstone and disturbed the langur habitats. Jodhpur city population (human) has increased from 3,17,612 in 1971 (inhabiting in an area of about 50 km^2 to about ten lacs in 2001 (expanded area up to 180 km^2). The problem we might face in next 10–15 years is that langurs' number may go beyond the carrying capacity, which has already been destructed and inhabited by human being. And these langurs would be taken as menace in the same way in which we hear from many of the other Indian cities.

Discussion

The langur population around Jodhpur has increased during last 35 years from about 870 langurs in 1971 (Mohnot, 1974; Mohnot *et al.*, 1987; Rajpurohit, 1987) and about 1500 in 1991–92 (Rajpurohit *et al.*, 1995; Srivastava and Mohnot, 1992) and then about 1800 langurs in 2001 (Rajpurohit *et al.*, 2001). The present study suggests 1829 langurs (in 2002–03) in this geographical isolated population.

The sex ratio found balanced at births, which turn male, biased at juvenile stage (*i.e.* 1.21 times) due to more female infant mortality. Further, with advancing age, male mortality increases dramatically with 1.4 times more females than males among subadults and then 5.7 times more females than males among adults. The proximate explanation for the greater female mortality among infants remains particularly unclear. Langurs resemble other old world monkeys in this respect, which generally show more serving juvenile males than females (see Dittus, 1979). Increased mortality in sub-adult and adult males, on the contrary is better understood at both proximate and ultimate levels of causation-maturing males leave their natal troop and join male bands, which move over large and unfamiliar areas where the risk of accidental death therefore increases (Rajpurohit, 1987; Rajpurohit and Sommer, 1991; 1993). Male migration has been identified as a main factor of higher male mortality in several other primate species such as toque macaque (Dittus, 1979), rhesus macaques (Wilson, 1970) or capuchin monkeys (Robinson, 1988). In langurs at Jodhpur the all-male band structure and invasions into home range of bisexual troops are direct and indirect consequences, respectively of male competition for access to females.

The study area Jodhpur is under severe pressure from variety of factors such as expansion of city area, agriculture, mining and tree cutting. Various development activities are also fragmenting the habitat. Looking to these human and nonhuman primate conflicts for habitat and conservation strategies we have to plan for management of Jodhpur langur population.

The conservation prospects for the Hanuman langur depend upon several aspects of India's ecology and economy-forest conservation, agricultural production, education and the public attitudes. Fortunately, India is a country with a substantial scientific and educational base and a strong development in the areas of conservation. India is also a country with a heritage and sensitivity to the values of its animal life. If these qualities are not overloaded by the demands of human population growth and the realistic of economics, primate conservation may have a bright future.

Acknowledgements

We are thankful to U.S. Fish and Wildlife Service (USFWS) for financial support and to Prof. S.M. Mohnot, Chairman, Primate Research Centre (PRC), Jodhpur for administrative support during the study period. Thanks are due to Dr. S.P. Singh, Head, Department of Zoology, J.N.V. University for logistic support. AKC has CSIR-RAship, RSR was getting UGC-University fellowship, DSR has CSFR-SRFship and GS is UGC Project Fellow.

References

Agoramoorthy, G. and Mohnot, S.M., 1988. Infanticide and juvenilicide in Hanuman langur (*Presbytis entellus*) around Jodhpur, India. *Hum. Ecol.*, 1: 279–296.

Altmann, J., 1974. Observation study of behaviour: sampling methods. *Behaviour*, 49: 227–267.

Chhangani, A.K., 2000. Ecobehavioural diversity of langurs, *Presbytis entellus* living in different ecosystems. Unpublished *Ph.D. Thesis*, J.N.V. University, Jodhpur.

Dittus, W.P.J., 1979. The evolution of behaviours regulating density and age specific sex ratios in a primate population. *Behaviour*, 69: 265–302.

Hrdy, S.B., 1977. *The Langurs of Abu*. Cambridge, MA: Harvard Univ. Press.

Mohnot, S.M., 1971a. Ecology and behaviour of the Hanuman langur, *Presbytis entellus* (Primates : Cercopithecidae) invading fields, gardens and orchards around Jodhpur, western India. *Trop. Ecol.*, 12: 237–249.

Mohnot, S.M., 1971b. Some aspects of social change and infant killing in the Hanuman langur, *Presbytis entellus* (Primates : Cercopithecidae) in western India. *Mammalia*, 35: 175–198.

Mohnot, S.M., 1974. Ecology and behaviour of the common Indian langur, *Presbytis entellus. Ph.D. Thesis*, University of Jodhpur, Jodhpur.

Mohnot, S.M., 1984. Observation on all male bands of the Hanuman langur (*Presbytis entellus*). In: *Current Primate Researches*, (Eds.) M.L. Roonwal, S.M. Mohnot and N.S. Rathore. University of Jodhpur, pp. 343–356.

Mohnot, S.M., Gadgil, M. and Makwana, S.C., 1981. The dynamics of the Hanuman langur population of Jodhpur, Rajasthan, India. *Primates*, 22: 182–191.

Mohnot, S.M., Agoramoorthy, G., Rajpurohit, L.S. and Srivastava, A., 1987. Ecobehavioural studies of Hanuman langur, *Presbytis entellus. Technical Report* (1983–87), DOE, Govt. of India, New Delhi, pp. ix+89.

Oppenheimer, J.R., 1977. *Presbytis entellus*, the Hanuman langur. In: *Primates Conservation*, (Eds.) H.S.H. Prince Rainier and G.H. Bourne. Academic Press, New York, pp. 469–512.

Rajpurohit, L.S., 1987. Male social organization in Hanuman langur, *Presbytis entellus. Ph.D. Thesis*, University of Jodhpur, Jodhpur.

Rajpurohit, L.S. and Mohnot, S.M., 1988. Fate of ousted male residents of one-male bisexual troops of Hanuman langur (*Presbytis entellus*) at Jodhpur, Rajasthan (India). *Hum. Evol.*, 3: 309–318.

Rajpurohit, L.S. and Sommer, V., 1991. Sex differences in mortality among langurs (*Presbytis entellus*) of Jodhpur, Rajasthan. *Folia Primatol.*, 56: 17–27.

Rajpurohit, L.S., Sommer, V. and Mohnot, S.M., 1995. Wanderers between harems and bachelor bands: male Hanuman langurs (*Presbytis entellus*) at Jodhpur in Rajasthan. *Behaviour*, 132: 255–299.

Rajpurohit, L.S., Mohnot, S.M., Srivastava, A. and Chhangani, A.K., 2001. Population dynamics of Jodhpur langurs. *Advances in Ethology*, 36: 245 (Blackwell, Vienna).

Roonwal, M.L. and Mohnot, S.M., 1977. *Primates of South Asia*. Cambridge (Mass), Harvard Univ. Press.

Robinson, J.G., 1988. Male migration in capuchin monkeys. *Behaviour*, 104: 202–232.

Sommer, V., 1987. Infanticide among free-ranging langurs (*Presbytis entellus*) at Jodhpur (Rajasthan, India): Recent observation and a reconsideration of hypotheses. *Primates*, 28: 163–197.

Srivastava, A., 1989. Feeding, ecology and behaviour of Hanuman langur, *Presbytis entellus. Ph.D. Thesis*, University of Jodhpur, Jodhpur.

Srivastava, A. and Mohnot, S.M., 1992. Existence of multimale troop and their transformation into unimale troops in Hanuman langur (*Presbytis entellus*). *Primate Report*, 34: 71–75.

Winkler, P., 1981. Zur oko-ethologie freilebender Hanuman langur (*Presbytis entellus*, Dufresne, 1797) in Jodhpur (Rajasthan), India. *Ph.D. Dissertation*, beorg-August Universitat, Gottingen.

Wilson, E.O., 1970. Competitive behaviour. Smithsonian Third International Symposium, Man and Beast: Comparative Social Behaviour. Washington, D.C. May 13–16, 1969, Smithsonian Institution.

Chapter 50

Phycological Aspects and Water Quality Assessment in the Rivers of Andhra Pradesh, India

P. Manikya Reddy and V. Venkateswarlu

Department of Botany, Osmania University, Hyderabad – 500 007, India

ABSTRACT

Phycological aspects in the rivers has been discussed based upon the data collected on algae from a number of rivers in Andhra Pradesh. Diatoms constituted the dominant group in benthic flora. Their generic and species sequential analysis gave some interesting results. The diatom flora appears to be more or less uniform at unpolluted stations of different results. It is concluded that benthic diatoms serve as good indicators of water quality and pollution in flowing water.

Keywords: *Phycological, Diatoms, Rivers, Generic and species sequential analysis indicators.*

Introduction

The study of river ecosystems / lotic habitats is quite different in many ways when compared to that of lentic environments, perhaps, due to (1) the fickle and unstable nature of the medium and substratum, (2) the insecurity of an organism in water current, (3) continuous nutrient supply and (4) considerable physiological and mechanical stress (Blum, 1960). The floristic and ecology of river algae have been reviewed by Blum (1956), Hynes (1970a) and the examples of benthic and epiphytic algae widespread in flowing waters are given by Whitton (1975). Algae in lotic habitats could be epilithic, epipelic, epizoic, epiphytic and planktonic. The lotic species exhibit very effective attaching devices. They show structural adaptations and Cedergen (1938) pointed out 4 such types. According to Symoens (1951) the communities of algae in rivers exhibit no constant composition of species from place to place and frequently their composition changes rapidly.

Venkateswarlu (1986) and Manikya Reddy and Venkateswarlu (1992) have collected lot of data on the algae of the rivers of Andhra Pradesh and the present paper aims at incorporating the various aspects concerning river algae. It also deals with the critical analysis of different species present in different situations. The species which serve as indicators of river water quality and pollution have been identified.

Materials and Methods

The river Moosi has its source in Ananthagiri hills near Vikarabad in the Ranga Reddy district of Andhra Pradesh. It runs through the city of Hyderabad and after traversing a course of about 250 km, it joins the river Krishna at Wazeerabad. In the thickly populated areas of the city, the river receives domestic wastes. The river Manjira originates in Balaghat range hills in Maharashtra State. After flowing through the states of Maharashtra and Karnataka it enters the Andhra Pradesh State in Medak district and finally joins the river Godavari. The river Thungabhadra rises in the Western ghats on the border of Karnataka State. It is composed of the rivers Tunga and Bhadra. The river after traversing a course of about 640 km joins the river Krishna at Sangameshwar near Alampur in the Mahabubnagar district of Andhra Pradesh. The river Tungabhadra receives effluents from Rayalaseema paper mills near Kurnool.

The river Kagna originates near Vikarabad in the Ranga Reddy district of Andhra Pradesh and after flowing in westernly direction through the states of Andhra Pradesh and Karnataka it joins the river Bhima near Wadi. The river Godavari is the largest river of the peninsula and has a total length 1500 km. It rises in the Western ghats in the Nasik district of Maharashtra state and receives a number of tributaries. It receives treated effluents form Bhadrachalam paper board mills near Bhadrachalam and from a fertilizer factory at Ramagundam. The river Krishna rises near Mahabaleshwar in the western ghats. Its length is about 1300 km. It flows through the Maharashtra and Andhra Pradesh and receives a number of tributaries.

Methods

Four stations were selected in each river for the collection of water and algal samples. Surface water samples were collected from all the rivers at monthly intervals and analyzed for various chemical, parameters by following standard methods (APHA, 1971). Simultaneously benthic algae were collected by the technique of Blum (1956) and Venkateswarlu (1969). For determining the frequency and percentage of different groups and different species of algae, the numbers of individuals per 120 high power fields of the microscope were recorded and percentages of different groups were calculated.

Results

The percentages of different groups of benthic algae, total algae, species number and number of diatom species recorded in all the rivers are given in Tables 50.1a,b, and 50.2a,b. The average concentrations of some important nutrients present at unpolluted and polluted stations of different rivers are given in Tables 50.3 and 50.4.

In all the rivers diatoms dominated the algal populations except in the river Manjira where Cyanophyceae are dominant. Even in the sequence of algae diatoms occupy first position followed by Cyanophyceae and Chlorophyceae. Diatoms constitute the major bulk of algal populations in the river benthos. In general, Pennales are dominant over Centrales (*Cyclotella* only).

Table 50.1a: Percentages of Different Lithophytic Algal Groups in the Rivers of Andhra Pradesh Unpolluted Stations

Algal Groups	*Moosi*	*Manjira*	*Tungabhadra*	*Kagna*	*Tungabhadra*	*Godavari*	*Krishna*
	1961–63	*1979–81*	*1981–83*	*1982–84*	*1984–85*	*1984–85*	*1985–86*
Bacillariophyceae	90.8	46.8	70.9	74.5	74.9	61.2	77.4
Chlorophyceae	4.83	9.5	6.6	6.5	6.5	6.2	2.3
Cyanophyceae	4.3	39.4	22.3	19.2	18.5	30.6	19.9
Euglenophyceae	0.06	3.5	–	0.12	–	22	0.5

Table 50.1b: Percentages of Different Lithophytic Algal Groups in the Rivers of Andhra Pradesh at Polluted Stations

Algal Groups	*Moosi*			*Tungabhadra*	*Tungabhadra*	*Godavari*
	1961–63	*1973–74*	*1981–82*	*1981–83*	*1984–85*	*1984–85*
Bacillariophyceae	81.55	15.4	8.1	67.9	59.7	24.9
Chlorophyceae	2.95	75.9	8.01	2.0	5.9	4.5
Cyanophyceae	11.7	5.5	6.4	38.5	31.8	68.1
Euglenophyceae	3.8	3.4	5.3	1.3	2.5	2.3

Table 50.2a: Total Algae, Total Number of Species and Total Number of Diatoms Species in the Rivers of Andhra Pradesh at Unpolluted Stations

Algal Groups	*Moosi*	*Manjira*	*Tungabhadra*	*Kagna*	*Tungabhadra*	*Godavari*	*Krishna*
	1961–63	*1979–81*	*1981–83*	*1982–84*	*1984–85*	*1984–85*	*1985–86*
Total algae	5464	18445	15225	10455	15268	13565	5620
Total number of species	58	56	57	64	62	62	65
Total number of diatom species	25	26	32	37	29	'36	42

Table 50.2b: Total Algae, Total Number of Species and Total Number of Diatoms Species in the Rivers of Andhra Pradesh at Polluted Stations

Algal Groups	*Moosi*		*Tungabhadra*	*Tungabhadra*	*Godavari*
	1961–63	*1973–74*	*1981–83*	*1984–85*	*1984–85*
Total algae	4602	9231	16165	18189	15692
Total number of species	45	23	42	41	47
Total number of diatom species	15	8	26	23	22

Table 50.3: The Average Concentrations of Some Important Nutrients in the Rivers of Andhra Pradesh at Unpolluted Stations (in mg/l except pH)

Nutrient Parameters	*Moosi 1961–63*	*Manjira 1979–81*	*Tungabhadra 1981–83*	*Kagna 1982–84*	*Tungabhadra 1984–85*	*Godavari 1984–85*
pH	8.3	8.4	8.4	8.5	8.5	8.5
CO_3	9.99	22.76	25.7	30.45	17.13	15.95
HCO_3	232.95	186.79	197.2	252.36	200.0	171.54
Cl	18.1	27.98	86.48	19.54	88.91	34.62
DO	7.9	7.6	7.3	7.6	7.5	7.8
Org. matter	1.23	0.88	1.95	1.22	1.22	0.81
F.NH_3	0.64	0.93	0.75	0.72	1.20	0.94
NO_3	1.12	1.59	1.29	1.13	0.29	0.48
NO_2	0.01	0.01	0.02	0.004	0.003	0.002
PO_4	0.001	0.01	0.015	0.067	0.007	traces
SiO_2	16.95	7.5	6.15	9.8	2.7	3.8

Table 50.4: The Average Concentrations of Some Important Nutrients in the Rivers of Andhra Pradesh at Polluted Stations (in mg/l except pH)

Chemical Parameters	*Moosi*			*Tungabhadra*	*Tungabhadra*	*Godavari*
	1961–63	*1973–74*	*1981–82*	*1981–83*	*1984–85*	*1984–85*
pH	7.9	7.2	7.4	8.0	8.2	8.1
CO_3	2.8	–	–	11.92	2.14	3.75
HCO_3	331.15	–	447.4	202.4	206.1	236.6
Cl	40.75	146.4	208.0	149.0	154.6	104.1
DO	3.0	0.6	nil	4.3	3.2	3.9
Org. matter	5.0	2.5	30.5	7.74	9.70	9.79
F.NH_3	2.62	14.7	20.1	1.09	2.2	1.09
NO_3	0.39	0.13	0.7	0.97	1.12	1.37
NO_2	0.05	0.06	–	0.002	0.001	0.001
PO_4	0.85	0.32	11.5	0.2	0.06	0.22
SiO_2	16.7	8.7	15.6	9.6	5.7	7.7

The generic and species sequential analysis of diatoms exhibited interesting phenomena and are given in the following:

Generic Sequential Analysis

1. River Moosi (1961–63): *Achnanthes* > *Cymbella* > *Synedra* > *Anomoeoneis* > *Navicula* > *Caloneis.*
2. River Moosi (1973–74): *Nitzschia* > *Cymbella* > *Synedra* > *Navicula.*
3. River Manjira: *Synedra* > *Cymbella* > *Navicula* > *Achnanthes.*

4. River Tungabhadra (1981–83): *Cymbella > Navicula > Synedra > Nitzschia > Mastogloia > Caloneis.*
5. River Tungabhadra (1984–86): *Cymbella > Synedra > Navicula > Nitzschia.*
6. River Godavari: *Synedra > Cymbella > Achnanthes > Navicula > Caloneis > Nitzschia.*
7. River Krishna: *Synedra > Cymbella > Gomphonema > Nitzschia > Achnanthes > Cocconeis.*
8. River Kagna: *Cymbella > Synedra > Achnanthes > Navicula > Gomphonema > Nitzschia.*

Species Sequential Analysis

River Moosi

Unpolluted

(1961–63): *Achnanthes microcephala > Cymbella microcephala > Synedra ulna > A. minutissima v. cryptocephala.*

1973–1974: *Nitzschia amphibia > Cymbella affinis > Synedra tabulala > Navicula cryptocephala v. veneta.*

Polluted

(1961–63): *Nitzschia palea > Achnanthes exigua > Cyclotella meneghiniana > Navicula pupula. f capitata > Pinnularia biceps > Navicula pigmaea.*

1973–74: *Nitzschia palea > Gomphonema sphaerophorum > Pinnularia biceps > Achnanthes exigua > Cyclotella meneghiniana.*

River Tungabhadra

Unpolluted

Cymbella affinis > Navicula cryptocephala > Synedra ulna > Nitzschia denticula v. curta > Mastogloia smithii V. amphicephala > Caloneis silicula > Achnanthes microcephala.

Polluted

Nitzschia palea > Gomphonema parvulum > Achnanthes exigua > Gomphonema lanceolatum > Nitzschia hungarica.

River Manjira

Synedra ulna v. aequalis > Cymbella cymbiformis v. jimboi > Cymbella affinis > Nitzschia denticula v. curta > Gomphonema constrictum v. capitata > Achnanthes minutissima v. cryptocephala.

River Kagna

Cymbella affinis > C. Cymbiformis > Synedra ulna > Achnanthes minutissima > Navicula rhynchocephala > Nitzschia denticula v. curta.

River Krishna

Synedra ulna v. aequalis > Cymbella aspera > Achnanthes minutissima v. cryptocephala > Cocconeis placentula.

River Godavari

Unpolluted

Synedra ulna v. aequalis > *Cymbella aspera* > *Achnanthes minulissima v. cryptocephala* > *Navicula bacillum* > *Caloneis silicula v. trancatula* > *Nitzschia denticula v. curta.*

Polluted

Nitzschia obtuse v. scalpelliformis > *Syndera ulna* > *Achnanthes exigua* > *Nitzschia hungarica.*

Discussion

From the literature survey and also from the data on different rivers of Andhra Pradesh, it is quite evident that diatoms constitute the major bulk of benthic populations. Perhaps, the substratum, the composition of the medium and the effective attaching mechanism of various species are the possible reasons for their dominance.

Hustedt (1957), Cholnoky (1968), Schoeman (1973) and Lowe (1974) have used diatoms as indicators of water quality and pollution. Prasad and Singh (1982) made a preliminary attempt on diatoms as indicators of water pollution. Trivedy (1986) surveyed the literature on the use of algae in biomonitoring of water pollution. However, these authors could not make a critical analysis of diatom species which represent the dominant flora in rivers and streams.

The generic analysis reveals that *Cymbella, Syndera, Navicula* and *Achnanthes* are the most predominant algae in the benthic communities at unpolluted stations of the rivers. Besides, *Epithemia, Rhopalodia* and Amphora are also found only in unpolluted waters. The genus *Cymbella* appears to occupy either the first or second place in most of the rivers investigated. Even the species analysis shows that the populations include a number of species belonging to this diatom. *Cymbella cymbiformis v. jimboi, C. affinis, C. microcephala, C. aspera, Synedra ulna v. aequalis, Achnanthes microcephala, Nitzschia denticula v. curta* and *N. amphibia* are the most dominant/commcn species in almost all the rivers investigated. Certain genera like *Achnanthes, Naavicula, Pinnularia, Gomphonema* and *Nitzschia* occur both in unpolluted and polluted habitats.

The diatom species sequential analysis in order of their abundance indicates that although certain species are common in different rivers investigated, their order of sequence is not uniform, especially at the unpolluted stations. However, in the polluted areas there appears to be some uniformity in their sequence. In the former it is mainly due to the large variety of species, the abundance of which depends upon the composition of the medium, type of substratum and physical conditions. On the other hand in polluted waters the species number is limited and perhaps, the less the number of species the more the uniformity in their sequence.

In polluted habitats species belonging to Araphidae group are not present as they can not exhibit any movement due to the absence of raphe. On the other hand the species with raphe can evade the areas which are toxic by way of movement within the habitat. So the species with raphe have advantage over rapheless forms specially in the polluted environments.

Benthic Algae as Indicators of River Water Quality

It is quite evident from the data certain species of algae play a role in assessing the quality of water. They are given in the following:

Name of the Species	*Water Quality*
Achnanthes microcephala	
A. minutissima v. cryptocephala	
Cymbella cymbiformis v. jimboi	
Navicula pupula f. rectangularis	Clean, dissolved oxygen > 5.0 mg/l.,
Rhopalodia gibba	low BOD, organic matter, ammonia
Nitzschia amphibia	chlorides and phosphates.
N. denticula v. curta	More number of species
Spirogryra adnata	
Coelosphaerium kuetzingianum	
Achnanthes exigua	
Navicula phymaea	
N. pupula f. capitata	Highly polluted (organically)
N. muralis	High organic matter,
Pinnularia biceps v. amphicephala	BOD, ammonia, chlorides and
Nitzschia palea	phosphates and very low D.O.
Cyclotella meneghiniana	or nil, less number of species,
Stigeoclonium tenue	more number of individuals.
Chlorella vulgaris	
Oscillatoria chalybea	
Euglena acus	
Large Populations of	
Nitzschia kutzingiana	High concentration of organic
N. palea	nitrogenous materials
N. thermalis	(Cholnoky, 1968)
Abundance of	
Nitzschia thermalis and	Anaerobic conditions
Disappearance of	(Schoeman, 1973)
Achanthes minutissima	
Spirogyra adnata	
Euastrum bombayense	
Navicula pupula f. rectangularis	
Anomoeoneis exilis f. lanceolala	Clean, fast flowing water,
Stigeoclonium tenue .	Polluted turbulent water.

Name of the Species	*Water Quality*
Large growths of *Cladophora glomerata*	Relatively high nutrient levels (Whitton, 1970) Heavy metals in low levels.
Abundant growth of *Stigeoclonium tenue*	High levels of heavy metals With relatively high nutrient levels.
Chlamydomonas botryopara	
Chlamydomonas subcaudata	
Carteria ovata	Chemical effluents from
S. obliquas	pharmaceutical industries mixed
Sphatrenopsis fluviatilis	with sewage water.
Scenedesmus incrassulatus	
Lepocinclis salina f. pachyderma	

Important Environmental Parameters

Water current is one of the most important factor in river ecology. For the growth of certain algae water movement is desirable. The benthic algae are some how benefited by moderate current whereas the fast currents cause mechanical danger to plankton (Blum, 1956). The size surface texture and chemical composition of the substratum may influence the distribution of algae in rivers. The natural substrata provide more reliable data than the artificial ones (Venkateswarlu and Manikya Reddy, 1985).

In lotic environments the nutrient supply will be continuous, because no nutrient gradient is established due to continuous agitation of water. The climatic conditions such as rainfall, temperature and light play an important role in river ecology. The distribution of the flora depends upon fluctuations in temperature, light regime and age of the water. Temperature determines the nutrient load because of its influence on evapotranspiration. In most of the rivers the fluctuations in temperature are minimum. The benthic flora gets mostly the reflected and refracted light (Witford and Schumacher, 1963).

Conclusions

On the basis of the literature survey and the critical analysis of the data on the rivers of Andhra Pradesh it can be concluded that benthic algae are abundant than plankton. In the former diatoms are dominant, may be due to the effective attaching adaptations shown by a number of species. Water current is the most important parameter in river ecology along with the substratum and climatic conditions. The diatom generic and species analysis reveal interesting phenomena. The abundance of diatoms frustules may give some clue to explore the oil deposits in different river basins.

References

APHA, AWWA and WPCF, 1971. *Standard Methods for the Examination of Water and Wastewater*, 13th ed. Washington U.S.A.

Blum, J.L., 1956. Ecology of river algae. *Bot. Rev.*, 22(5): 291–341.

Blum, J.L., 1960. Algal populations in flowing waters in the ecology of algae. *The Pymatu Symposia in Ecology. Spl.* Publ. No. 2, Univ. Pittasburgh, pp. 11–21.

Cedergren, G.R., 1938. Reofila eller det rinnands vattnets algamhallen, Sveensk. Bot. Tidskr., 32: 363–373.

Cholnoky, B.J., 1968. Die Okologie der Diatomeen in Sinnengewassem. T. Cramer Publishers, pp. 699.

Hustedt, F., 1957. Die Diatomeenflora des Fluss system a der weser in Gebiet der Hansestadt Bremen. Abh. Naturw. Ver. Bremen., 34: 181–440.

Hynes, H.B.N., 1970. *The Ecology of Running Waters*. Liverpool Univ., Toronto Press, Canada.

Lowe, R.L., 1974. Environmental requirements and pollution tolerance of fresh water diatoms. *Envi. Monit. Ser.,* EPA 670/4-74-005. Nat. Env. Res. Centre. Res. and Dev., U.S. Env. Prot. Agency, Cincinnati, Ohio, 334 pp.

Manikya Reddy, P. Venkateswarlu, V., 1992. The impact of paper mill effluents on the algal flora of the river Tungabhadra. *J. Indian. Bot. Soc.*, 71: 109–114.

Prasad, B.N. and Singh, Y., 1982. On diatoms as indicators of water pollution. *J. Indian Bot. Soc.*, 61: 326–336.

Schoeman, F.R., 1973. A systematic and ecological study of the diatom flora of Lesotho with special reference to the water quality. V& H Printeres, Pretoria, 365 pp.

Symeoens, J.J., 1951 Esquise s'um systeme des associations algales s'ean douce. Verh. Inst. *Ver. Theoret.* Angew. Lin. 11: 395–408.

Trivedy, R.C., 1986. Role of algae in biomonitoring of water pollution. *Asian Environment*, 8(3): 31–42.

Venkateswarlu, V., 1969. An ecological study of the algae of the River Moosi, Hyderabad (India) with special reference to water pollution. I. Physico-chemical complexes. *Hydrobiologia*, 33(1): 117–143.

Venkateswarlu, V., 1986. Ecological studies on the rivers of Andhra Pradesh with special reference to water quality and pollution. In: *Proc. Indian Acad. Sci. (Pl. Sci.)*, 96(6): 495–508.

Venkateswarlu, V. and Manikya Reddy, P., 1985. Algae as biomonitors in river ecology. In: *Symp. Biomonitoring State Environ. Poll.*, pp. 183–189.

Whitford, L.A. and Schumacher, G.J., 1963. Communities of algae in North Carolina streams and their seasonal relations. *Hydrobiologia*, 22(1–2): 137–167.

Whitton, B.A. (Ed.), 1975. *River Ecology*. Blackwell Scientific Publ., London.

Whitton, B.A., 1970. Toxicity of heavy metals to Chlorophyta from flowing waters. *Arch. Microbial.*, 72: 353–360.

Whitton, B.A., 1984 Algae as monitors of heavy metals in freshwaters. In: *Algae as Ecological Indicators*, (Ed.) L. Elliot Shubert. Academic Press Inc., London Ltd., pp. 257–280.

Chapter 51

Biocontrol of House Fly, *Musca domestica* L. (Diptera : Muscidae) by Hymenopteran Pupal Parasitoid *Spalangia cameroni* P. (Hymenoptera : Pteromalidae)

J. Muruheswari, N. Krishnaveni and Sarojini Sukumar
Department of Life Sciences, Avinashilingam Deemed University, Coimbatore – 641 043

ABSTRACT

Spalangia cameroni is an effective biocontrol agent in suppressing the house fly population. The two-day old housefly pupae were selected and stuck to a card and then exposed to the *S. cameroni* for parasitization. *S. cameroni* were reared in clear plastic bottles (15 cm × 7.5 cm diameter) containing house fly pupae at 28±2°C and 80 RH. Approximately 1,000, two-day old *Musca domestica* pupae were introduced three times a week held in the cage for 18–24 hours and then removed and placed in bottles (4 litres) and after 3–5 days the empty puparia and emerged adult house flies were removed and remaining parasitized pupae were kept in containers for 19–23 days required for the parasites to complete development. Then the newly emerged adult parasites were transferred to the colony. Each *S. cameroni* is found to control 20 *M. domestica* pupae after 12 hours of exposure.

Keywords: Biocontrol, Musca domestica, Spalangia cameroni.

Introduction

Housefly, *Musca domestica* widely distributed and closely associated with man and his environment, poses serious threat to human and animal health. Its public health importance can be understood from the fact that over 100 species of disease causing pathogens including bacteria, viruses, fungi, protozoans and helminthes have been recorded from house flies which are only mechanical carriers and its possible role in their transmission has been elucidated.

Current fly control methods heavily rely on the usage of increasing amounts of insecticides. Because of the extensive use of chemical insecticides fly population have become resistant to registered compounds. Extensive use of chemical insecticides and larvicides for a prolonged period has resulted in these undesirable managerial problems. This increase the alternative housefly control strategies. Evaluation of biological control agents provides a promising tool in offering excellent scope for fly management.

In a biological control sense, one could effectively employ pupal parasites as this stage often represents a residue of less than one per cent of the total eggs deposited by the previous generation of flies. Therefore, any biotic agent destroying the pupal stage can exert a vastly greater natural control effect per individual than by destroying any previous stage. Among the pteromalid pupal parasitoids, the parasitic species belonging to the genus Spalangia appear to be of practical value in the natural control of house flies. The present investigation was an attempt to study the biology and biocontrol potential of *S. cameroni* on the house fly pupae.

Materials and Methods

Mass Rearing of Housefly, *M. domestica*

The cyclic colonies of *M. domestica* were established in the laboratory by the adult houseflies collected by the sweep net from the surrounding areas of Coimbatore. They were maintained on skimmed milk powder and mango in the cage (30 cm × 30 cm × 30 cm) to which a stockinette attached at one side of the cage, kept under the controlled temperature 28±2°C and 80 RH. A cotton wool soaked in water was kept in the cage for maintaining humidity (Balakrishnan, 1992). Styrofoam pieces were placed inside the cage and they act as landing platforms to reduce adult mortality due to drowning. (Singh and Moore, 1985). The ovitrap kept inside the cage constituted a mixture of soaked groundnut oil cake and wheat bran in 1 : 1 ratio in 100 ml plastic container. This is the larval breeding medium.

Eggs

Eggs deposited in the ovitraps were isolated soon after oviposition and placed in a 500 ml beaker containing 250 ml of tap water. The egg masses were broken up by gentle stirring. The individual sunken viable eggs were collected and the floating dead eggs were removed. The eggs collected were placed on a black paper and the incubation time was recorded (Moore, 1954).

Larvae

The hatched larvae were transferred to the breeding medium in glass bottles (1 litre) covered with nylon nettings.

Pupae

The third instar larvae of the flies migrated to the peripheral areas and pupated close to the surface of the medium. The medium containing 2 day old pupae was transferred gently to a floatation container with 10 litres of water and stirred continuously. The medium absorbed moisture and sunk to the bottom of the container. Then the pupae were dried at room temperature and can be used for study purpose (Moore, 1954; Singh and Moore, 1985 and Balakrishnan, 1992).

Mass Rearing of *S. cameroni*

The pupae of *S. cameroni* was supplied by the Biocontrol Research Laboratories, Bangalore. They were reared in clear plastic bottles (15 cm × 7.5 cm diameter). The adult emerged was provided with water and droplets of honey daily as a source of carbohydrates (Halland Fisher, 1988.)

Introduction of Host Pupae for Parasitization

Approximately, 1000 uniform sized (6±0.5 mm) two-day old house fly pupae were selected and stuck to a card and introduced three times a week for 18–24 hours. Mated female parasitoids were allowed to oviposit on fly pupal hosts, at a ratio of one female per 20 pupae. The card was then removed and placed in bottles (4 litres). After 19–25 days, the newly emerged adult parasites were transferred to the colony (Morgan *et al.*, 1975).

Dissection Techniques

Dissection technique was performed daily, about 20 pupae per day for three weeks. The outer shell of the house fly puparium was cut all the way around at its midpoint. One end was pulled off and the fly pupae eased out of the other half of the shell. Measurements of the egg, larva, pupa were made using micrometers and were drawn with the help of camera lucida. The representative specimens were preserved in 70 per cent ethyl alcohol.

Results and Discussion

Oviposition

The mated female parasitoid, after finding a house fly pupae, was observed to make a thorough examination of it. It walked slowly over the puparium while tapping it with the antennae and then rubbing the tip of the abdomen on the host for several seconds to get a desired spot for oviposition and drilled through the puparium by fixing itself firmly on the host pupa. The duration of the oviposition ranged between 10.30–16.40 minutes (X = 13.472±2.138). This change in the oviposition time was due to the reduced thickness of the puparium.

Characteristics of Parasitized Pupa

It was observed to be darker in colour than normal pupa, but became discoloured at the drill site. Similar changes had been observed by Wylie (1972) in *S. cameroni* and Pinkus (1913) in *S. muscidarum*.

The results of the morphometric analysis of different stages of life cycle of *S. cameroni* is depicted in Table 51.1.

Table 51.1: Morphometric Analysis of Different Stages of Life Cycle of *S. cameroni* Measurements Based on the Average of 20 Specimens

Sl.No.	*Stage*	*Length (mm)*		*Width (mm)*		*Larval Period (in days)*	
		Range	*Mean±SD*	*Range*	*Mean±SD*	*Range*	*Mean±SD*
1.	Egg	0.36–0.48	0.422±0.040	0.10–0.22	0.156±0.043	22–36 (hours)	27.40±4.652
2.	First instar larva	0.43–0.67	0.554±0.073	0.15–0.30	0.237±0.056	3–4 (hours)	3.52±0.331
3.	Second instar larva	0.70–2.20	1.434±0.519	0.35–0.95	0.637±0.204	2–3 (hours)	2.60±0.385
4.	Third instar larva	1.80–2.80	2.540±0.504	1.20–1.70	1.450±0.150	3–4 (hours)	3.50±0.369
5.	Pre-pupa	1.90–3.00	2.470±0.269	0.90–1.40	1.216±0.166	24 hours	One day
6.	Pupa	2.30–3.20	2.780±0.282	0.98–1.20	1.130±0.081	7–9 (hours)	7.64±0.739

Egg

Egg is elongate, oval, opaque, white, usually convex dorsally and faintly convex or flat ventrally. The eggs were usually laid between pupal capsule and the integument. Eclosion occurred through the anterior tapering transparent region of the egg.

First Instar Larva

Once eclosed, the larva stood erect with its hind segments on the host pupa and continuously moved its head and thorax for a few seconds before crawling. This translucent larva has a wide anterior position and a tapering posterior portion.

Second Instar Larva

The gut region of the larva has increased in size and occupied the major part of the larva. Nine pairs of spiracles are present.

Third Instar Larva

The body is widest at the middle region and narrow at both the ends. The body is grayish white in colour, strongly arched dorsally. This larva has 2 longitudinal row of 11 tubercles situated laterodorsally. Tubercles are turgid and conspicuous and the size and shape vary from each other. Tubercles located at the middle of the body are larger in size. There are nine pairs of spiracles located anterior of the border of the tubercles. The larva ceased feeding after three days.

Prepupa

The prepupa is white in colour and is differentiated into two distinct regions namely, a narrow anterior region and a wider posterior region corresponding to the abdomen. There was an average decrease in length and width as compared with the third instar and a deposition of dark brown to grayish semi-liquid larval meconium enclosed in a peritrophic membrane is seen.

Pupa

The future adult appendages are visible in the young pupa. The newly formed pupa is white, then become cream to light than at a latter period. The eyes and ocelli became distinct in the latter pupal stage. The pupae could be sexed through comparison of the lengths of their antennae. The female antennae extended only to the posterior part of the thorax, while those of the males reached the posterior part of the abdomen. The day prior to adult emergence the movements of the adults appendages were seen through the transparent pupal covering.

Pre-adult

After emerging from the pupal seen, the adult has the habit of remaining within the puparia of the host for sometime. During this period, they bore more of less regular, circular holes through the puparia for emergence. The characters and behaviour that has been observed is more or less similar to the study by Arellano and Reuda (1988) in *S. endius*.

Acknowledgement

The authors are grateful and express their gratitude to the authorities of Avinashilingam Deemed University for the facilities provided.

References

Arellano, G.M. and Rueda, L.M., 1988. Biological study of the housefly pupal parasitoid *Spalangia endius* (Walker) Hymenoptera : Pteromalidae, Philipp. *Ent.*, 7(4): 329–350.

Balakrishnan, N., 1992. Studies on hymenopteron pupal parasitoids of house fly (*Musca domestica* L.) in Pondicherry with special reference to the bionomics and host parasitoid interactions of *Dirhinus himalayanus* (Westwood, 1836) and *Pachycerepoideus vidnemmiae* (Rondaini, 1875). *M.Sc. Thesis.*, VCRC, Pondicherry.

Hall, R.D. and Fischer, F.J., 1988. Laboratory studies on the biology of *Spalangia nigra* (Hymenoptera : Pteromalidae). *Entomophaga*, 33(4): 495–504.

Moore, I., 1954. An efficient method of collecting dung beetles. The Pan-Pacific, *Entomol.*, 30(3): 208.

Morgan, P.B., Patterson, R., Labreque, G.C., Weidhaas, D.E., Benton, A. and Whitfield, T., 1975. Rearing and release of the house fly pupal parasite *Spalangia endius*, Walker. *Environ. Entomol.*, 4(4): 609–611.

Pinkus, H., 1913. The life history and habits of *Spalangia muscidarum* Richardson. Parasite of the Stable fly. *Psyche*, 20: 149–158.

Singh, P. and Moore, 1985. *Handbook of Insect Rearing*, Vol. II. Elsevier Science Publishers, B.V., pp. 129–134.

Wylie, H.G., 1972. Oviposition restraint of *Spalangia cameroni* (Hymenoptera : Pteromalidae) on parasitized house fly pupae Can. *Entomol.*, 104: 209–214.

Index

Y

Z